Jürgen Beyerer, Raphael Hagmanns, Daniel Stadler
Pattern Recognition

Also of interest

Mathematics of Deep Learning
An Introduction
Leonid Berlyand, Pierre-Emmanuel Jabin, 2023
ISBN 978-3-11-102431-8, e-ISBN (PDF) 978-3-11-102555-1

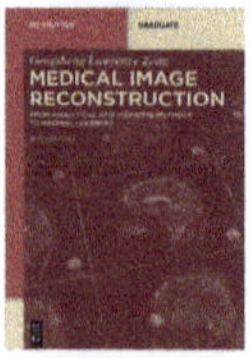

Medical Image Reconstruction
From Analytical and Iterative Methods to Machine Learning
Gengsheng Lawrence Zeng, 2023
ISBN 978-3-11-105503-9, e-ISBN (PDF) 978-3-11-105540-4

Category Theory
Invariances and Symmetries in Computer Science
Edited by Siddhartha Bhattacharyya, Vaclav Snasel, Aboul Ella Hassanien
Zoran Majkic, 2023
ISBN 978-3-11-108056-7, e-ISBN (PDF) 978-3-11-108167-0

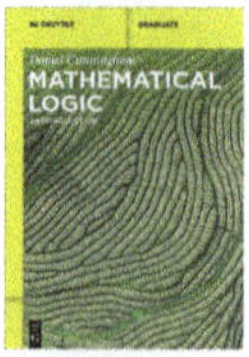

Mathematical Logic
An Introduction
Daniel Cunningham, 2023
ISBN 978-3-11-078201-1, e-ISBN (PDF) 978-3-11-078207-3

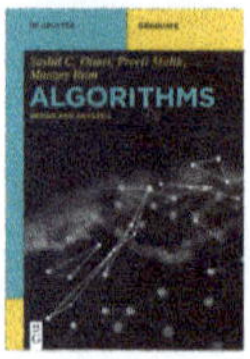

Algorithms
Design and Analysis
Sushil C. Dimri, Preeti Malik, Mangey Ram, 2021
ISBN 978-3-11-069341-6, e-ISBN (PDF) 978-3-11-069360-7

Jürgen Beyerer, Raphael Hagmanns, Daniel Stadler

Pattern Recognition

Introduction, Features, Classifiers and Principles

2nd Edition

DE GRUYTER
OLDENBOURG

Authors
Prof. Dr.-Ing. habil. Jürgen Beyerer
Fraunhofer Institute of Optronics, System Technologies and Image Exploitation (IOSB)
Fraunhoferstr. 1
76131 Karlsruhe
juergen.beyerer@iosb.fraunhofer.de
-and-
Institute for Anthropomatics and Robotics, Chair IES
Karlsruhe Institute of Technology (KIT)
Haid-und-Neu-Str. 7
76131 Karlsruhe

Raphael Hagmanns
Institute for Anthropomatics and Robotics, Chair IES
Karlsruhe Institute of Technology (KIT)
Haid-und-Neu-Str. 7
76131 Karlsruhe
raphael.hagmanns@kit.edu

Daniel Stadler
Institute for Anthropomatics and Robotics, Chair IES
Karlsruhe Institute of Technology (KIT)
Haid-und-Neu-Str. 7
76131 Karlsruhe
daniel.stadler@kit.edu

ISBN 978-3-11-133919-1
e-ISBN (PDF) 978-3-11-133920-7
e-ISBN (EPUB) 978-3-11-133941-2

Library of Congress Control Number: 2024931594

Bibliographic information published by the Deutsche Nationalbibliothek
The Deutsche Nationalbibliothek lists this publication in the Deutsche Nationalbibliografie; detailed bibliographic data are available on the Internet at http://dnb.dnb.de.

Printing and binding: CPI books GmbH, Leck
Cover image: Narong KHUEANKAEW / iStock / Getty Images Plus
∞ Printed on acid-free paper
Printed in Germany

www.degruyter.com

Preface

PATTERN RECOGNITION ⊂ MACHINE LEARNING ⊂ ARTIFICIAL INTELLIGENCE: This relation could give the impression that pattern recognition is only a tiny, very specialized topic. That, however, is misleading. Pattern recognition is a very important field of machine learning and artificial intelligence with its own rich structure and many interesting principles and challenges. For humans, and also for animals, their natural abilities to recognize patterns are essential for navigating the physical world which they perceive with their naturally given senses. Pattern recognition here performs an important abstraction from sensory signals to categories: on the most basic level, it enables the classification of objects into "Eatable" or "Not eatable" or, e.g., into "Friend" or "Foe". These categories (or, synonymously, classes) do not always have a tangible character. Examples of non-material classes are, e.g., "secure situation" or "dangerous situation". Such classes may even shift depending on the context, for example, when deciding whether an action is socially acceptable or not. Therefore, everybody is very much acquainted, at least at an intuitive level, with what pattern recognition means to our daily life. This fact is surely one reason why pattern recognition as a technical subdiscipline is a source of so much inspiration for scientists and engineers. In order to implement pattern recognition capabilities in technical systems, it is necessary to formalize it in such a way, that the designer of a pattern recognition system can systematically engineer the algorithms and devices necessary for a technical realization. This textbook summarizes a lecture course about pattern recognition that one of the authors (Jürgen Beyerer) has been giving for students of technical and natural sciences at the Karlsruhe Institute of Technology (KIT) since 2005. The aim of this book is to introduce the essential principles, concepts and challenges of pattern recognition in a comprehensive and illuminating presentation. We will try to explain all aspects of pattern recognition in a well understandable, self-contained fashion. Facts are explained with a mixture of a sufficiently deep mathematical treatment, but without going into the very last technical details of a mathematical proof. The given explanations will aid readers to understand the essential ideas and to comprehend their interrelations. Above all, readers will gain the big picture that underlies all of pattern recognition.

The authors would like to thank their peers and colleagues for their support:

Special thanks are owed to Dr. Ioana Gheţa who was very engaged during the early phases of the lecture "Pattern Recognition" at the KIT. She prepared most of the many slides and accompanied the course along many lecture periods.

Thanks as well to Dr. Martin Grafmüller and to Dr. Miro Taphanel for supporting the lecture Pattern Recognition with great dedication.

Moreover, many thanks to Prof. Michael Heizmann and Prof. Fernando Puente León for inspiring discussions, which have positively influenced to the evolution of the lecture.

Thanks to Christian Hermann and Lars Sommer for providing additional figures and examples of deep learning. Our gratitude also to our friends and colleagues Alexey

https://doi.org/10.1515/9783111339207-202

Pak, Ankush Meshram, Chengchao Qu, Christian Hermann, Ding Luo, Julius Pfrommer, Julius Krause, Johannes Meyer, Lars Sommer, Mahsa Mohammadikaji, Mathias Anneken, Mathias Ziebarth, Miro Taphanel, Patrick Philipp, and Zheng Li for providing valuable input and corrections for the preparation of this manuscript.

Lastly, we thank DeGruyter for their support and collaboration in this project.

Karlsruhe, Summer 2017

Jürgen Beyerer
Matthias Richter
Matthias Nagel

Preface of 2nd edition

In recent years, "Pattern Recognition" has emerged as one of the fastest evolving areas in machine learning. In a landscape characterized by rapid advances, our commitment to providing a comprehensive and contemporary resource is reflected by various content modernizations. Several new classifiers are introduced to align with current problem formulations. In addition, different fundamental sections have been added to provide further insight into important topics such as Gaussian mixture models and their parameter estimation, hidden Markov models, or transformer models. The updates include a thorough unification of the notation, along with fontification, aiming for a more cohesive reading experience. All changes align with a major update of the "Pattern Recognition" lecture at the Karlsruhe Institute of Technology (KIT).

We would like to thank Stefan Wolf for accompanying this lecture with great dedication. Special thanks go to Robert Zimmermann and Dr. Christian Frese for proofreading various parts of the second edition. Finally, we would like to thank the team at DeGruyter for their continuous and inexhaustible efforts in overcoming the challenges of applying the new book template.

Karlsruhe, Spring 2024

Jürgen Beyerer
Raphael Hagmanns
Daniel Stadler

https://doi.org/10.1515/9783111339207-203

Preface of 2nd Edition

Contents

List of Tables

https://doi.org/10.1515/9783111339207-205

List of Figures

https://doi.org/10.1515/9783111339207-206

Notation

General identifiers

$a, \dots, z$	Scalar, function mapping to a scalar, or a realization of a random variable
$\underline{a}, \dots, \underline{z}$	Random variable (scalar)
$\mathbf{a}, \dots, \mathbf{z}$	Vector, function mapping to a vector, or realization of a vectorial random variable
$\underline{\mathbf{a}}, \dots, \underline{\mathbf{z}}$	Random variable (vectorial)
$\hat{\mathbf{a}}, \dots, \hat{\mathbf{z}}$	Realized estimator of denoted variable
$\underline{\hat{\mathbf{a}}}, \dots, \underline{\hat{\mathbf{z}}}$	Estimator of denoted variable as random variable itself
$\mathbf{A}, \dots, \mathbf{Z}$	Matrix
$\underline{\mathbf{A}}, \dots, \underline{\mathbf{Z}}$	Matrix as random variable
$\mathcal{A}, \dots, \mathcal{Z}$	Set
$\mathfrak{A}, \dots, \mathfrak{Z}$	System of sets

Special identifiers

c	Number of classes
d	Dimension of feature space
$\mathcal{D}$	Set of training samples
i, j, k	Indices along the dimension, i.e., $i, j, k \in \{1, \dots, d\}$, or along the number of samples, i.e., $i, j, k \in \{1, \dots, N\}$
$\mathbf{I}$	Identity matrix
j	Imaginary unit, $\mathrm{j}^2 = -1$
$\mathbf{J}$	Fisher information matrix
$k(\cdot,\cdot)$	Kernel function
$k(\cdot)$	Decision function
$\mathcal{K}$	Decision space
l	Cost function $l : \Omega^0/{\sim} \times \Omega/{\sim} \longrightarrow \mathbb{R}$
L	Cost matrix $\in \mathbb{R}^{(c+1)\times c}$
$\mathbf{m}$	Feature vector
$\mathbf{m}_i$	Feature vector of the i-th sample
m_{ij}	The j-th component of the i-th feature vector
M_{ij}	The component at the i-th row and j-th column of the matrix $\mathbf{M}$
$\mathcal{M}$	Feature space
N	Number of samples
o	Object
ω	Class of objects, i.e., $\omega \subseteq \Omega$
ω_0	Rejection class
Ω	Set of objects (the relevant part of the world) $\Omega = \{o_1, \dots, o_N\}$
$\Omega/{\sim}$	The domain factorized w.r.t. the classes, i.e., the set of classes $\Omega/{\sim} = \{\omega_1, \dots, \omega_c\}$
$\Omega^0/{\sim}$	The set of classes including the rejection class, $\Omega^0/{\sim} = \Omega/{\sim} \cup \{\omega_0\}$

https://doi.org/10.1515/9783111339207-207

$p(m)$	Probability density function for random variable $\underline{m}$ evaluated at m
$P(\omega)$	Probability mass function for (discrete) random variable $\underline{\omega}$ evaluated at ω
$\Pr(e)$	Probability of an event e
$\mathfrak{P}(\mathcal{A})$	Power set, i.e., the set of all subsets of $\mathcal{A}$
$\mathcal{S}$	Set of all samples, $\mathcal{S} = \mathcal{D} \uplus \mathcal{T} \uplus \mathcal{V}$
$\mathcal{T}$	Set of test samples
$\mathcal{V}$	Set of validation samples
$\mathbf{U}$	Unit matrix, i.e., the matrix all of whose entries are 1
$\boldsymbol{\theta}$	Parameter vector
Θ	Parameter space

General sets

$\mathbb{C}$	Set of complex numbers
$\mathbb{H}$	Poincaré half plane
$\mathbb{N}$	Set of natural numbers (without zero)
$\mathbb{N}_0$	Set of natural numbers (including zero)
$\mathbb{Q}$	Set of rational numbers
$\mathbb{Q}^{>0}$, $\mathbb{Q}^{<0}$	Set of positive, negative rational numbers
$\mathbb{R}$	Set of real numbers
$\mathbb{R}^{>0}$, $\mathbb{R}^{<0}$	Set of positive, negative real numbers
$\mathbb{Z}$	Set of integer numbers

Special symbols

$\lvert\mathbf{A}\rvert$	Determinant of matrix $\mathbf{A}$
$\lvert\mathcal{A}\rvert$	Cardinality of set $\mathcal{A}$
$\propto$	"Proportional to"-relation
$\xrightarrow{\text{P}}$	Convergence in probability
$\xrightarrow{\text{w}}$	Weak convergence
$\rightsquigarrow$	Leads to (not necessarily in a strict mathematical sense)
$\uplus$	Disjoint union of sets, i.e., $\mathcal{C} = \mathcal{A} \uplus \mathcal{B} \Leftrightarrow \mathcal{C} = \mathcal{A} \cup \mathcal{B}$ and $\mathcal{A} \cap \mathcal{B} = \emptyset$
$\langle \cdot, \cdot \rangle$	Scalar product
∇, $\nabla_{\mathbf{e}}$	Gradient, gradient w.r.t. $\mathbf{e}$
$\mathrm{Cov}\{\cdot\}$	Covariance
$\det(\cdot)$	Determinant
$\frac{\partial a}{\partial b}$	Derivative of a w.r.t. b
δ_i^j	Kronecker delta/symbol; $\delta_i^j = 1$ iff $i = j$, else $\delta_i^j = 0$
$\delta_{[\cdot]}$	Generalized Kronecker symbol, i.e., $\delta_{[\Pi]} = 1$ iff Π is true and $\delta_{[\Pi]} = 0$ otherwise
$\mathrm{E}\{\cdot\}$	Expected value

$\mathcal{N}(\mu, \sigma^2)$	Normal/Gaussian distribution with expectation μ and variance σ^2
$\mathcal{N}(m; \mu, \sigma^2)$	Normal/Gaussian distribution with expectation μ, variance σ^2, and explicit nomination of the random variable
$\mathcal{N}(\boldsymbol{\mu}, \boldsymbol{\Sigma})$	Multivariate normal/Gaussian distribution with expectation $\boldsymbol{\mu}$ and covariance matrix $\boldsymbol{\Sigma}$
$\mathcal{N}(\mathbf{m}; \boldsymbol{\mu}, \boldsymbol{\Sigma})$	Multivariate normal/Gaussian distribution with expectation $\boldsymbol{\mu}$, covariance matrix $\boldsymbol{\Sigma}$, and explicit nomination of the random variable
$\operatorname{tr} \mathbf{A}$	Trace of the matrix $\mathbf{A}$
$\operatorname{Var}\{\cdot\}$	Variance

Abbreviations

cf.	"confer" (latin: compare)
iff	if and only if
i.i.d.	independent and identically distributed
n.b.	"nota bene" (latin: note well, take note)
w.r.t.	with respect to

Introduction

The overall goal of pattern recognition is to develop systems that can distinguish and classify objects. The range of possible objects is vast. Objects can be physical things existing in the real world, like banknotes, as well as non-material entities, e.g., e-mails, or abstract concepts such as actions or situations. The objects can be of natural origin or artificially created. Examples of objects in pattern recognition tasks are shown in Figure 1.

On the basis of recorded patterns, the task is to classify the objects into previously assigned classes by defining and extracting suitable features. The type as well as the number of classes is given by the classification task. For example, banknotes (see Figure 1b) could be classified according to their monetary value or the goal could be to discriminate between real and counterfeited banknotes. For now, we will refrain from defining what we mean by the terms *pattern*, *feature*, and *class*. Instead, we will rely on an intuitive understanding of these concepts. A precise definition will be given in the next chapter.

From this short description, the fundamental elements of a pattern recognition task and the challenges to be encountered at each step can be identified even without a precise definition of the concepts pattern, feature, and class:

Pattern acquisition, sensing, measuring In the first step, suitable properties of the objects to be classified have to be gathered and put into computable representations. Although *pattern* might suggest that this (necessary) step is part of the actual pattern recognition task, it is not. However, this process has to be considered so far as to

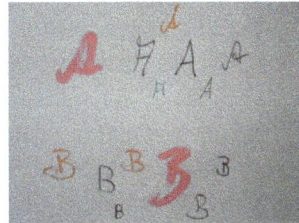

(a) Screws **(b)** Banknotes **(c)** Handwriting **(d)** Plant seeds **(e)** Plant leaves

Fig. 1: Examples of artificial and natural objects.

https://doi.org/10.1515/9783111339207-208

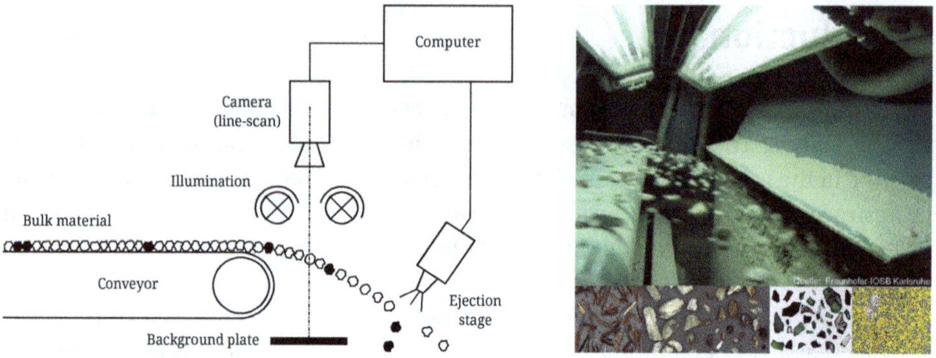

(a) Schematic overview

(b) Inspection and ejection stage

Fig. 2: Industrial bulk material sorting system.

provide an awareness of any possible complications it may cause in the subsequent steps. Measurements of any kind are usually affected by random noise and other disturbances that, depending on the application, can not be mitigated by methods of metrology alone: for example, changes of lighting conditions in uncontrolled and uncontrollable environments. A pattern recognition system has to be designed so that it is capable of solving the classification task regardless of such factors.

Feature definition, feature acquisition Suitable features have to be selected based on the available patterns and methods for extracting these features from the patterns have to be defined. The general aim is to find the smallest set of the most informative and discriminative features. A feature is discriminative if it varies little with objects within a single class, but varies significantly with objects from different classes.

Design of the classifier After the features have been determined, rules to assign a class to an object have to be established. The underlying mathematical model has to be selected so that it is powerful enough to discern all given classes and thus solve the classification task. On the other hand, it should not be more complicated than it needs to be. Determining a given classifier's parameters is a typical learning problem and is therefore also affected by the problems pertaining to this field. These topics will be discussed in greater detail in Chapter 1.

These lecture notes on pattern recognition are mainly concerned with the last two issues. The complete process of designing a pattern recognition system will be covered in its entirety and the underlying mathematical background of the required building blocks will be given in depth.

Pattern recognition systems are generally parts of larger systems, in which pattern recognition is used to derive decisions from the result of the classification. Industrial sorting systems are typical of this (see Figure 2). Here, products are processed differently depending on their class memberships.

Tab. 1: Capabilities of humans and machines in relation to pattern recognition.

	Association & cognition	Combinatorics & precision
Human	very good	poor
Machine	medium	very good

Hence, as a pattern recognition system is not an end in itself, the design of such a system has to consider the consequences of a bad decision caused by a misclassification. This puts pattern recognition between human and machine. The main advantage of automatic pattern recognition is that it can execute recurring classification tasks with great speed and without fatigue. However, an automatic classifier can only discern the classes that were considered in the design phase and it can only use those features that were defined in advance. A pattern recognition system to tell apples from oranges may label a pear as an apple and a lemon as an orange if lemons and pears were not known in the design phase. The features used for classification might be chosen poorly and not be discriminative enough. Different environmental conditions (e.g., lighting) in the laboratory and in the field that were not considered beforehand might impair the classification performance, too. Humans, on the other hand, can use their associative and cognitive capabilities to achieve good classification performance even in adverse conditions. In addition, humans are capable of undertaking further actions if they are unsure about a decision. The contrasting abilities of humans and machines in relation to pattern recognition are compared in Table 1. In many cases one will choose to build a hybrid system: easy classification tasks will be processed automatically, ambiguous cases require human intervention, which may be aided by the machine, e.g., by providing a selection of the most probable classes.

1 Fundamentals and definitions

The aim of this chapter is to describe the general structure of a pattern recognition system and properly define the fundamental terms and concepts that were partially used in the Introduction already. A description of the generic process of designing a pattern recognizer will be given and the challenges at each step will be stated more precisely.

1.1 Goals of pattern recognition

The purpose of pattern recognition is to assign classes to objects according to some similarity properties. Before delving deeper, we must first define what is meant by *class* and *object*. For this, two mathematical concepts are needed: equivalence relations and partitions.

Definition 1.1 (Equivalence relation). Let Ω be a set of elements with some relation $\sim$. Suppose further that $o, o_1, o_2, o_3 \in \Omega$ are arbitrary. The relation $\sim$ is said to be an *equivalence relation* if it fulfills the following conditions:

1. Reflexivity: $o \sim o$.
2. Symmetry: $o_1 \sim o_2 \Leftrightarrow o_2 \sim o_1$.
3. Transitivity: $o_1 \sim o_2$ and $o_2 \sim o_3 \Rightarrow o_1 \sim o_3$.

Two elements o_1, o_2 with $o_1 \sim o_2$ are said to be *equivalent*. We further write $[o]_\sim \subseteq \Omega$ to denote the subset

$$[o]_\sim = \left\{o' \in \Omega \,\middle|\, o' \sim o\right\} \tag{1.1}$$

of all elements that are equivalent to o. The object o is also called a *representative* of the set $[o]_\sim$ (equivalence class). In the context of pattern recognition, each $o \in \Omega$ denotes an object and each $[o]_\sim$ denotes a class. A different approach to classifying every element of a set is given by partitioning the set:

Definition 1.2 (Partition, class). Let Ω be a set and $\omega_1, \omega_2, \omega_3, \ldots \subseteq \Omega$ be a system of subsets. This system of subsets is called a *partition of* Ω if the following conditions are met:

1. $\omega_i \cap \omega_j = \emptyset$ for all $i \neq j$, i.e., the subsets are pairwise disjoint and
2. $\bigcup_i \omega_i = \Omega$, i.e., the system is exhaustive.

Every subset ω is called a *class* (of the partition).

It is easy to see that *equivalence relations* and *partitions* describe synonymous concepts: every equivalence relation induces a partition and every partition induces an equivalence relation.

The underlying principle of all pattern recognition is illustrated in Figure 1.1. On the left it shows—in abstract terms—the world and a (sub)set Ω of objects that live within

https://doi.org/10.1515/9783111339207-001

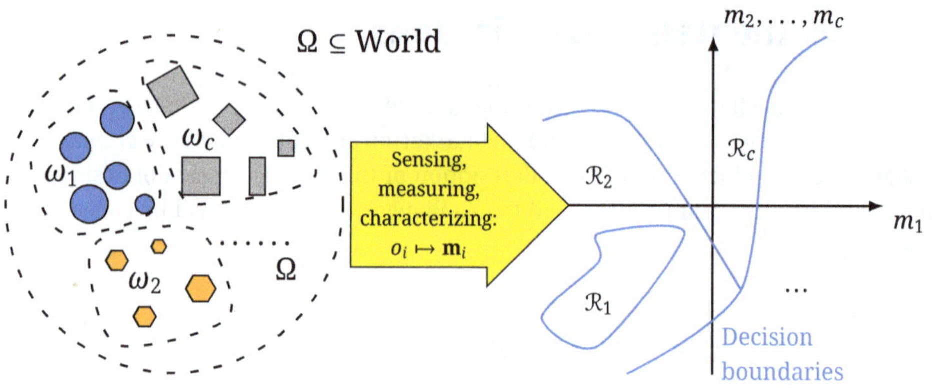

Fig. 1.1: Transformation of the domain Ω into the feature space $\mathcal{M}$.

the world. The set Ω is given by the pattern recognition task and is also called the *domain*. Only the objects in the domain are relevant to the task; this is the so called *closed world assumption*. The task also partitions the domain into classes $\omega_1, \omega_2, \omega_3, \ldots \subseteq \Omega$. A suitable mapping associates every object o_i to a feature vector $\mathbf{m}_i \in \mathcal{M}$ inside the *feature space* $\mathcal{M}$. The goal is now to find rules that partition $\mathcal{M}$ along *decision boundaries* so that the classes of $\mathcal{M}$ match the classes of the domain. Hence, the rule for classifying an object o is

$$\hat{\omega}(o) := \omega_i \quad \text{if} \quad \mathbf{m}(o) \in \mathcal{R}_i. \tag{1.2}$$

This means that the estimated class $\hat{\omega}(o)$ of object o is set to the class ω_i if the feature vector $\mathbf{m}(o)$ falls inside the region $\mathcal{R}_i$. For this reason, the $\mathcal{R}_i$ are also called decision regions. The concept of a *classifier* can now be stated more precisely:

Definition 1.3 (classifier). A classifier is a collection of rules that state how to evaluate feature vectors in order to sort objects into classes. Equivalently, a classifier is a system of decision boundaries in the feature space.

Readers experienced in machine learning will find these concepts very familiar. In fact, machine learning and pattern recognition are closely intertwined: pattern recognition is (mostly) *supervised learning*, as the classes are known in advance. This topic will be picked up again later in this chapter.

1.2 Structure of a pattern recognition system

In the previous section it was already mentioned that a pattern recognition system maps objects onto feature vectors (see Figure 1.1) and that the classification is carried out in the feature space. This section focuses on the steps involved and defines the terms *pattern* and *feature*.

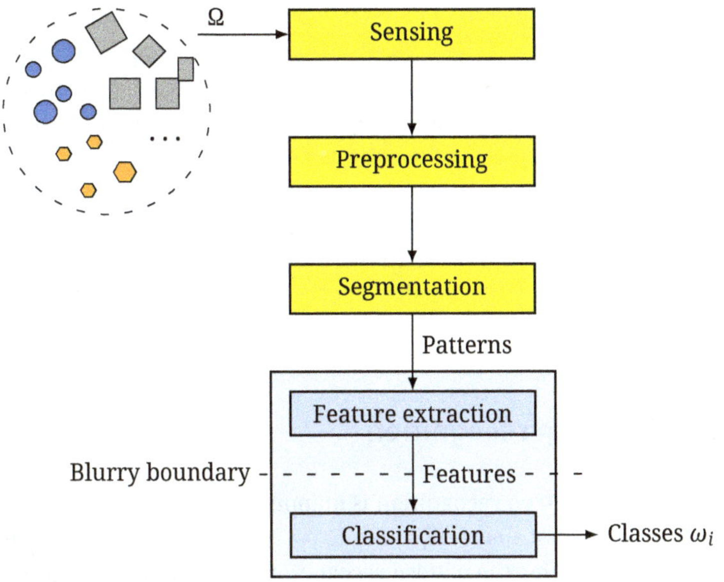

Fig. 1.2: Processing pipeline of a pattern recognition system.

Figure 1.2 shows the processing pipeline of a pattern recognition system. In the first steps, the relevant properties of the objects from Ω must be put into a machine readable interpretation. These first steps (yellow boxes in Figure 1.2) are usually performed by methods of sensor engineering, signal processing, or metrology, and are not directly part of the pattern recognition system. The result of these operations is the *pattern* of the object under inspection.

Definition 1.4 (Pattern). A *pattern* is the collection of the observed or measured properties of a single object.

The most prominent pattern is the image, but patterns can also be (text) documents, audio recordings, seismograms, or indeed any other signal or data. The pattern of an object is the input to the actual pattern recognition, which is itself composed of two major steps (gray boxes in Figure 1.2): previously defined features are extracted from the pattern and the resulting feature vector is passed to the classifier, which then outputs an equivalence class according to Equation (1.2).

Definition 1.5 (feature). A *feature* is an obtainable, characteristic property, which will be the basis for distinguishing between patterns and therefore also between the underlying classes.

A feature is any quality or quantity that can be derived from the pattern, for example, the area of a region in an image, the count of occurrences of a key word within a text, or the position of a peak in an audio signal.

As an example, consider the task of classifying cubical objects as either “small cube” or “big cube” with the aid of a camera system. The *pattern* of an object is the camera image, i.e., the pixel representation of the image. By using suitable image processing algorithms, the pixels that belong to the cube can be separated from the pixels that show the background and the length of the edge of the cube can be determined. Here, “edge length” is the *feature* that is used to classify the object into the classes “big” or “small”.

Note that the boundary between the individual steps is not clearly defined, especially between feature extraction and classification. Often there is the possibility of using simple features in conjunction with a powerful classifier or of combining elaborate features with a simple classifier.

1.3 Abstract view of pattern recognition

From an abstract point of view, pattern recognition is mapping the set of objects to be classified Ω to the equivalence classes $\omega \in \Omega/\sim$, i.e., $\Omega \to \Omega/\sim$ or $o \mapsto \omega$. In some cases, this view is sufficient for treating the pattern recognition task. For example, if the objects are e-mails and the task is to classify the e-mails as either “ham” $\hat{=} \omega_1$ or “spam” $\hat{=} \omega_2$, this view is sufficient for deriving the following simple classifier: The body of an incoming e-mail is matched against a list of forbidden words. If it contains more than S of these words, it is marked as spam, otherwise it is marked as ham.

For a more complicated classification system, as well as for many other pattern recognition problems, it is helpful and can provide additional insights to break up the mapping $\Omega \to \Omega/\sim$ into several intermediate steps. In this book, the pattern recognition process is subdivided into the following steps: observation, sensing, measurement; feature extraction; decision preparation; and classification. This subdivision is outlined in Figure 1.3.

To come back to the example mentioned above, an e-mail is already digital data, hence it does not need to be sensed. It can be further seen as an object, a pattern, and a feature vector, all at once. A spam classification application that takes the e-mail as input and accomplishes the desired assignment to one of the two categories could be considered as a black box that performs the mapping $\Omega \to \Omega/\sim$ directly.

In many other cases, especially if objects of the physical world are to be classified, the intermediate steps of $\Omega \to \mathcal{P} \to \mathcal{M} \to \mathcal{K} \to \Omega/\sim$ will help to better analyze and understand the internal mechanisms, challenges and problems of object classification. It also supports engineering a better pattern recognition system. The concept of the pattern space $\mathcal{P}$ is especially helpful if the raw data acquired about an object has a very high dimension, e.g., if an image of an object is taken as the pattern. Explicit use of $\mathcal{P}$ will be made in Section 2.4.8, where the tangent distance is discussed, and in Section 2.6.3, where invariant features are considered. The concept of the decision space $\mathcal{K}$ helps to generalize classifiers and is especially useful to treat the rejection problem in Section 9.5. Lastly, the concept of the feature space $\mathcal{M}$ is fundamental to pattern recognition and

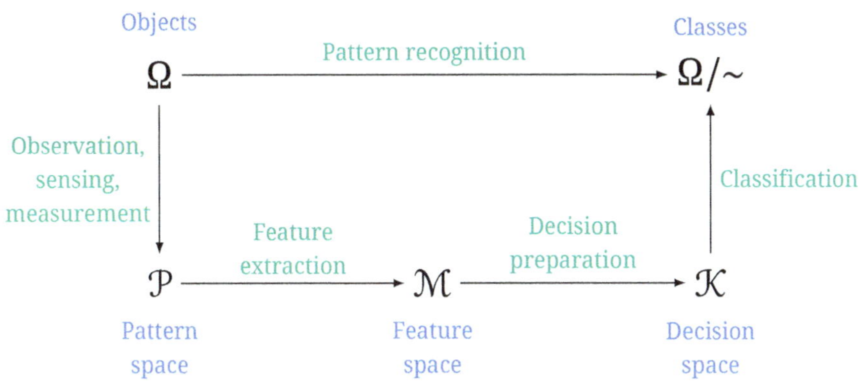

Fig. 1.3: Subdividing the pattern recognition process allows deeper insights and helps to better understand important concepts such as: the curse of dimensionality, overfitting, and rejection.

permeates the whole textbook. Features can be seen as a concentrated extract from the pattern, which essentially carries the information about the object which is relevant for the classification task.

Overall, any pattern recognition task can be formally defined by a quintuple $(\Omega, \sim, \omega_0, l, \mathcal{S})$, where Ω is the set of objects to be classified, $\sim$ is an equivalence relation that defines the classes in Ω, ω_0 is the rejection class (see Section 9.5), l is a cost function that assesses the classification decision $\hat{\omega}$ compared to the true class ω (see Section 3.3), and $\mathcal{S}$ is the set of examples with known class memberships. Note that the rejection class ω_0 is not always needed and may be empty. Similarly, the cost function l may be omitted, in which case it is assumed that incorrect classification creates the same costs independently of the class and no cost is incurred by a correct classification (*0–1 loss*).

These concepts will be further developed and refined in the following chapters. For now, we will return to a more concrete discussion of how to design systems that can solve a pattern recognition task.

1.4 Design of a pattern recognition system

Figure 1.4 shows the principal steps involved in designing a pattern recognition system: data gathering, selection of features, definition of the classifier, training of the classifier, and evaluation. Every step is prone to making different types of errors, but the sources of these errors can broadly be sorted into four categories:

1. too small a dataset,
2. a non-representative dataset,
3. inappropriate, non-discriminative features, and
4. an unsuitable or ineffective mathematical model of the classifier.

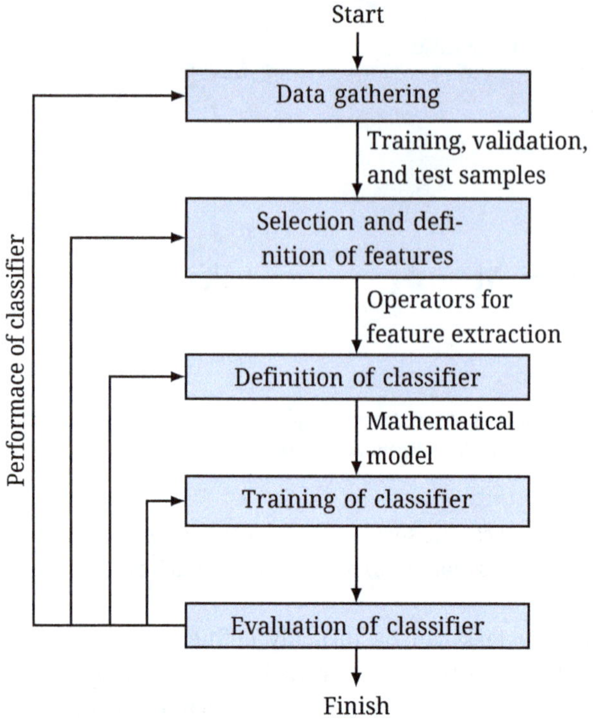

Fig. 1.4: Design phases of a pattern recognition system.

The following section will describe the different steps in detail, highlighting the challenges faced and pointing out possible sources of error.

The first step is always to gather samples of the objects to be classified. The resulting dataset is labeled $\mathcal{S}$ and consists of patterns of objects where the corresponding classes are known a priori, for example because the objects have been labeled by a domain expert. As the class of each sample is known, deriving a classifier from $\mathcal{S}$ constitutes *supervised learning*. The complement to supervised learning is *unsupervised learning*, where the class of the objects in $\mathcal{S}$ is not known and the goal is to uncover some latent structure in the data. In the context of pattern recognition, however, unsupervised learning is only of minor interest.

A common mistake when gathering the dataset is to pick pathological, characteristic samples from each class. At first glance, this simplifies the following steps, because it seems easier to determine the discriminative features. Unfortunately, these seemingly discriminative features are often useless in practice. Furthermore, in many situations, the most informative samples are those that represent edge cases. Consider a system where the goal is to pick out defective products. If the dataset only consists of the most perfect samples and the most defective samples, it is easy to find highly discriminative

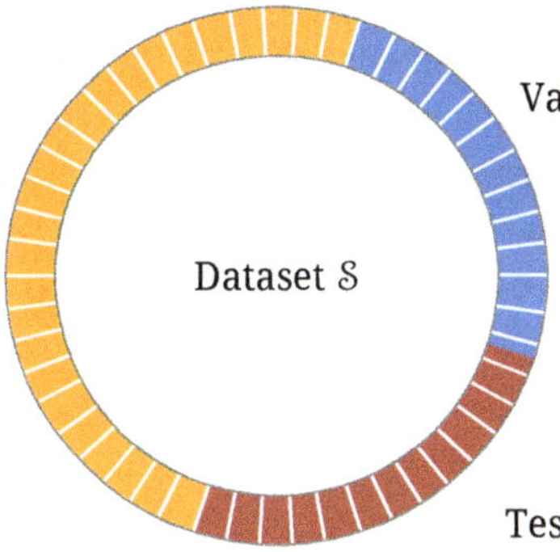

Fig. 1.5: Rule of thumb to partition the dataset into training, validation, and test sets.

features and one will assume that the classifier will perform with high accuracy. Yet in practice, imperfect, but acceptable products may be picked out or products with a subtle, but serious defect may be missed. A good dataset contains both extreme and common cases. More generally, the challenge is to obtain a dataset that is representative of the underlying distribution of classes. However, an unrepresentative dataset is often intentional or practically impossible to avoid when one of the classes is very sparsely populated but representatives of all classes are needed. In the above example of picking out defective products, it is conceivable that on average only one in a thousand products has a defect. In practice, one will select an approximately equal number of defective and intact products to build the dataset $\mathcal{S}$. This means that the so called *a priori* distribution of classes must not be determined from $\mathcal{S}$, but has to be obtained elsewhere.

The dataset $\mathcal{S}$ is further partitioned into a training set $\mathcal{D}$, a validation set $\mathcal{V}$, and a test set $\mathcal{T}$. A rule of thumb is to use 50 % of $\mathcal{S}$ for $\mathcal{D}$, 25 % of $\mathcal{S}$ for $\mathcal{V}$, and the remaining 25 % of $\mathcal{S}$ for $\mathcal{T}$ (see Figure 1.5). The test set $\mathcal{T}$ is held back and not considered during most of the design process. It is only used once to evaluate the classifier in the last design step (see Figure 1.4). The distinction between training and validation set is not always necessary. The validation set $\mathcal{V}$ is needed if the classifier in question is governed not only by parameters that are estimated from the training set $\mathcal{D}$, but also depends on so called design parameters or *hyperparameters*. The optimal design parameters are determined using the validation set.

A general issue is that the available dataset is often too small. The reason is that obtaining and (manually) pre-classifying a dataset is typically very time consuming and thus costly. In some cases, the number of samples is naturally limited, e.g., when the goal is to classify earthquakes. The partition into training, test and validation sets further reduces the number of available samples, sometimes to a point where carrying out the remaining design phases is no longer reasonable. Chapter 9 will suggest methods for dealing with small datasets.

The second step of the design process (see Figure 1.4) is concerned with choosing suitable features. Different types of features and their characteristics will be covered in Chapter 2 and will not be discussed at this point. However, two general design principles should be considered when choosing features:

1. Simple, comprehensible features should be preferred. Features that correspond to immediate (physical) properties of the objects or features which are otherwise meaningful, allow understanding and optimizing the decisions of the classifier.
2. The selection should contain a small number of highly discriminative features. The features should show little deviation within classes, but vary greatly between classes.

The latter principle is especially important to avoid the so called *curse of dimensionality* (sometimes also called the *Hughes effect*): a higher dimensional feature space means that a classifier operating in this feature space will depend on more parameters. Determining the appropriate parameters is a typical estimation problem. The more parameters need to be estimated, the more samples are needed to adhere to a given error bound. Chapter 6 will give more details on this topic.

The third design step is the definition of a suitable classifier (see Figure 1.4). The boundary between feature extraction and classifier is arbitrary and was already called "blurry" in Figure 1.2. In the example in Figure 2.4c, one has the option to either stick with the features and choose a more powerful classifier that can represent curved decision boundaries or to transform the features and choose a simple classifier that only allows linear decision boundaries. It is also possible to take the output of one classifier as input for a higher order classifier. For example, the first classifier could classify each pixel of an image into one of several categories. The second classifier would then operate on the features derived from the intermediate image. Ultimately, it is mostly a question of personal preference where to put the boundary and whether feature transformation is part of the feature extraction or belongs to the classifier.

After one has decided on a classifier, the fourth design step (see Figure 1.2) is to train it. Using the training and validation sets $\mathcal{D}$ and $\mathcal{V}$, the (hyper-)parameters of the classifier are estimated so that the classification is in some sense as accurate as possible. In many cases, this is achieved by defining a loss function that punishes misclassification, then optimizing this loss function w.r.t. the classifier parameters. As the dataset can be considered as a (finite) realization of a stochastic process, the parameters are subject to statistical estimation errors. These errors will become smaller the more samples are available.

An edge case occurs when the sample size is so small and the classifier has so many parameters that the estimation problem is under-determined. It is then possible to choose the parameters in such a way that the classifier classifies all training samples correctly. Yet novel, unseen samples will most probably not be classified correctly, i.e., the classifier does not generalize well. This phenomenon is called *overfitting* and will be revisited in Chapter 6.

In the fifth and last step of the design process (see Figure 1.2), the classifier is evaluated using the test set $\mathcal{T}$, which was previously held back. In particular, this step is important to detect whether the classifier generalizes well or whether it has been overfitted. If the classifier does not perform as needed, any of the previous steps—in particular the choice of features and classifier—can be revisited and adjusted. Strictly speaking, the test set $\mathcal{T}$ is already depleted and must not be used in a second run. Instead, each separate run should use a different test set, which has not yet been seen in the previous design steps. However, in many cases it is not possible to gather new samples. Again, Chapter 9 will suggest methods for dealing with such situations.

1.5 Exercises

(1.1) Let $\mathcal{S}$ be the set of all computer science students at the KIT. For $x,y \in \mathcal{S}$, let $x \sim y$ be true iff x and y are attending the same class. Is $x \sim y$ an equivalence relation?

(1.2) Let $\mathcal{S}$ be as above. Let $x \sim y$ be true iff x and y share a grandparent. Is $x \sim y$ an equivalence relation?

(1.3) Let $\mathbf{x}, \mathbf{y} \in \mathbb{R}^d$. Is $\mathbf{x} \sim \mathbf{y} \Leftrightarrow \mathbf{x}^\mathsf{T}\mathbf{y} = 0$ an equivalence relation?

(1.4) Let $\mathbf{x}, \mathbf{y} \in \mathbb{R}^d$. Is $\mathbf{x} \sim \mathbf{y} \Leftrightarrow \mathbf{x}^\mathsf{T}\mathbf{y} \geq 0$ an equivalence relation?

(1.5) Let $x,y \in \mathbb{N}$ and $f : \mathbb{N} \mapsto \mathbb{N}$ be a function on the natural numbers. Is the relation $x \sim y \Leftrightarrow f(x) \leq f(y)$ an equivalence relation?

(1.6) Let $\mathcal{A}$ be a set of algorithms and for each $X \in \mathcal{A}$ let $r(X,n)$ be the runtime of that algorithm for an input of length n. Is the following relation an equivalence relation?

$$X \sim Y \quad \Leftrightarrow r(X,n) \in \mathcal{O}\,(r(Y,n)) \qquad \text{for } X,Y \in \mathcal{A}.$$

Note: The Landau symbol $\mathcal{O}$ ("big O notation") is defined by

$$\mathcal{O}\,(f(n)) := \{g(n) \mid \exists a > 0\, \exists n_0 > 0\, \forall n \geq n_0 : |g(n)| \leq a|f(n)|\}\,,$$

i.e., $\mathcal{O}\,(f(n))$ is the set of all functions of n that are asymptotically bounded below by $f(n)$.

2 Features

A good understanding of features is fundamental for designing a proper pattern recognition system. Thus this chapter deals with all aspects of this concept, beginning with a mere classification of the kinds of features, up to the methods for reducing the dimensionality of the feature space. A typical beginner's mistake is to apply mathematical operations to the numeric representation of a feature, just because it is syntactically possible, albeit these operations have no meaning whatsoever for the underlying problem. Therefore, the first section elaborates on the different types of possible features and their traits.

2.1 Types of features and their traits

In empiricism, the *scale of measurement* (also: *level of measurement*) is an important characteristic of a feature or variable. In short, the scale defines the allowed transformations that can be applied to the variable without adding more meaning to it than it had before. Roughly speaking, the scale of measurement is a classification of the expressive power of a variable. A transformation of a variable from one domain to another is possible if and only if the transformation preserves the structure of the original domain.

Table 2.1 shows five scales of measurement in conjunction with their characteristics as well as some examples. The first four categories—the nominal scale, the ordinal scale, the interval scale, and the ratio scale—were proposed by Stevens [1946]. Lastly, we also consider the absolute scale. The first two scales of measurement can be further subsumed under the term qualitative features, whereas the other three scales represent quantitative features. The order of appearance of the scales in the table follows the cardinality of the set of allowed feature transformations. The transformation of a nominal variable can be any function f that represents an unambiguous relabeling of the features, that is, the only requirement on f is injectivity. At the other end, the only allowed transformation of an absolute variable is the identity.

2.1.1 Nominal scale

The nominal scale is made up of pure labels. The only meaningful question to ask is whether two variables have the same value: the nominal scale only allows to compare two values w.r.t. equivalence. There is no meaningful transformation besides relabeling. No empirical operation is permissible, i.e., there is no mathematical operation of nominal features that is also meaningful in the material world.

A typical example is the sex of a human. The two possible values can be either written as "f" vs. "m", "female" vs. "male", or be denoted by the special symbols ♀ vs. ♂. The labels are different, but the meaning is the same. Although nominal values are

https://doi.org/10.1515/9783111339207-002

Tab. 2.1: Taxonomy of scales of measurement. Empirical relations are mathematical relations that emerge from experiments, e.g., comparing the volume of two objects by measuring how much water they displace. Likewise, empirical operations are mathematical operations that can be carried out in an experiment, e.g., adding the mass of two objects by putting them together, or taking the ratio of two masses by putting them on a balance scale and noting the point of the fulcrum when the scale is balanced.

Trait	**Qualitative**		**Quantitative**		
	Nominal scale	Ordinal scale	Interval scale	Ratio scale	Absolute scale
Empirical relation	Equivalence $\sim$	Equivalence $\sim$ Ordering $\prec$	Equivalence $\sim$ Ordering $\prec$	Equivalence $\sim$ Ordering $\prec$	Equivalence $\sim$ Ordering $\prec$
Empirical operation			Addition $\oplus$	Addition $\oplus$ Multiplication $\otimes$	Addition $\oplus$ Multiplication $\otimes$
Allowed transformation	$m' = f(m)$ injective	$m' = f(m)$ strictly increasing	$m' = am + b$ with $a > 0$	$m' = am$ with $a > 0$	$m' = m$
Typical domain	Integers, names, Symbols	Integers	Real numbers	Real numbers	Natural numbers
Expressiveness	Very low	Low	Medium	High	Very high
Examples	Telephone numbers, Postal codes, Gender, Scale name	School grades, Degree of hardening, Wind intensity, Scale expressiveness	Temperature in °F, Calendar time, Geographic altitude	Temperature in K, Electric current, Bank account balance, Edge length	Electron count, Euler characteristic, Number of test failures

sometimes represented by digits, one must not interpret them as numbers. For example, the postal codes used in Germany are digits, but there is no meaning in, e.g., adding two postal codes. Similarly, nominal features do not have an ordering, i.e., the postal code 12345 is not "smaller" than the postal code 56789. Of course, most of the time there are options for how to introduce some kind of lexicographic sorting scheme, but this is purely artificial and has no meaning for the underlying objects.

With respect to statistics, the permissible average is not the mean (since summation is not allowed) or the median (since there is no ordering), but the mode, i.e., the most common value in the dataset.

2.1.2 Ordinal scale

The next higher scale is made of values on an ordinal scale. The ordinal scale allows comparing values w.r.t. equivalence and rank. Any transformation of the domain must preserve the order, which means that the transformation must be strictly increasing. But there is still no way to add an offset to one value in order to obtain a new value or to take the difference between two values.

Probably the best known example is school grades. In the German grading system, the grade 1 ("excellent") is better than 2 ("good"), which is better than 3 ("satisfactory") and so on. But quite surely the difference in a student's skills is not the same between the grades 1 and 2 as between 2 and 3, although the "difference" in the grades is unity in both cases. In addition, teachers often report the arithmetic mean of the grades in an exam, even though the arithmetic mean does not exist on the ordinal scale. In consequence, it is *syntactically* possible to compute the mean, even though the result, e.g., 2.47 has no place on the grading scale, other than it being "closer" to a 2 than a 3. The Anglo-Saxon grading system, which uses the letters "A" to "F", is somewhat immune to this confusion.

The correct average involving an ordinal scale is obtained by the median: the value that separates the lower half of the sample from the upper half. In other words, 50 % of the sample is smaller and 50 % is larger than the median. One can also measure the scatter of a dataset using the quantile distance. The p-quantile of a dataset is the value that separates the lower $p \cdot 100\,\%$ from the upper $(1-p) \cdot 100\,\%$ of the dataset (the median is the 0.5-quantile). The p-quantile distance is the distance (number of values) between the p and $(1-p)$-quantile. Common values for p are $p = 0$, which results in the range of the data set, and $p = 0.25$, which results in the inter-quartile range.

2.1.3 Interval scale

The interval scale allows adding an offset to one value to obtain a new one or to calculate the difference between two values—hence the name. However, the interval scale lacks a naturally defined zero. Values from the interval scale are typically represented using real numbers, which contains the symbol “0”, but this symbol has no special meaning and its position on the scale is arbitrary. For this reason, the scalar multiplication of two values from the interval scale is meaningless. Permissible transformations preserve the order, but may shift the position of the zero.

A prominent example is the (relative) temperature in °F and °C. The conversion from Celsius to Fahrenheit is given by $T_F = \frac{9\,°F}{5\,°C} T_C + 32\,°F$. The temperatures 10 °C and 20 °C on the Celsius scale correspond to 50 °F and 68 °F on the Fahrenheit scale. Hence, one cannot say that 20 °C is twice as warm as 10 °C: this statement does not hold w.r.t. the Fahrenheit scale.

The interval scale is the first of the discussed scales that allows computing the arithmetic mean and standard deviation.

2.1.4 Ratio scale and absolute scale

The ratio scale has a well defined, non-arbitrary zero and therefore allows calculating ratios of two values. This implies that there is a scalar multiplication and that any transformation must preserve the zero. Many features from the field of physics belong to this category and any transformation is merely a change of units. Note that although there is a semantically meaningful zero, this does not mean that features from this scale may not attain negative values. An example is one’s account balance, which has a defined zero (no money in the account), but may also become negative (open liabilities).

The absolute scale shares these properties, but is equipped with a natural unit and features of this scale can not be negative. In other words, features of the absolute scale represent counts of some quantities. Therefore, the only allowed transformation is the identity.

2.2 Feature space inspection

For a well working system, the question, how to find “good”, i.e., distinguishing features of objects, needs to be answered. The primary course of action is to visually inspect the feature space for good candidates.

In order to find discriminative features, one needs to get an idea about the structure of the feature space. In the one- or two-dimensional case, this can be easily done by looking at a visual representation of the dataset in question, e.g., a histogram or a scatter plot. Even with three dimensions, a perspective view of the data might suffice. However, this approach becomes problematic when the number of dimensions is larger than three.

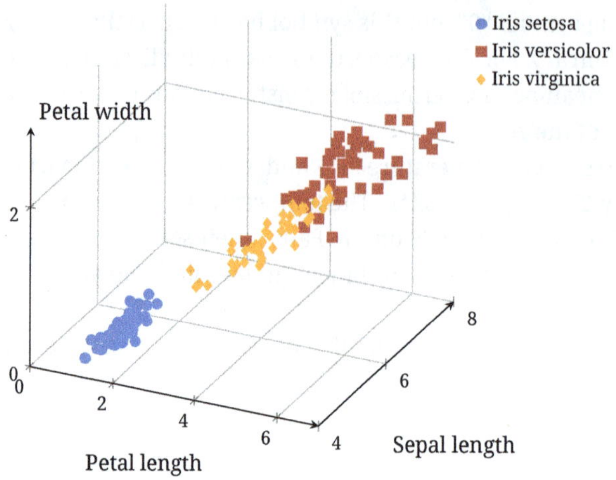

(a) Three-dimensional feature space

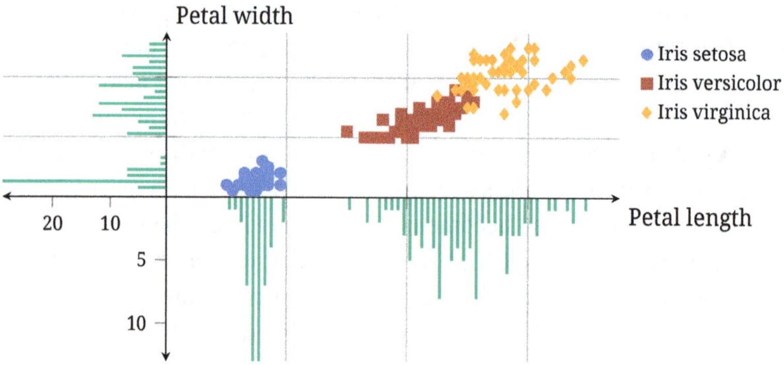

(b) Two-dimensional projection

Fig. 2.1: Iris flower dataset as an example of how projection helps the inspection of the feature space.

2.2.1 Projections

The simplest approach to inspecting high-dimensional feature spaces is to visualize every pair of dimensions of the dataset. More formally, the dataset is visualized by projecting the data onto a plane defined by pairs of basis vectors of the feature space. This approach works well if the data is rather cooperative. Figure 2.1 illustrates Fisher's Iris flower dataset, which quantifies the morphological variation of Iris flowers of three related species.

Figure 2.1a depicts a perspective drawing of the three features petal width, petal length and sepal length. Figure 2.1b shows a two-dimensional projection and two aligned histograms of the same data by omitting the sepal length. The latter clearly shows that the features petal length and petal width are already sufficient to distinguish the species

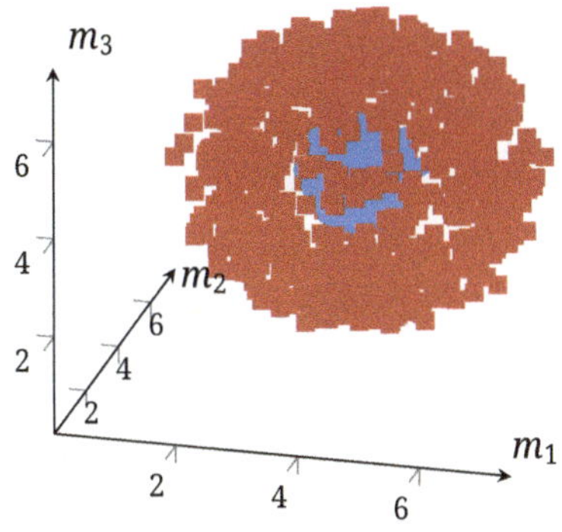

(a) 3D scatter plot of all samples

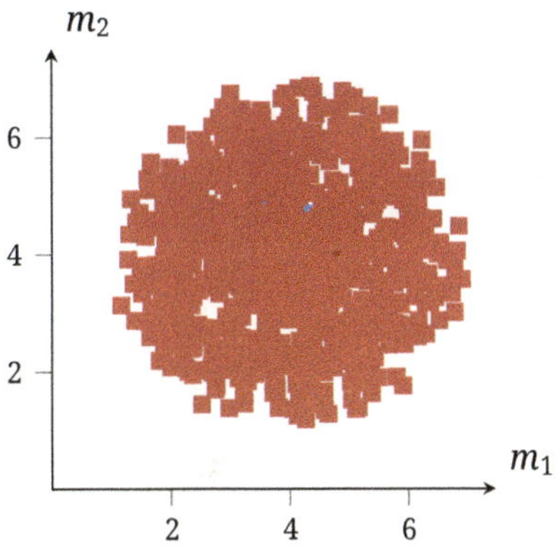

(b) Projection of all samples onto the m_1,m_2-plane

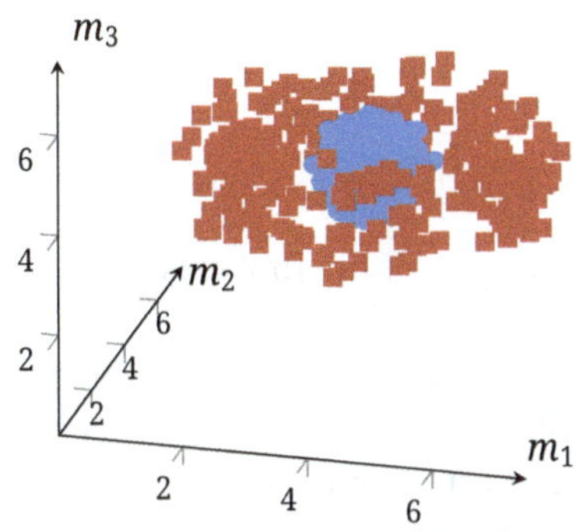

(c) 3D scatter plot of samples in a slice

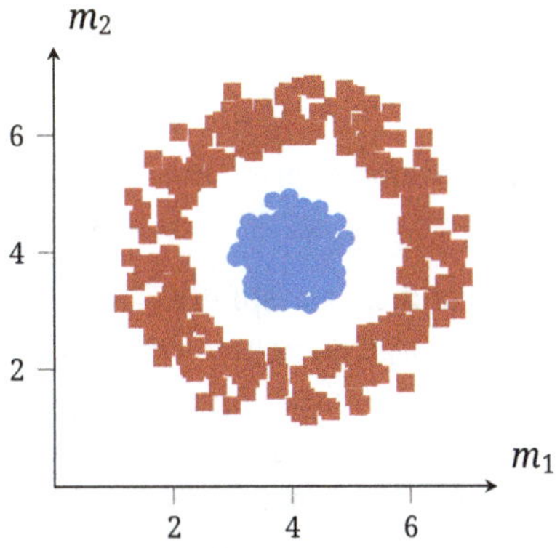

(d) Projection of samples in the slice onto the m_1,m_2-plane

Fig. 2.2: Difference between the full projection and the slice projection techniques.

Iris setosa from the others. Further two-dimensional projections might show that Iris versicolor and Iris virginica can also be easily separated from each other.

2.2.2 Intersections and slices

If the distribution of the samples in the feature space is more complex, simple projections might fail. Even worse, this approach might lead to the wrong conclusion that the samples of two different classes cannot be separated by the features in question even though they can be. Figure 2.2 shows this issue using artificial data. The objects of the first class are all distributed within a solid sphere. The samples of the second class lie close to the surface

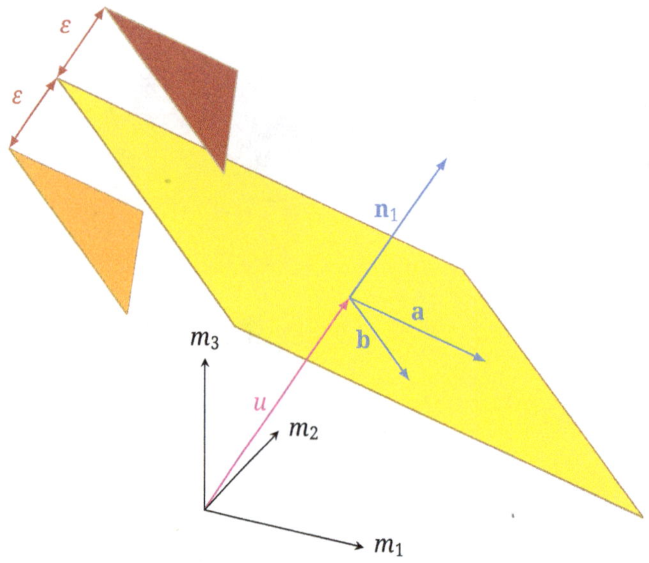

Fig. 2.3: Construction of two-dimensional slices.

of a second, larger sphere. This sphere encloses the samples of the first class, but the radius is large enough to separate the classes.

The initial situation is depicted in Figure 2.2a. Even though the samples can be separated, any projection to a two-dimensional subspace will suggest that the classes overlap each other, as shown in Figure 2.2b. However, if one projects slices of the data instead all of it at once, the structure becomes apparent. Figure 2.2c shows the result of such a slice in the three dimensional space and Figure 2.2d shows the corresponding projection. The latter clearly shows that one class only encloses the other but can be distinguished nonetheless.

The principal idea of the construction is illustrated in Figure 2.3. The slice is defined by its mean plane (yellow) and a bound ε that defines half of the thickness of the slice. Any sample that is located at a distance less than this bound is projected onto the plane. The mean plane itself is given by its two directional vectors **a**, **b** and its oriented distance u from the origin. The mean plane on its own, i.e., a slice with zero thickness ($\varepsilon = 0$), does not normally suffice to “catch” any sample points: If the samples are continuously distributed, the probability that a sample is intersected by the mean plane is zero.

Let $d \in \mathbb{N}$ be the dimension of the feature space. A two-dimensional plane is defined either by its two directional vectors **a** and **b** or as the intersection of $d - 2$ linearly independent hyperplanes. Hence, let

$$\{\mathbf{a}, \mathbf{b}, \mathbf{n}_1, \ldots, \mathbf{n}_{d-2}\} \tag{2.1}$$

denote an orthonormal basis of the feature space, where each $\mathbf{n}_j$ is the normal vector of a hyperplane. Let $u_1, \ldots, u_{d-2}$ be the oriented distances of the hyperplanes from the origin.

The two-dimensional plane is defined by the solution of the system of linear equations

$$\begin{aligned} \mathbf{n}_1^\mathsf{T}\mathbf{m} - u_1 &= 0 \\ &\vdots \\ \mathbf{n}_{d-2}^\mathsf{T}\mathbf{m} - u_{d-2} &= 0. \end{aligned} \tag{2.2}$$

Let $\mathbf{m} = (m_1, \ldots, m_d)^\mathsf{T}$ be an arbitrary point of the feature space. The distance of $\mathbf{m}$ from the plane in the direction of $\mathbf{n}_j$ is given by $\mathbf{n}_j^\mathsf{T}\mathbf{m} - u_j$, hence the total Euclidean distance of $\mathbf{m}$ from the plane is

$$v = \sqrt{\sum_{j=1}^{d-2} \left(\mathbf{n}_j^\mathsf{T}\mathbf{m} - u_j\right)^2}. \tag{2.3}$$

Let $\mathbf{m}_1, \ldots, \mathbf{m}_N$ be the feature vectors of the dataset. Then the two-dimensional projection of the feature vectors within the slice to the plane is given by

$$\left\{ \begin{pmatrix} \mathbf{a}^\mathsf{T}\mathbf{m}_i \\ \mathbf{b}^\mathsf{T}\mathbf{m}_i \end{pmatrix} \middle| \sqrt{\sum_{j=1}^{d-2} \left(\mathbf{n}_j^\mathsf{T}\mathbf{m}_i - u_j\right)^2} < \varepsilon \right\}. \tag{2.4}$$

2.3 Transformations of the feature space

Because the sample size is limited, it is usually advisable to restrict the number of features used. Apart from limiting the selection, this can also be achieved by a suitable transformation of the feature space (see Figure 2.4). In Figure 2.4a it is possible to separate the two classes using the feature m_1 alone. Hence, the feature m_2 is not needed and can be omitted. In Figure 2.4b, both features are needed, but the classes are separable by a straight line. Alternatively, the feature space could be rotated in such a way that the new feature m_2' is sufficient to discriminate between the classes. The annular classes in Figure 2.4c are not linearly separable, but a nonlinear transformation into polar coordinates shows that the classes can be separated by the radial component. Section 2.7 will present methods for automating such transformations to some degree. Especially the principal component analysis will play a central role.

2.4 Measurement of distances in the feature space

As will be shown in later chapters, many classifiers need to calculate some kind of distance between feature vectors. A very simple, yet surprisingly well-performing classifier is the so-called nearest neighbor classifier: Given a dataset with known points in the feature space and known class memberships for each point, a new point with unknown membership is assigned to the same class as the nearest known point. Obviously, the concept "being nearest to" requires a measure of distance.

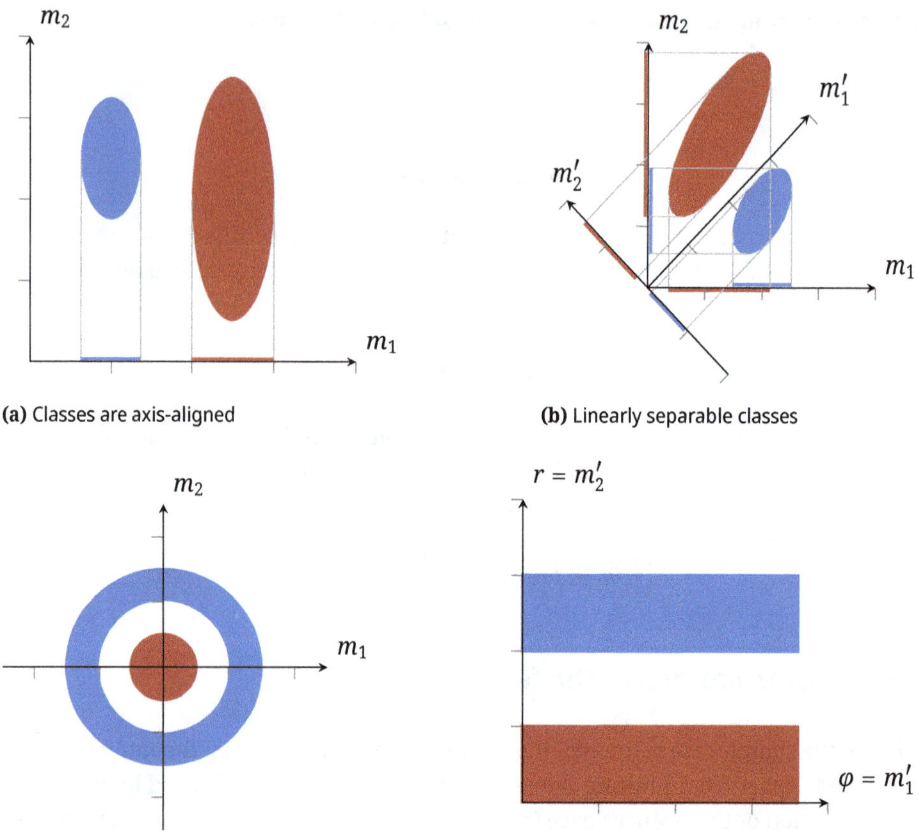

(a) Classes are axis-aligned

(b) Linearly separable classes

(c) Annular classes in Cartesian coordinates

(d) Annular classes in polar coordinates

Fig. 2.4: Feature transformation for dimensionality reduction.

If the feature vector was an element of a standard Euclidean vector space, one could use the well known Euclidean distance

$$\|\mathbf{m} - \mathbf{m}'\| = \sqrt{\sum_{i=1}^{d} |m_i - m'_i|^2}, \tag{2.5}$$

but this approach relies on some assumptions that are generally not true for real-world applications. The cause of this can be summarized by the *heterogeneity of the components of the feature space*, meaning

- features on different scales of measurement,
- features with different (physical) units,
- features with different meanings, and
- features with differences in magnitude.

Above all, Equation (2.5) requires that all components $m_i, m'_i, i = 1, \dots, d$ are at least on an interval scale. In practice, the components are often a mixture of real numbers, ordinal values and nominal values. In these cases, the Euclidean distance in Equation (2.5) does not make sense; even worse, it is syntactically incorrect.

In cases where all the components are real numbers, there is still the problem of different scales or units. For example, the same (physical) feature, "length", can be given in "inches" or "miles". The problem gets even worse if the components stem from different physical magnitudes, e.g., if the first component is a mass and the second component is a length. A simple solution to this problem is a weighted sum of the individual component distances, i.e.,

$$D\left(\mathbf{m}, \mathbf{m}'\right) = \sum_{i=1}^{d} \alpha_i D_i\left(m_i, m'_i\right) \quad \text{for } \alpha_i > 0 \text{ and } \sum_{i=1}^{d} \alpha_i = 1. \tag{2.6}$$

The coefficients $\alpha_1, \dots, \alpha_d$ handle the different units by containing the inverse of the component's unit, so that each summand becomes a dimensionless quantity. Nonetheless, the question of the difference in size is still an open problem and affords many free design parameters that must be carefully chosen.

Finally, the sum of squares (see Equation (2.5)) is not the only way to merge the different components into one distance value. The first section of this chapter introduces the more general Minkowski norms and metrics. The choice of metric can also influence a classifier's performance.

2.4.1 Basic definitions

To discuss the oncoming concepts, we must first define the terms that will be used.

Definition 2.1 (metric, metric space). Let $\mathcal{M}$ be a set and $\mathbf{m}$, $\mathbf{m}'$, and $\mathbf{m}'' \in \mathcal{M}$. A function $D : \mathcal{M} \times \mathcal{M} \to \mathbb{R}^{\geq 0}$ is called a *metric* iff

1. $D\left(\mathbf{m}, \mathbf{m}'\right) \geq 0$ (non-negativity)
2. $D\left(\mathbf{m}, \mathbf{m}'\right) = 0 \Leftrightarrow \mathbf{m} = \mathbf{m}'$ (reflexivity, coincidence)
3. $D\left(\mathbf{m}, \mathbf{m}'\right) = D\left(\mathbf{m}', \mathbf{m}\right)$ (symmetry)
4. $D\left(\mathbf{m}, \mathbf{m}''\right) \leq D\left(\mathbf{m}, \mathbf{m}'\right) + D\left(\mathbf{m}', \mathbf{m}''\right)$ (triangle inequality)

A set $\mathcal{M}$ equipped with a metric D is called a *metric space.*

With respect to real-world applications, having a metric feature space is an ideal, but unrealistic situation. Luckily, fewer requirements will often suffice. As will be seen in Section 2.4.6, the Kullback–Leibler divergence is not a metric because it lacks the symmetry property and violates the triangle inequality, but it is quite useful nonetheless. Those functions that fulfill some, but not all of the above requirements are usually called distance functions, discrepancys or divergences. None of these terms is precisely defined. Moreover, "distance function" is also used as a synonym for metric and should be avoided to prevent confusion. "Divergence" is generally only used for functions that quantify the difference between probability distributions, i.e., the term is used in a very specific context. Another important concept is given by the term (vector) norm:

Definition 2.2 (Norm, normed vector space). Let $\mathcal{M}$ be a vector space over the real numbers and let $\mathbf{m}, \mathbf{m}' \in \mathcal{M}$. A function $\|\cdot\| : \mathcal{M} \to \mathbb{R}^{\geq 0}$ is called a *norm* iff
1. $\|\mathbf{m}\| \geq 0$ and $\|\mathbf{m}\| = 0 \Leftrightarrow \mathbf{m} = 0$ (positive definiteness)
2. $\|\alpha\mathbf{m}\| = |\alpha|\, \|\mathbf{m}\|$ with $\alpha \in \mathbb{R}$ (homogeneity)
3. $\|\mathbf{m} + \mathbf{m}'\| \leq \|\mathbf{m}\| + \|\mathbf{m}'\|$ (triangle inequality)

A vector space $\mathcal{M}$ equipped with a norm $\|\cdot\|$ is called a *normed vector space.*

Due to the prerequisite of the definition, a normed vector space can only be applied to features that are at least on a ratio scale. A norm can be used to construct a metric, which means that every normed vector space is a metric space, too.

Definition 2.3 (Induced metric). Let $\mathcal{M}$ be a normed vector space and $\|\cdot\|$ its norm and let $\mathbf{m}, \mathbf{m}' \in \mathcal{M}$. Then

$$D\left(\mathbf{m}, \mathbf{m}'\right) := \left\|\mathbf{m} - \mathbf{m}'\right\| \tag{2.7}$$

defines an *induced metric* on $\mathcal{M}$.

Note that because of the homogeneity property, Definition 2.2 requires the value to be on a ratio scale; otherwise the scalar multiplication would not be well defined. However, the induced metric from Definition 2.3 can be applied to an interval scale, too, because the proof does not need the scalar multiplication. Of course, one must not say that the metric $D(\mathbf{m}, \mathbf{m}') = \|\mathbf{m} - \mathbf{m}'\|$ stems from a norm, because there is no such thing as a norm on an interval scale.

2.4.2 Elementary norms and metrics

Inarguably, the most familiar example of a norm is the Euclidean norm. But this norm is just a special embodiment of a whole family of vector norms that can be used to quantify the distance of features on a ratio scale. The norms of this family are called Minkowski norms or p-norms.

Definition 2.4 (Minkowski norm, p-norm). Let $\mathcal{M}$ denote a real vector space of finite dimension d and let $r \in \mathbb{Z} \cup \{\infty\}$ be a constant parameter. Then

$$\|\mathbf{m}\|_r = \begin{cases} \left(\sum_{i=1}^{d} |m_i|^r\right)^{\frac{1}{r}} & \text{if } r < \infty \\ \max_{i=1}^{d} |m_i| & \text{if } r = \infty \end{cases} \tag{2.8}$$

is a norm on $\mathcal{M}$.

The name "p-norm" comes from the fact that the parameter is traditionally denoted by p and not r as seen here. This book uses r to avoid a clash of names, because p is already used to denote a probability density function.

Although r can be any integer or infinity, only a few choices are of greater importance. For $r = 2$

$$\|\mathbf{m}\|_e = \|\mathbf{m}\|_2 = \sqrt{\sum_{i=1}^{d} |m_i|^2} \tag{2.9}$$

is the Euclidean norm. Likewise, $r = 1$ yields the absolute norm

$$\|\mathbf{m}\|_1 = \sum_{i=1}^{d} |m_i|. \tag{2.10}$$

This norm—or more precisely: the induced metric—is also known as taxicab metric or Manhattan metric. One can visualize this metric as the distance that a car must go between two points of a city with a rectilinear grid of streets like in Manhattan. For $r = \infty$ the resulting norm

$$\|\mathbf{m}\|_T = \|\mathbf{m}\|_\infty = \max_{i=1}^{d} |m_i| \tag{2.11}$$

is called maximum norm or Chebyshev norm. Figure 2.5 depicts the unit circles for different choices of r in the upper right quadrant of the two-dimensional Euclidean space. Furthermore, the Mahalanobis norm is another common metric for real vector spaces:

Definition 2.5 (Mahalanobis norm). Let $\mathcal{M}$ denote a real vector space of finite dimension d and let $\mathbf{A} \in \mathbb{R}^{d \times d}$ be a positive definite matrix. Then

$$\|\mathbf{m}\|_m = \sqrt{\mathbf{m}^\top \mathbf{A} \mathbf{m}} \tag{2.12}$$

is a norm on $\mathcal{M}$.

To a certain degree, the Mahalanobis norm is another way to generalize the Euclidean norm: they coincide for $\mathbf{A} = \mathbf{I}_d$. More generally, elements A_{ii} on the diagonal of $\mathbf{A}$ can be thought of as scaling the corresponding dimension i, while off-diagonal elements $A_{ij}, i \neq j$ assess the dependence between the dimension i and j. The Mahalanobis also appears in the multivariate normal distribution (see Definition 3.4), where the matrix $\mathbf{A}$ is the inverse of the covariance $\boldsymbol{\Sigma}$ of the data.

So far only norms and their induced metrics that require at least an interval scale were considered. The metrics handle all quantitative scales of Table 2.1. The next sections will introduce metrics for features on other scales.

2.4.3 A metric for sets

Lets assume one has a finite set $\mathcal{U}$ and the features in question are subsets of $\mathcal{U}$. In other words, the feature space $\mathcal{M}$ is the power set $\mathfrak{P}(\mathcal{U})$ of $\mathcal{U}$. On the one hand the features are clearly not ordinal, because the relation "$\subseteq$" induces only a partial order. Of course, it is possible to artificially define an ad hoc total order because $\mathcal{M}$ is finite, but the focus shall

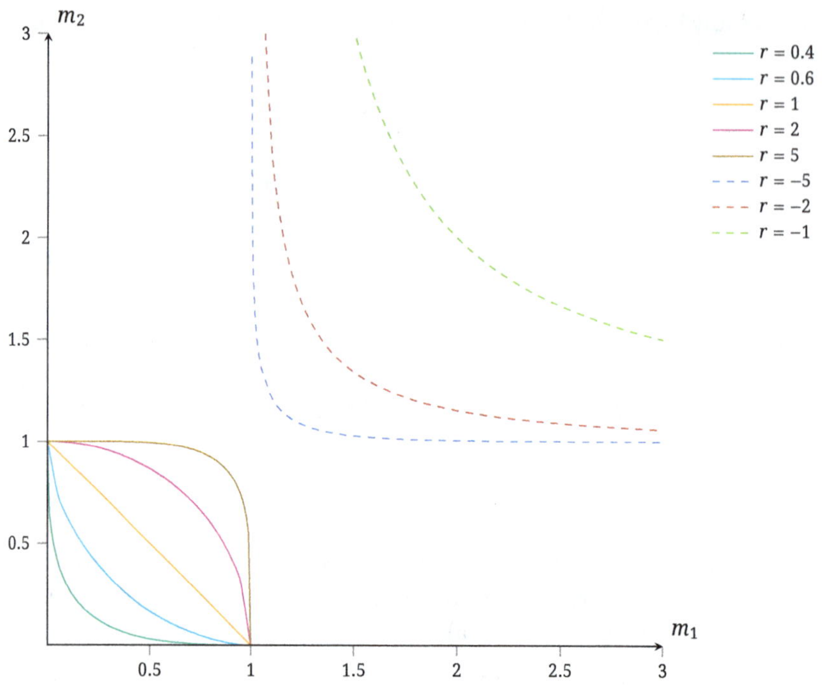

Fig. 2.5: Unit circles for Minkowski norms with different choices of r. Only the upper right quadrant of the two-dimensional Euclidean space is shown.

remain on generally meaningful metrics. On the other hand, a mere nominal feature only allows to state if two values (here: two sets) are equal or not. However, two sets can also be said to be "nearly equal" when both the intersection and the set difference is non-empty (i.e., they share some, but not all elements). The Tanimoto metric reflects these situations.

Definition 2.6 (Tanimoto metric). Let $\mathcal{U}$ be a finite set, $\mathcal{M} = \mathfrak{P}(\mathcal{U})$ and $\mathcal{S}_1, \mathcal{S}_2 \in \mathcal{M}$, i.e., $\mathcal{S}_1, \mathcal{S}_2 \subseteq \mathcal{U}$. Then

$$D_{\text{Tanimoto}}(\mathcal{S}_1, \mathcal{S}_2) = \frac{|\mathcal{S}_1| + |\mathcal{S}_2| - 2\,|\mathcal{S}_1 \cap \mathcal{S}_2|}{|\mathcal{S}_1| + |\mathcal{S}_2| - |\mathcal{S}_1 \cap \mathcal{S}_2|} \in [0, 1] \tag{2.13}$$

defines a metric on $\mathcal{M}$.

Here, we will omit the proof that D_{Tanimoto} is indeed a metric (interested readers are referred to, e.g., the proof of Lipkus [1999]) and instead investigate its properties. If $\mathcal{S}_1$ and $\mathcal{S}_2$ denote the same set, then $|\mathcal{S}_1| = |\mathcal{S}_2| = |\mathcal{S}_1 \cap \mathcal{S}_2|$ and therefore $D_{\text{Tanimoto}}(\mathcal{S}_1, \mathcal{S}_2) = 0$. Contrary, if $\mathcal{S}_1$ and $\mathcal{S}_2$ do not have any element in common, $|\mathcal{S}_1 \cap \mathcal{S}_2| = 0$ holds and $D_{\text{Tanimoto}}(\mathcal{S}_1, \mathcal{S}_2) = 1$. Altogether, the Tanimoto metric varies on the interval from 0 (identical) to 1 (completely different).

Moreover, the Tanimoto metric takes the overall number of elements into account. Two sets that differ in one element are judged to be increasingly similar, as the number of shared elements increases. For example, let $\mathcal{U} = \{a, \ldots, z\}$, $\mathcal{S}_1 = \{a, b, c\}$, $\mathcal{S}_2 = \{a, b, d\}$, $\mathcal{S}'_1 = \{a, b, d, e, f\}$ and $\mathcal{S}'_2 = \{a, b, d, e, g\}$. It follows, that

$$D_{\text{Tanimoto}}(\mathcal{S}_1, \mathcal{S}_2) = \frac{3+3-4}{3+3-2} = \frac{1}{2} \quad \text{and} \tag{2.14}$$

$$D_{\text{Tanimoto}}\left(\mathcal{S}'_1, \mathcal{S}'_2\right) = \frac{5+5-8}{5+5-4} = \frac{1}{3}. \tag{2.15}$$

2.4.4 Metrics on the ordinal scale

It is not immediately clear how to define a meaningful metric for ordinal features, since there is no empirical addition on that scale. A possible solution is to consider the metric $D(m,m')$ of two ordinal features m, m' as the number of swaps of neighboring elements in order to reach m' from m.

Consider, for example, the set of characters in the English language {A, B, C, . . . , Z}, where the order corresponds to the position in the alphabet. The metric informally defined above would yield $D(\text{A},\text{C}) = 2$ and $D(\text{A},\text{A}) = 0$. This example can be generalized as follows:

Definition 2.7 (Permutation metric). Let $\mathcal{M}$ be a locally finite and totally ordered set with a unique successor function, i.e., for each element $x \in \mathcal{M}$ there is a unique next element $x' \in \mathcal{M}$. Then

$$D(x, x') = \text{smallest number of permutations of successive elements to get from } x \text{ to } x' \tag{2.16}$$

is a metric on $\mathcal{M}$.

Another way to look at Definition 2.7 is to homomorphically map $\mathcal{M}$ into the integers (i.e., successive elements of $\mathcal{M}$ are mapped to successive integers) and calculate the absolute difference of the numbers corresponding to the sets.

2.4.5 The cosine distance

Despite not being a metric because the triangle inequality is not fulfilled, the cosine distance is an important measure for vector differences. In contrast to considering the absolute values of vectors as in the Minkowski norms, the enclosing angle of two vectors is considered:

Definition 2.8 (Cosine distance). Let $\mathcal{M}$ denote a real vector space of finite dimension d with vectors $\mathbf{m}, \mathbf{m}' \in \mathcal{M}$ and let φ be the angle enclosed by $\mathbf{m}$ and $\mathbf{m}'$. Then

$$D_{\text{cosine}}\left(\mathbf{m}, \mathbf{m}'\right) = 1 - \cos(\varphi) = 1 - \left(\frac{\mathbf{m}^\top \mathbf{m}'}{\|\mathbf{m}\|_{\mathrm{e}} \|\mathbf{m}'\|_{\mathrm{e}}}\right) \tag{2.17}$$

is the cosine distance between $\mathbf{m}$ and $\mathbf{m}'$.

Note that the term $\left(\frac{\mathbf{m}^\top \mathbf{m}'}{\|\mathbf{m}\|_{\mathrm{e}} \|\mathbf{m}'\|_{\mathrm{e}}}\right)$ is often referred to as cosine similarity, which is zero when $\mathbf{m}$ and $\mathbf{m}'$ are orthogonal to each other. The cosine similarity reaches its maximum value of 1 if $\mathbf{m}$ and $\mathbf{m}'$ point to the same direction and its minimum value of -1 if they point in the opposite direction. Consequently, the cosine distance is restricted to $[0,2]$. As we will see in Section 6.1, the cosine distance is especially a good choice for measuring differences in high-dimensional vector spaces.

2.4.6 The Kullback–Leibler divergence

The Kullback–Leibler divergence (KL divergence) does not directly quantify a difference between features $\mathbf{m}$, but between probability distributions (characterized by the probability mass function or the probability density) over the features. It is often used as a meta metric to compare objects o_i, $i = 1, \ldots, N$ that are in turn characterized by a set of features $\mathcal{O}_i = \{\mathbf{m}_j \mid j = 1, \ldots, M_i\}$. To this extent, the features in $\mathcal{O}_i$ are used to estimate the probability mass $\hat{P}(\mathbf{m} \mid o_i)$ or probability density $\hat{p}(\mathbf{m} \mid o_i)$ for each object o_i. The KL divergence is then used to compute the distance between two object-dependent distributions and by proxy the distance between two objects. Here, the $\hat{P}(\mathbf{m} \mid o_i)$ or $\hat{p}(\mathbf{m} \mid o_i)$ can themselves be interpreted as features for o_i. An extended example of this approach is given below.

Definition 2.9 (Kullback–Leibler divergence).

1. Let P, P' be two probability mass functions on the same space $\mathcal{M}$. The *Kullback–Leibler divergence* of P' with respect to P is given by

$$D\left(P \| P'\right) = \sum_{\mathbf{m} \in \operatorname{supp} P} P(\mathbf{m}) \ln \frac{P(\mathbf{m})}{P'(\mathbf{m})}. \tag{2.18}$$

2. Let p, p' be two probability density functions on the same space $\mathcal{M}$. The *Kullback–Leibler* divergence of p' with respect to p is given by

$$D\left(p \| p'\right) = \int \cdots \int_{\operatorname{supp} p} p(\mathbf{m}) \ln \frac{p(\mathbf{m})}{p'(\mathbf{m})} \, \mathrm{d}\mathbf{m}. \tag{2.19}$$

The Kullback–Leibler divergence is not a metric: As can be seen from definition, it is not symmetric and does not obey the triangle inequality. If $P'(\mathbf{m}) = 0$ and $P(\mathbf{m}) \neq 0$, one sets $D(P\|P') = \infty$, hence the range of the KL divergence is $[0, \infty]$. The Kullback–Leibler divergence is zero if the distributions are equal almost everywhere.

Despite the fact that the Kullback–Leibler divergence is missing important properties of a metric, it is still very useful, because it dominates the difference of the probability density functions with respect to the absolute norm (Minkowski norm with $r = 1$) by

$$D(p\|p') \geq \frac{1}{2\log_2 \mathrm{e}} \left(\int_{\mathcal{M}} \cdots \int \left| p(\mathbf{m}) - p'(\mathbf{m}) \right| \mathrm{d}\mathbf{m} \right)^2 = \frac{1}{2\log_2 \mathrm{e}} \left\| p - p' \right\|_1^2 . \tag{2.20}$$

The Kullback–Leibler divergence can be interpreted as the expectation of the logarithm of the so called likelihood ratio $\frac{p(\mathbf{m})}{p'(\mathbf{m})}$ based on features that are distributed according to $p(\mathbf{m})$,

$$D(p\|p') = \mathrm{E}\left\{ \ln \frac{p(\underline{\mathbf{m}})}{p'(\underline{\mathbf{m}})} \right\} . \tag{2.21}$$

The likelihood ratio is the crucial quantity of optimal statistical tests to decide between two competing hypotheses $H_1 : \underline{\mathbf{m}} \sim p(\mathbf{m})$ and $H_2 : \underline{\mathbf{m}} \sim p'(\mathbf{m})$ (Neyman and Pearson [1992]). In other words, the Kullback–Leibler divergence measures the mean discriminability between H_1 and H_2.

To become accustomed to the Kullback–Leibler divergence, we will now discuss some simple examples. First, consider the family of Bernoulli distributions parametrized by the probability of success $\tau \in [0,1]$. That is,

$$\Pr(\text{Success}) = \tau \quad \text{and} \quad \Pr(\text{Failure}) = 1 - \tau . \tag{2.22}$$

The Kullback–Leibler divergence between two different Bernoulli distributions is therefore given by

$$D(P_a\|P_b) = a \ln \frac{a}{b} + (1 - a) \ln \frac{1 - a}{1 - b} . \tag{2.23}$$

Figure 2.6 depicts the value of the Kullback–Leibler divergence as a function of the success probability b of the second distribution while keeping the success probability of the first distribution fixed at $a = 0.3$.

Now, as an example of the continuous case, consider the Kullback–Leibler divergence of two univariate Gaussian distributions. The Gaussian density function is

$$p(m \mid \mu,\sigma) = \frac{1}{\sqrt{2\pi}\sigma} \exp\left(-\frac{1}{2\sigma^2} (m - \mu)^2 \right) \tag{2.24}$$

and hence for two Gaussian distributions with parameters μ_1,σ_1 and μ_2,σ_2, one obtains

$$D(p_1\|p_2) = \int_{\mathbb{R}} p(m \mid \mu_1,\sigma_1) \ln \frac{\left(2\pi\sigma_1^2\right)^{-\frac{1}{2}} \exp\left(-\frac{1}{2\sigma_1^2} (m - \mu_1)^2 \right)}{\left(2\pi\sigma_2^2\right)^{-\frac{1}{2}} \exp\left(-\frac{1}{2\sigma_2^2} (m - \mu_2)^2 \right)} \, dm$$

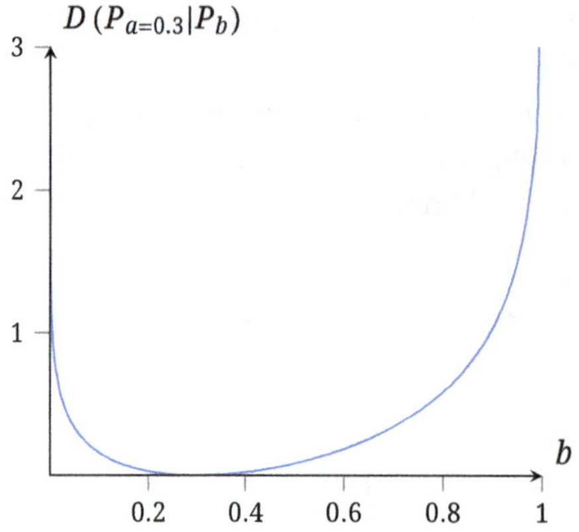

Fig. 2.6: Kullback–Leibler divergence between two Bernoulli distributions with fixed success probability $a = 0.3$ for the first distribution.

$$\begin{aligned}
&= -\frac{1}{2}\ln\frac{\sigma_1^2}{\sigma_2^2} - \frac{1}{2\sigma_1^2}\int_{\mathbb{R}} p(m\,|\,\mu_1,\sigma_1)(m-\mu_1)^2\,\mathrm{d}m \\
&\qquad + \frac{1}{2\sigma_2^2}\int_{\mathbb{R}} p(m\,|\,\mu_1,\sigma_1)(m-\mu_2)^2\,\mathrm{d}m \\
&= -\frac{1}{2}\ln\frac{\sigma_1^2}{\sigma_2^2} - \frac{\sigma_1^2}{2\sigma_1^2} + \frac{\sigma_1^2+(\mu_1-\mu_2)^2}{2\sigma_2^2} \\
&= \frac{1}{2}\left(\frac{\sigma_1^2}{\sigma_2^2} - \ln\frac{\sigma_1^2}{\sigma_2^2} - 1\right) + \frac{(\mu_1-\mu_2)^2}{2\sigma_2^2}.
\end{aligned} \tag{2.25}$$

It is interesting to note that the Kullback–Leibler divergence becomes symmetric if the family of distributions is further restricted to Gaussian distributions with equal variances, i.e., $\sigma_1^2 = \sigma_2^2 = \sigma^2$, i.e., when $p_1(m\,|\,\mu_1,\sigma)$, $p_2(m\,|\,\mu_2,\sigma)$, one has that $D(p_1\|p_2) = D(p_2\|p_1)$. Figure 2.7 shows two pairs of Gaussian distributions with equal variance. The corresponding Kullback–Leibler divergence is noted in the diagram. In order to illustrate the asymmetric behavior of the Kullback–Leibler divergence, Figure 2.8 depicts a pair of Gaussian distributions where p_{μ_1,σ_1} and p_{μ_2,σ_2} have swapped roles between the diagrams. Once again, the Kullback–Leibler divergence is given in the diagram. As a last example, a pair of rectangle-like densities with unequal support is considered. This example illustrates that the Kullback–Leibler divergence can even take on the value infinity depending on the order of its arguments (see Figure 2.9).

Extended example: Grading of honing textures

The Kullback–Leibler divergence can be used to derive robust distance measures between features. Consider, for example, the problem of grading honing textures in the cylinder bores of combustion engines, shown in Figure 2.10. This honing texture is the result of a grinding tool rotating around its axis while oscillating inside the cylinder hole. The resulting grooves' function is to retain lubricant while the pistons move inside

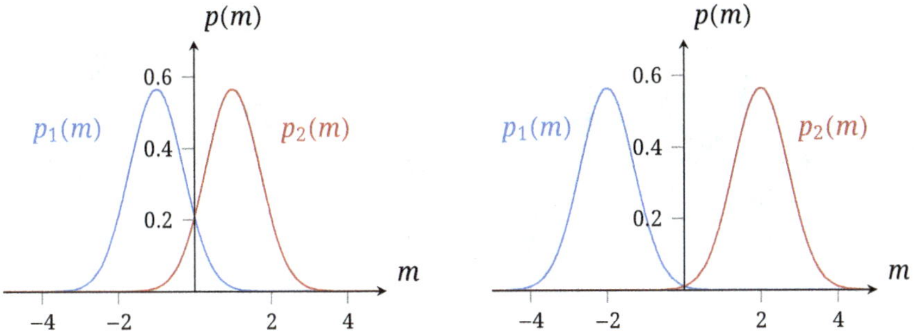

(a) $\mu_1 = -1$, $\mu_2 = 1$, $\sigma^2 = 0.5$ and $D_{\mathrm{KL}} = 4$

(b) $\mu_1 = -2$, $\mu_2 = 2$, $\sigma^2 = 0.5$ and $D_{\mathrm{KL}} = 16$

Fig. 2.7: Pairs of Gaussian distributions with equal variance $\sigma^2 = 0.5$ and their KL divergences.

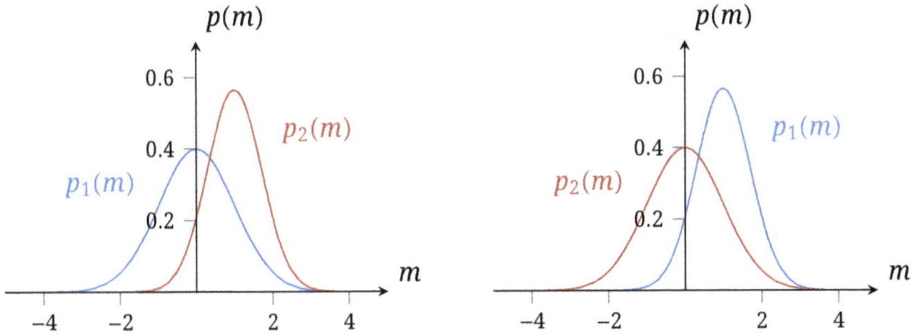

(a) $\mu_1 = 0$, $\sigma_1^2 = 1$, $\mu_2 = 1$, $\sigma_2^2 = 0.5$ and $D_{\mathrm{KL}} = 1.153$

(b) $\mu_1 = 1$, $\sigma_1^2 = 0.5$, $\mu_2 = 0$, $\sigma_2^2 = 1$ and $D_{\mathrm{KL}} = 0.597$

Fig. 2.8: Pairs of Gaussian distribution with different variances and their KL divergences $D(p_1||p_2)$.

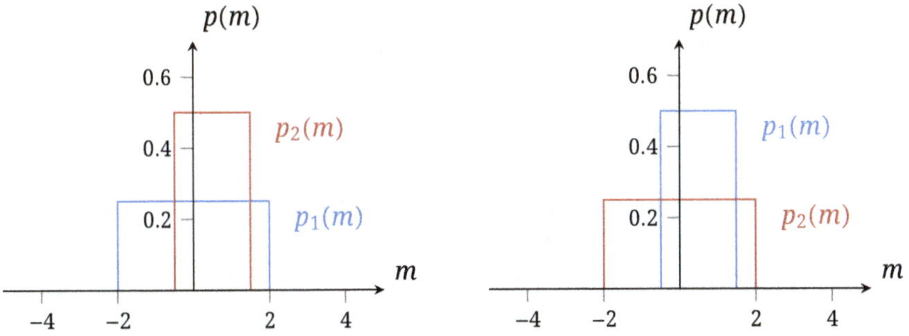

(a) $D_{\mathrm{KL}} = \infty$

(b) $D_{\mathrm{KL}} < \infty$

Fig. 2.9: Pairs of rectangle-like densities and their KL divergences $D(p_1||p_2)$.

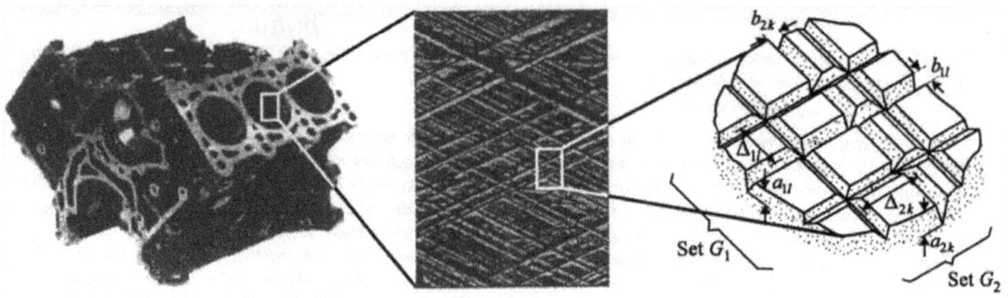

Fig. 2.10: Combustion engine, microscopic image of bore texture, and detail with texture model with groove parameters. Source: Krahe and Beyerer [1997].

the motor block. A single groove in the honing texture is characterized by its amplitude (depth) a and width b. Let further Δ denote the distance between two given adjacent grooves. From Figure 2.10 it can be seen that the grooves can be divided into two sets, depending on their orientation. The question is: how similar are these sets of grooves?

One possible approach is to model the parameters $\mathbf{u} := (a,b)^\mathsf{T}$ and Δ of the two groove sets stochastically. In particular, let us assume that Δ and $\mathbf{u}$ are independent. It is known that the stride Δ follows an exponential distribution and it is sensible to assume that the groove depth or amplitude a and width b are jointly normally distributed. This yields the overall density

$$\begin{aligned} p_i(\mathbf{u},\Delta) &= p_{\Delta i}(\Delta) \cdot p_{ui}(\mathbf{u}) \\ &= \lambda_i \exp(-\lambda_i \Delta) \frac{1}{2\pi\sqrt{\det(\mathbf{C}_i)}} \exp\left(-\frac{(\mathbf{u}-\boldsymbol{\mu}_i)^\mathsf{T}\mathbf{C}_i^{-1}(\mathbf{u}-\boldsymbol{\mu}_i)}{2}\right), \end{aligned} \tag{2.26}$$

where $i = 1, 2$ indicates the first or second set. Here, $\boldsymbol{\mu}_i$ is the expected value of $\mathbf{u}$ in the ith groove set,

$$\boldsymbol{\mu}_i = \mathrm{E}_i\{\underline{\mathbf{u}}\} = \iiint \mathbf{u} p_i(\mathbf{u},\Delta)\,\mathrm{d}\mathbf{u}\,\mathrm{d}\Delta = \begin{pmatrix} \mu_{a_i} \\ \mu_{b_i} \end{pmatrix}. \tag{2.27}$$

With ρ_i denoting the correlation coefficient between a_i and b_i,

$$\rho_i = \frac{\mathrm{E}\{(\underline{a}_{ij} - \mu_{a_i})(\underline{b}_{ij} - \mu_{b_i})\}}{\sigma_{a_i}\sigma_{b_i}}. \tag{2.28}$$

Now let $\mathbf{C}_i$ be the covariance matrix of $\mathbf{u}$ in the ith groove set:

$$\mathbf{C}_i = \mathrm{E}\left\{(\underline{\mathbf{u}} - \boldsymbol{\mu}_i)(\underline{\mathbf{u}} - \boldsymbol{\mu}_i)^\mathsf{T}\right\} = \begin{pmatrix} \sigma_{a_i}^2 & \rho_i\sigma_{a_i}\sigma_{b_i} \\ \rho_i\sigma_{a_i}\sigma_{b_i} & \sigma_{b_i}^2 \end{pmatrix}. \tag{2.29}$$

The parameter λ_i denotes the groove density in the ith set, i.e.,

$$\lambda_i = \frac{1}{\mathrm{E}\{\underline{\Delta}_{ij}\}}. \tag{2.30}$$

This model will be used to construct a measure of the distance between two groove sets. Recall from Definition 2.9 that the KL divergence between p_1 and p_2 is asymmetric. In order to derive a symmetric measure of the distance between two groove sets, we simply take the sum of the KL divergence between p_1 and p_2, and the KL divergence with these arguments transposed:

$$\begin{aligned} D_{ab\Delta} &:= D(p_1 \| p_2) + D(p_2 \| p_1) \\ &= \iiint p_1(\mathbf{u},\Delta) \ln \frac{p_1(\mathbf{u},\Delta)}{p_2(\mathbf{u},\Delta)} \, \mathrm{d}\mathbf{u} \, \mathrm{d}\Delta + \iiint p_2(\mathbf{u},\Delta) \ln \frac{p_2(\mathbf{u},\Delta)}{p_1(\mathbf{u},\Delta)} \, \mathrm{d}\mathbf{u} \, \mathrm{d}\Delta. \end{aligned} \tag{2.31}$$

After a fair bit of algebra, this expression simplifies to the sum of three terms:

$$\begin{aligned} D_{ab\Delta} = &\frac{1}{2}(\boldsymbol{\mu}_1 - \boldsymbol{\mu}_2)^\mathsf{T} \left(\mathbf{C}_1^{-1} + \mathbf{C}_2^{-1}\right)(\boldsymbol{\mu}_1 - \boldsymbol{\mu}_2) && \text{(Mahalanobis distance of means)} \\ &+ \frac{1}{2} \operatorname{tr}\left(\mathbf{C}_1^{-1}\mathbf{C}_2 + \mathbf{C}_2^{-2}\mathbf{C}_1\right) - 2 && \text{(distance of covariances)} \\ &+ \frac{\lambda_1}{\lambda_2} + \frac{\lambda_2}{\lambda_1} - 2. && \text{(distance of groove densities)} \end{aligned} \tag{2.32}$$

This metric can be used to compare two sets of grooves as follows. First, measure the amplitudes a, depths b, and distances Δ of the grooves in the set. Second, estimate the parameters $\boldsymbol{\mu}_i$, $\mathbf{C}_i$ and λ_i, $i = 1,2$ for both sets according to Equations (2.27), (2.29) and (2.30). Third, compute the distance between the two sets by using Equation (2.32). If the distance is below a given threshold, the honing texture passes the inspection. Otherwise, it is rejected as of insufficient quality.

2.4.7 The *t*-distributed stochastic neighbor embedding

The graphical representation of high-dimensional data for human visual perception is restricted to dimensionalities less or equal three. Human cognition and intuition based on experiences in low-dimensional spaces cannot be generalized to high-dimensional spaces and can lead to misleading imagination and wrong conclusions. Another phenomenon is the *curse of dimensionality* (cf. Section 6.1), which generally describes the problem of rapid sparsification of data in higher dimensions. Therefore, we will learn a number of methods to reduce the dimensionality of feature spaces in order to overcome these visualization limitations. However, the Kullback–Leibler divergence can be used directly as main driver for the visualization of high-dimensional data. The corresponding method is called *t-distributed stochastic neighbor embedding* (t-SNE) (van der Maaten and Hinton [2008]). The objective is to create a distribution P' in a lower dimensional space $\mathcal{M}' \subseteq \mathbb{R}^{d'}$ where the proximity of low-dimensional data points reflects their similarities in the original high-dimensional space $\mathcal{M} \subseteq \mathbb{R}^d$ of the corresponding distribution P. Such a distribution P' can then be found by minimizing some objective function that measures the discrepancy between the two distributions P and P'.

In order to do that, t-SNE first uses a Gaussian representation for measuring similarities between data points $\mathbf{m}_i, \mathbf{m}_j$ in the high-dimensional space $\mathcal{M}$:

$$p_{i|j} = p(\mathbf{m}_i|\mathbf{m}_j) := \frac{\exp(-(\|\mathbf{m}_i - \mathbf{m}_j\|^2)/2\sigma_i^2}{\sum_{i\neq j}\exp(-(\|\mathbf{m}_i - \mathbf{m}_j\|^2)/2\sigma_i^2}. \tag{2.33}$$

The values $p_{i|j}$ are therefore probabilities which express the similarity between two data points. When two data points $\mathbf{m}_i, \mathbf{m}_j$ are close within the high-dimensional space, the value of $p_{i|j}$ will be high. Conversely, if they are located far apart, the value $p_{i|j}$ will be low. This is achieved by centering a Gaussian at each data point and measuring the density of all other points in its local neighborhood (numerator of Equation (2.33)). The denominator of Equation (2.33) basically normalizes the probability. To satisfy the conditions of a probability distribution, $\sum_i p_{i|j} = 1$ must hold. The bandwidth σ_i can be interpreted as design parameter, which controls the balance between preserving local or global structures in the original data. Also, the final pairs are symmetrized by summing and averaging the two involved conditionals

$$p_{ij} = \frac{p_{j|i} + p_{i|j}}{2N} \tag{2.34}$$

with N being the number of data points.

The approach used in the low-dimensional space $\mathcal{M}'$ is similar, but instead of using a Gaussian distribution, the t-distribution with one degree of freedom (which is equal to a Cauchy distribution) is utilized to implicitly model the data:

$$p'_{i|j} = p'(\mathbf{m}'_i|\mathbf{m}'_j) := \frac{(1 + \|\mathbf{m}'_i - \mathbf{m}'_j\|^2)^{-1}}{\sum_{i\neq j}(1 + \|\mathbf{m}'_i - \mathbf{m}'_j\|^2)^{-1}}. \tag{2.35}$$

The utilized student t-distribution, eponym of the approach, is heavy-tailed and is therefore capable of distributing the data points effectively. Note that the p'_{ij} cannot be calculated explicitly. Instead, the goal is to iteratively place the data points $\mathbf{m}'_i$ in a way that minimizes the distance between the p_{ij} and the p'_{ij}. This distance D_{KL} is expressed using the KL divergence between the two probability distributions P and P':

$$D_{\mathrm{KL}}(P\|P') = \sum_i \sum_j p_{ij} \log \frac{p_{ij}}{p'_{ij}}. \tag{2.36}$$

The KL divergence is an ideal measure for quantifying the dissimilarity between distributions in low- and high-dimensional feature spaces. It ensures that large values of p_{ij} in the high-dimensional feature space will correspond to large values of p'_{ij} in the low-dimensional feature space. However, due to the asymmetry of the KL divergence this does not necessarily hold for the other way: small values for p_{ij} mapped to larger values p'_{ij} will not result in large penalties in Equation (2.36). Therefore, t-SNE aims on maintaining the *local similarity structures* of the data. We will further explore the concrete algorithm of t-SNE and relevant examples in Section 2.7.5.

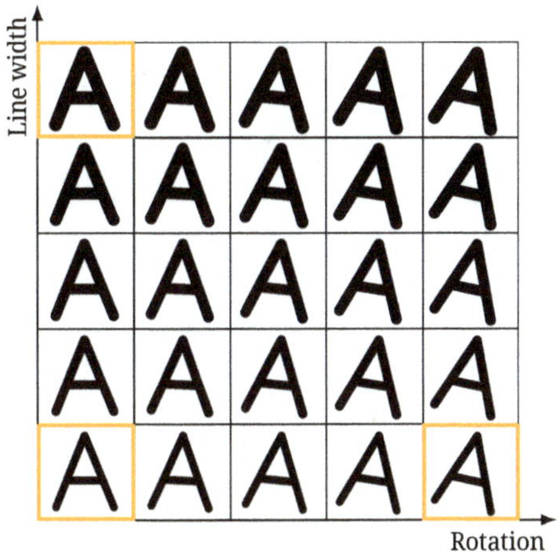

Fig. 2.11: Systematic variations in optical character recognition.

2.4.8 Tangential distance measure

To conclude this section on metrics, we will discuss the tangential distance measure. This method does not introduce a new metric, but rather builds on top of a given one and makes this metric more robust against small, systematic disturbances of the feature vectors that may be caused by varying lighting conditions, out of focus images, small rotations of the pattern, etc. The key is that these disturbances should be systematic and not due to random noise.

Consider, for example, the problem of optical character recognition. Figure 2.11 shows two possible systematic variations of a character (the pattern): rotation and line thickness. This variation causes differing patterns and will therefore result in different feature vectors. However, since the variations in the pattern are systematic, so will the variations in the feature vectors. More precisely, small variations of the pattern will move the feature vector within a small neighborhood of the feature vector of the original pattern (given that the feature mapping is smooth in the appropriate sense).

This observation leads the following assumption: the feature vectors $\mathcal{O}_i = \{\mathbf{m}_j \mid j = 1, \dots, M_i\}$ derived from the patterns of an object o_i lie on a topological manifold. The mathematical details and implications of this insight are quite profound and outside the scope of this book. Nonetheless, Appendix B gives a primer of the most important terms and concepts of the underlying theory. For the purposes of this section, it is sufficient to interpret "manifold" as a lower-dimensional hypersurface embedded in the feature space. In other words, the features $\mathbf{m}_j \in \mathcal{O}_i$ of the object o_i do not populate the feature space arbitrarily, but are restricted to some surface within the feature space.

An example is shown in Figure 2.12, where the black curves show the manifolds to which are restricted the features of two objects o_i and o_k. In the context of the optical character recognition (OCR) example, the two objects stand for different characters, e.g., o_i for the character "A" and o_k for the character "B". Formally, the manifolds are denoted

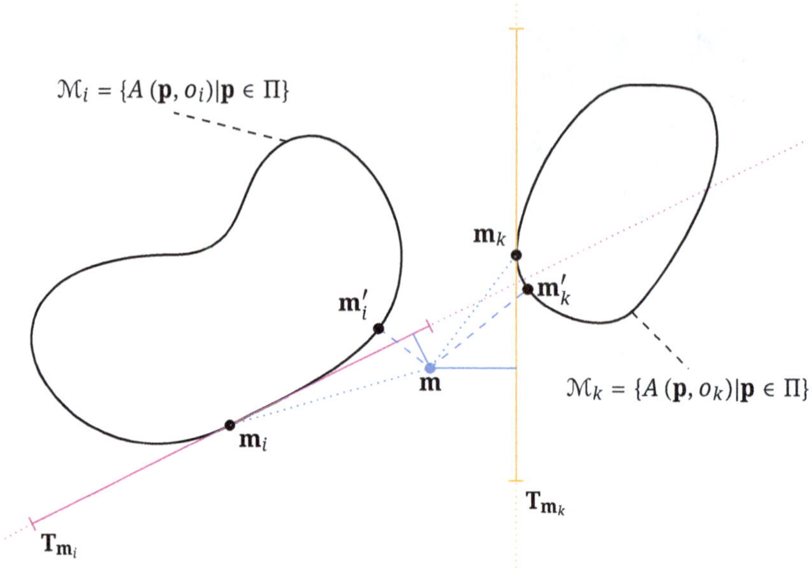

Fig. 2.12: Tangential distance measure: Improving the distance measure of two feature vectors by linear interpolation of the manifold corresponding to the underlying object.

by the set of all points given by an action A of some transformation $\mathbf{p}$ on the object o,

$$\mathcal{M}_i = \{A(\mathbf{p},o_i) \mid \mathbf{p} \in \Pi\}. \tag{2.37}$$

Again, the mathematical definitions of the terms are found in Appendix B. Here, it is sufficient to interpret $A(\mathbf{p},o_i)$ as the feature vector that is extracted from some systematic variation parametrized by $\mathbf{p}$.

The Figure 2.12 also highlights an important issue: given a feature vector $\mathbf{m}$ to classify and two feature vectors $\mathbf{m}_i$ and $\mathbf{m}_k$ derived from the objects o_i and o_k, respectively, computing the distance between the features might lead to the wrong conclusions. Here, $\mathbf{m}$ is closest to $\mathbf{m}_k$ and hence one could conclude that the object that produced $\mathbf{m}$ is more similar to o_k than to o_i. If, however, one considers the entire manifold of features of both objects, one arrives at a different picture: the closest point to $\mathbf{m}$ on $\mathcal{M}_i$ ($\mathbf{m}'_i$) is closer than the closest point on $\mathcal{M}_k$ ($\mathbf{m}'_k$). In consequence, $\mathbf{m}$ is closer to o_i than to o_k, which is the opposite of what was deduced from the distances of the given features. This motivates the following improved distance measure:

$$D_{\text{manifold}}(\mathbf{m}, \mathbf{m}_i) := \min_{\mathbf{m}'_i \in \mathcal{M}_i} \|\mathbf{m} - \mathbf{m}'_i\|, \quad \text{where } \mathcal{M}_i \ni \mathbf{m}_i. \tag{2.38}$$

Unfortunately, the manifold is generally not known and even if it were, computing the minimal distance is generally computationally infeasible. A solution to this is given by the tangential distance measure: Similar to a first order Taylor expansion, the true distance

$D_{\text{manifold}}(\mathbf{m}, \mathbf{m}_i)$ is approximated using a tangential (that is, linear) approximation at $\mathbf{m}_i$:

$$D_{\text{tangential}}(\mathbf{m}, \mathbf{m}_i) := \min_{\|\mathbf{a}\| \leq \varepsilon} \|\mathbf{m} - (\mathbf{m}_i + \mathbf{T}_{\mathbf{m}_i}\mathbf{a})\| \approx D_{\text{manifold}}(\mathbf{m}, \mathbf{m}_i). \tag{2.39}$$

Here, $\mathbf{T}_{\mathbf{m}_i}$ denotes the tangent space (or, more precisely, the projection onto the tangent space) of the manifold at $\mathbf{m}_i$. The search for the closest distance is further restricted to a small neighborhood $\{\mathbf{m}_i + \mathbf{T}_{\mathbf{m}_i}\mathbf{a} \mid \|\mathbf{a}\| < \varepsilon\}$ around $\mathbf{m}_i$. The reason is that, similar to the Taylor expansion, the linear approximation becomes more and more inaccurate the further one deviates from $\mathbf{m}_i$.

Figure 2.12 illustrates this approach: the feature vector $\mathbf{m}$ is closer to the tangent identified by $\mathbf{T}_{\mathbf{m}_i}$ (purple line) than to the tangent identified by $\mathbf{T}_{\mathbf{m}_k}$ (orange line). Hence, $\mathbf{m}$ is correctly assigned to the object o_i instead of o_k. Note that in the figure, the neighborhoods (denoted by the perpendicular stops on the tangents) are chosen to be very large. In consequence, the approximation to the manifold $\mathcal{M}_k$ does not hold. In practice, one would probably choose a smaller neighborhood. However, the neighborhood might also not be chosen too small, because then the distance to the tangent $\mathbf{T}_{\mathbf{m}_w}$ will not differ much from the distance to the original feature vector $\mathbf{m}_i$.

Note: If one chooses the Euclidean norm, the minimization w.r.t. $\mathbf{a}$ in Equation (2.39) reduces to a quadratic optimization problem, which can be solved using standard tools.

However, exactly computing the tangent space $\mathbf{T}_{\mathbf{m}}$ requires the evaluation of the gradient of $A(\mathbf{p}, o_k)$ at $\mathbf{m}$. Unfortunately, this information is rarely available in practice. However, the tangent space can be approximated by a secant $\hat{\mathbf{t}}(\mathbf{m})$:

$$\hat{\mathbf{t}}_j(\mathbf{m}_i) = \mathbf{m}_i - A(\mathbf{p} + \Delta\mathbf{p}_j; o_j), \tag{2.40}$$

where $\det\left[\Delta\mathbf{p}_1, \ldots, \Delta\mathbf{p}_q\right] \neq 0$, i.e., the $\Delta\mathbf{p}_j$ are linearly independent. The small disturbance $\Delta\mathbf{p}_j, j = 1, \ldots, q$, can be obtained by recording the objects under various conditions or (more commonly) by simulating in a software application these conditions from a small number of actual measurements.

An example of this is shown in Figure 2.13, where only the patterns in the orange boxes were obtained by measurement. The remaining variations were approximated using linear interpolation. The corresponding features lie on the secants between the features of the measured patterns. Comparing Figure 2.13 to the true variations in Figure 2.11 highlights another important trade-off when using this method: how many measurements, or sampling points of the manifold, should one obtain? Measuring the variations takes a considerable effort, but measuring too few variations will reduce the quality of the approximation.

Lastly, Figure 2.12 illustrates another drawback of the method: the manifolds $\mathcal{M}_k = \{A(\mathbf{p}, o_k) \mid \mathbf{p} \in \Pi\}$ and $\mathcal{M}_i = \{A(\mathbf{p}, o_i) \mid \mathbf{p} \in \Pi\}$ are drawn as closed curves. However, this is not necessarily true. Indeed, the manifold need not even be connected, but might consist of several, disconnected strips. As a result, the tangential approximation may significantly underestimate the actual distance to the manifold. Furthermore, the secant approximation may assume a manifold where there is none and therefore invalidate the whole method. However, these issues rarely occur in practice.

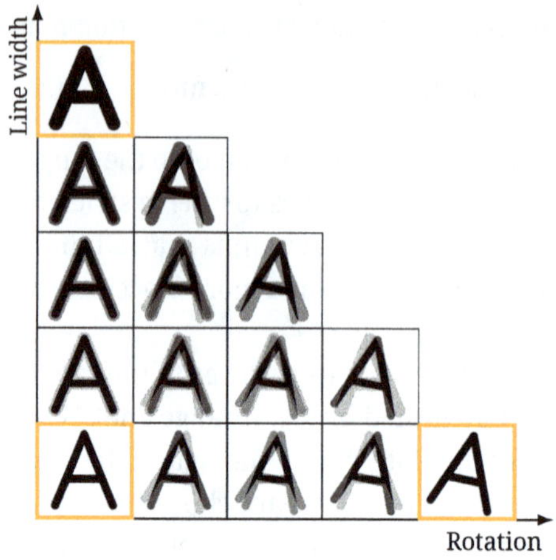

Fig. 2.13: Linear approximation of the variation in Figure 2.11.

2.5 Normalization

The previous section has already presented an approach to enhance the robustness of a distance measure. Normalization has a similar goal, but is applied at an earlier stage in the processing chain. Instead of improving a metric, normalization tries

- to eliminate extraneous disturbances of the patterns,
- to eliminate extraneous variations of the patterns, and
- to eliminate extraneous variations of the features.

If modifications are already avoided at the stage of the patterns, the deduced features become independent of those modifications. Surely, this task is highly domain specific, and requires a good understanding of the concrete pattern recognition system. For this reason, this section can only present some examples of what can be done in certain cases. These examples are:

1. planimetric adjustment of images;
2. lighting adjustment of images;
3. amplitude recovery of audio signals, e.g., by automatic gain adjustment;
4. distortion adjustment of images (due to lens aberrations);
5. alignment, elimination of physical dimension, and leveling of proportions; and
6. dynamic time warping.

2.5.1 Alignment, elimination of physical dimension, and leveling of proportions

Let us assume one has features with values on a interval scale at least and let $\underline{\mathbf{m}}$ be the feature vector modeled as a random variable. Then Item 5 can be realized by

$$\underline{\mathbf{m}}' = \left(\frac{\underline{m}_1 - \mathrm{E}\{\underline{m}_1\}}{\sqrt{\mathrm{Var}\{\underline{m}_1\}}}, \ldots, \frac{\underline{m}_d - \mathrm{E}\{\underline{m}_d\}}{\sqrt{\mathrm{Var}\{\underline{m}_d\}}} \right)^{\mathsf{T}} . \tag{2.41}$$

Although $\underline{\mathbf{m}}$ might be equipped with physical dimensions, $\underline{\mathbf{m}}'$ is not, because of the division by the standard deviation. Moreover,

$$\mathrm{E}\{\underline{\mathbf{m}}'\} = \mathbf{0}, \qquad \mathrm{Var}\{\underline{m}_j'\} = 1,\ j = 1,\ldots,d. \tag{2.42}$$

In practice, one normally approximates the unknown expectation and variance by the empirical mean and standard deviation of the learning samples $\mathbf{m}_i$, $j = i, \ldots, N$,

$$\overline{m}_j = \frac{1}{N} \sum_{i=1}^{N} m_{ij}, \tag{2.43}$$

$$s_j = \sqrt{\frac{1}{N-1} \sum_{i=1}^{N} (m_{ij} - \overline{m}_j)} \quad \text{and} \tag{2.44}$$

$$\mathbf{m}' = \left(\frac{m_1 - \overline{m}_1}{s_1}, \ldots, \frac{m_d - \overline{m}_d}{s_d} \right)^{\mathsf{T}} . \tag{2.45}$$

2.5.2 Lighting adjustment of images

The adjustment and normalization of the lighting conditions of images is a very broad field. As this textbook is about pattern recognition, this section can only touch on this topic. A more detailed discussion can be found in the relevant literature, e.g., in *Machine Vision* by Beyerer et al. [2016]. The examples shown here are taken from that book.

Chromaticity normalization

Here and in the discussions below, the color values of the pixels are assumed to be within the interval $[0, 1]$. Chromaticity normalization transforms each pixel so that all the pixels (except the black ones) have the same intensity. More formally, given the color components (r, g, b) of a pixel, the transformation maps to the color value

$$(r', g', b') = \begin{cases} \frac{1}{r+g+b}(r, g, b) & \text{if } r + g + b > 0 \\ (0,0,0) & \text{if } r + g + b = 0 \end{cases} . \tag{2.46}$$

Figure 2.14 shows an example.

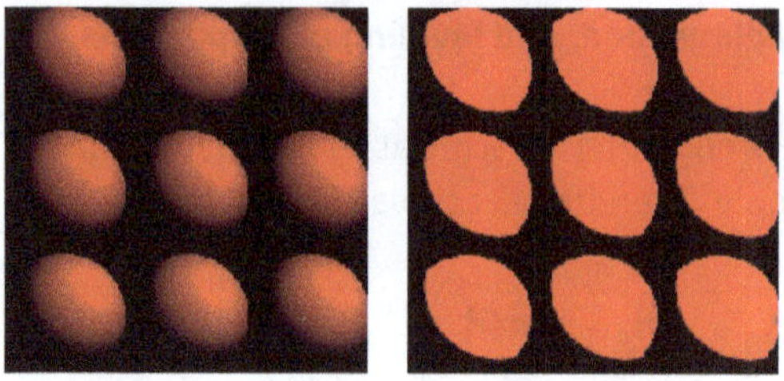

(a) Original image (b) Normalized image

Fig. 2.14: Chromaticity normalization of an image. Source: Beyerer et al. [2016].

Illumination normalization tries to equalize the overall proportions of the color channels in order to mitigate the effect of different light sources with varying color temperatures. Assume the image contains J pixels $(r_1, g_1, b_1), \ldots, (r_J, g_J, b_J)$ and let $(R, G, B) = \frac{3}{J}\sum_{j=1}^{J}(r_j, g_j, b_j)$ denote the overall sum for each channel. Then each pixel is transformed to $\left(r'_j, g'_j, b'_j\right) = \left(\frac{r_j}{R}, \frac{g_j}{G}, \frac{b_j}{B}\right)$. This means that $\sum_{j=1}^{J} r'_j = \sum_{j=1}^{J} g'_j = \sum_{j=1}^{J} b'_j = \frac{J}{3}$ for each color channel and the final image has equal proportions of each channel.

Let us consider both steps together. An RGB image with J pixels can be described by $3J$ variables. The lightning adjustment creates J linear constraints, i.e., one for each pixel. The illumination normalization contributes 3 additional constraints, one for each channel. This leads to the following system of linear equations

$$r'_1 + g'_1 + b'_1 = 1 \qquad \ldots \qquad r'_J + g'_J + b'_J = 1 \tag{2.47}$$

$$\sum_{j=1}^{J} r'_j = \frac{J}{3} \qquad \sum_{j=1}^{J} g'_j = \frac{J}{3} \qquad \sum_{j=1}^{J} b'_j = \frac{J}{3}. \tag{2.48}$$

These are $J + 2$ linearly independent equations, because

$$\frac{J}{3} = \sum_{j=1}^{J} r'_j = \sum_{j=1}^{J}\left(1 - g'_j - b'_j\right) = J - \sum_{j=1}^{J} g'_j - \sum_{j=1}^{J} b'_j = J - \frac{J}{3} - \frac{J}{3}. \tag{2.49}$$

In summary, the system has $3J - J - 2 = 2(J - 1)$ degrees of freedom. Normally, if the transformation $(r'_j, g'_j, b'_j) = \frac{1}{r_j+g_j+b_j}(r_j, g_j, b_j)$ or the transformation $\left(r'_j, g'_j, b'_j\right) = \left(\frac{r_j}{R}, \frac{g_j}{G}, \frac{b_j}{B}\right)$ is applied to an image, the result is normalized either in chromaticity or in illumination, but not both. However, if both steps are iteratively applied, the sequence of images converges. Figure 2.15 shows an example of such a comprehensive image normalization.

Signal theoretic approach

Another method of image normalization is to look at the image from a signal theoretic point of view and consider image (improving) operations as filters. For the sake of simplicity, assume that the image is a gray-scale image and is given by a two-dimensional function $g : \mathbb{R}^2 \to \mathbb{R}$. Though the digital representation of an image only defines the values at discrete points (at the pixels), it is assumed that the function can be continuously

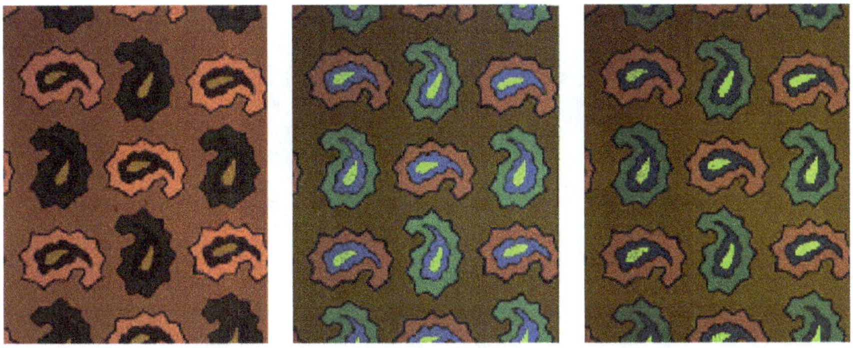

(a) High proportion of red **(b)** High proportion of green **(c)** High proportion of yellow

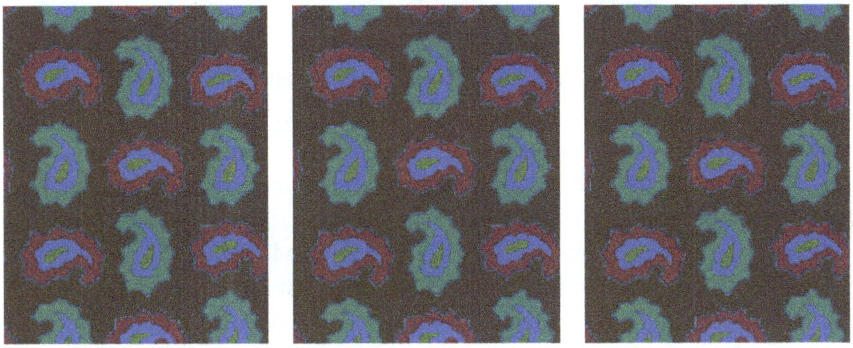

(d) Normalized image **(e)** Normalized image **(f)** Normalized image

Fig. 2.15: Normalization of lighting conditions by iterated normalization of the chromaticity and illumination. Source: Beyerer et al. [2016].

interpolated. Moreover, some of the following discussion requires that g be periodic and without discontinuities at the borders of the image. Such a representation can be obtained from the image by mirroring it along the borders, e.g., $g(x,y) = g(w - x,y)$ for $w \leq x \leq 2w$. From here on, we will assume that g fulfills all the necessary requirements (especially smoothness, periodicity, and integratability) without explicitly stating this. We further restrict ourselves to linear filters, i.e., filters T that satisfy

$$T(\alpha f + \beta g) = \alpha Tf + \beta Tg, \tag{2.50}$$

where f and g are signals and $\alpha, \beta \in \mathbb{R}$ are scalars. Informally, the above requires that the result of applying T to a combination of signals has to be the same as first applying T to each signal and then combining the results.

Let us now assume a very simple signal model for the image generation process, given by

$$g(\mathbf{x}) = s(\mathbf{x}) \bigtriangleup b(\mathbf{x}), \quad \mathbf{x} = (x,y)^{\mathsf{T}}. \tag{2.51}$$

(a) Original (b) After homomorphic filtering (c) After homogenization of degree 2

Fig. 2.16: Images of the surface of agglomerated cork. Source: Beyerer et al. [2016].

Here $g : \mathbb{R}^2 \to \mathbb{R}$ denotes the final image as seen by the pattern recognition system, $s : \mathbb{R}^2 \to \mathbb{R}$ the true underlying image we wish to recover, $b : \mathbb{R}^2 \to \mathbb{R}$ the disturbance, and $\triangle$ a binary operator on two signals.

If $\triangle$ is addition, then s is said to be subject to additive noise. Additive systems are often preferred because they can usually be treated by linear filters, which have been intensively studied and are well understood tools. If $\triangle$ is not addition, one can try to map the system equation into a different space so that the transformed system is additive. Generally, let T denote a linear filter and U such a transformation. Then the filter can be applied by

$$U^{-1}TUg = U^{-1}TU\,(s \triangle b) = U^{-1}\,(TUs + TUb)\,. \tag{2.52}$$

Such a transformed filter is called a "homomorphic" filter.

Assume a slowly varying disturbance $b(\mathbf{x})$ due to an inhomogeneous illumination of the scene. This disturbance will act by multiplication on the true signal,

$$g\,(\mathbf{x}) = s\,(\mathbf{x}) \cdot b\,(\mathbf{x})\,. \tag{2.53}$$

Furthermore, assume that $g, s, b > 0$ and that the support of the Fourier transforms of $\ln s$ and $\ln b$ is not identical. This implies that the logarithm of the disturbance $\ln b$ varies much more slowly than the logarithm $\ln s$ of the true signal. Under these assumptions, the image can be improved by a high pass filter (H) in combination with a logarithmic transformation. It follows that

$$\begin{aligned}(\exp H \ln)g &= (\exp H \ln)(s \cdot b) = \exp(\overbrace{H \ln s}^{\substack{\approx \ln s \text{ by} \\ \text{assumption}}} + \overbrace{H \ln b}^{\substack{\approx 0 \text{ by} \\ \text{assumption}}}) \\ &\approx \exp(\ln s + 0) = s.\end{aligned} \tag{2.54}$$

Figure 2.16b shows the result of such a homomorphic filter.

Lastly, we will discuss the homogenization of images by interpreting the image as a random process $\underline{g} : \mathbb{R}^2 \to [0,1]$. Details on random processes can be found in Appendix C.

Here, consider $\underline{g}$ as a random function that gives a value given a position $\mathbf{x}$. Since $\underline{g}$ is a random function, we can also give the expectation $\mu = \mathrm{E}\{\underline{g}\} : \mathbb{R}^2 \to [0,1]$ and variance $\sigma^2 = \mathrm{Var}\{\underline{g}\} : \mathbb{R}^2 \to \mathbb{R}_+$ (both functions themselves). In the stochastic signal model, the image $\underline{g}$ is modeled by

$$\underline{g}\,(\mathbf{x}) = \underline{s}\,(\mathbf{x}) \bigtriangleup \underline{b}\,(\mathbf{x})\,. \tag{2.55}$$

As before, $\underline{g}$ denotes the final image as seen by the system, $\underline{s}$ the true but unknown image, and $\underline{b}$ a disturbance. It is assumed that $\underline{s}$ is a homogeneous process—a reasonable assumption for regular texture like the cork surface in Figure 2.16—and that $\underline{b}$ destroys the homogeneity of $\underline{g}$. For example, $\underline{b}$ could represent differences in lighting that make some parts of the image appear brighter than others. With this in mind, one can estimate the underlying image using

$$\hat{\underline{s}}(\mathbf{x}) = \frac{\underline{g}(\mathbf{x}) - \mu(\mathbf{x})}{\sigma(\mathbf{x})}\,. \tag{2.56}$$

Although reasonable, this approach is problematic in practice, because it requires estimating the expectation and variance of $\underline{g}$. Neither can be estimated reliably, since only one realization of $\underline{g}$, the image g at hand, is given. However, under the assumption that $\underline{g}$ is ergodic (which due to $\underline{b}$ it generally is not), one can determine the unknown expectation and variance in Equation (2.56) by taking the average over all points of only one single realization g.

Further details of the method can be found in Beyerer et al. [2016]. An example of the result of the homogenization transformation is shown in Figure 2.16c.

2.5.3 Distortion adjustment of images

Let V denote a distortion of the image due to aberration or perspective deformation. Each (continuous) coordinate $(\tilde{x}, \tilde{y})^\mathsf{T}$ of the true world is mapped to the image coordinates $(x, y)^\mathsf{T}$ (see Figure 2.17)

$$\begin{pmatrix} x \\ y \end{pmatrix} = V \begin{pmatrix} \tilde{x} \\ \tilde{y} \end{pmatrix}. \tag{2.57}$$

If V is known, the adjusted image γ can be reconstructed from g. As both γ and g are discrete, the final coordinates (ξ, η) are used as the starting point and the value of the induced coordinates $(x, y) = V(\xi, \eta)$ is copied:

$$\gamma(\xi, \eta) = g\,(V(\xi, \eta))\,. \tag{2.58}$$

Usually, $(x, y) = V(\xi, \eta)$ does not denote a valid lattice point of the preliminary image, hence $g\,(V(\xi, \eta))$ must be interpolated. In practice, these three methods are customary:

- nearest-neighbor interpolation,
- bilinear interpolation, and
- bicubic interpolation.

For more details, again see Beyerer et al. [2016].

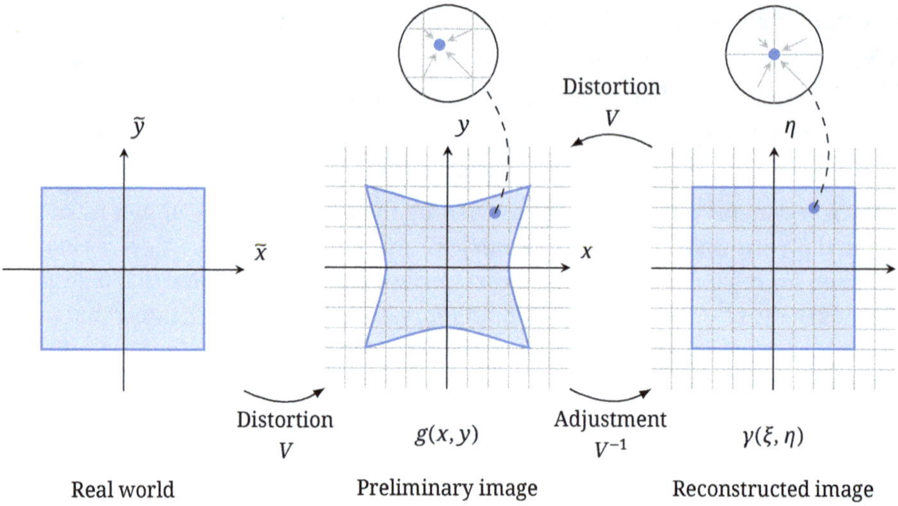

Fig. 2.17: Adjustment of geometric distortions.

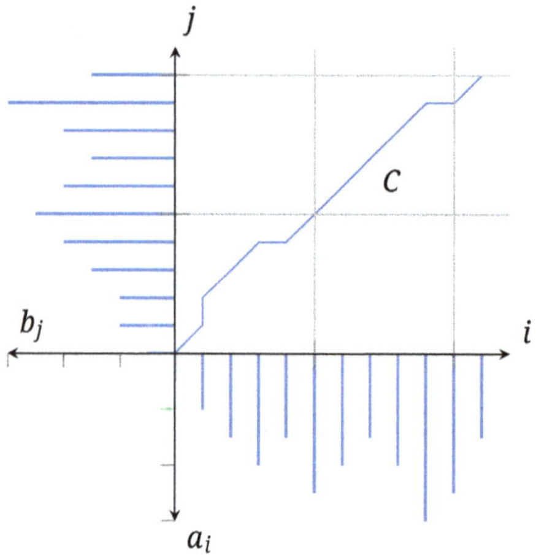

Fig. 2.18: Adjustment of temporal distortions using dynamic time warping.

2.5.4 Dynamic time warping

Dynamic time warping is necessary if one has to match two signals of different lengths. A typical example is a pair of audio recordings with different speed profiles. The goal is to find the best mapping between the two signals that equalizes the different temporal speed courses (see Figure 2.18). Let

$$\mathbf{A} = (\mathbf{a}_1, \ldots, \mathbf{a}_M) \tag{2.59}$$

$$\mathbf{B} = (\mathbf{b}_1, \ldots, \mathbf{b}_L) \tag{2.60}$$

be two discrete signals **A** and **B** with lengths M and L, respectively. The goal is to find a sequence of pairs of indices

$$\mathbf{C} = (\mathbf{c}_1, \ldots, \mathbf{c}_K) \qquad \text{with} \qquad \mathbf{c}_k = (i_k, j_k)\,, k = 1, \ldots, K \tag{2.61}$$

that obey the constraints

$$i_1 = j_1 = 1 \qquad \text{(starting points must match)} \tag{2.62}$$

$$i_K = M, j_K = L \qquad \text{(end points must match)} \tag{2.63}$$

$$i_k \leq i_{k+1}, j_k \leq j_{k+1} \qquad \text{(monotonicity)} \tag{2.64}$$

$$i_k - i_{k-1} \leq 1, j_k - j_{k-1} \leq 1 \qquad \text{(no point must be skipped)} \tag{2.65}$$

such that

$$C_{\text{opt}} = \arg\min_{C} \sum_{k=1}^{K} \|\mathbf{a}_{i_k} - \mathbf{b}_{j_k}\| \tag{2.66}$$

is minimized.

2.6 Selection and construction of features

Though the whole chapter has already been about features, it has been implicitly assumed that those features are already present. Of course, every section came up with some examples of features, where it was helpful to discuss the concepts.

The first section classified features according to their scale of measurement, the third section illustrated some transformations of the feature space, the fourth section dealt with distance measures, and the previous section gave some examples of feature normalization. In summary, all the sections relied on the fact that there were already available features that could be handled, modified, or transformed. Only the second section focused a little bit on how to obtain the features. But even that section assumed that there already was a pool of features from which to select.

In order to fill this gap, this section will put the focus on the question of how to initially find good features. The first subsection will give some examples of descriptive features and why descriptive features should be preferred. The second subsection is about features derived from a model of the generation process of the object. The third subsection will present a way of systematically constructing invariant features and is closely related to Section 2.4.8.

2.6.1 Descriptive features

The most straightforward approach is to select standard descriptive features that characterize obvious traits of the object's class and that carry a natural interpretation. Despite

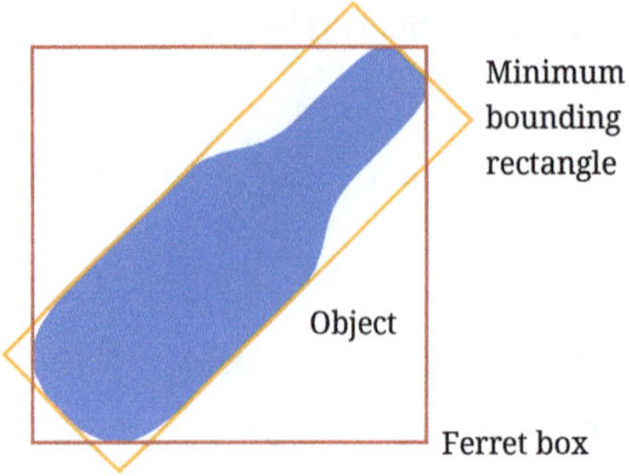

Fig. 2.19: Different bounding boxes around an object.

Fig. 2.20: The convex hull around a concave object.

the simplicity of this heuristic method, it is often quite successful. Moreover, descriptive features have the distinct advantage that they can easily be understood by the system designer, something which simplifies the debugging process for when the pattern recognition system fails.

Geometric features

If the border of an object can be identified, the object's area is computable as well. The *degree of filling* m is defined as

$$m = \frac{\text{area of the object}}{\text{area of the bounding rectangle}} \in [0,1]. \tag{2.67}$$

Usually there are two options for how a minimum bounding rectangle can be defined (see Figure 2.19). The minimum bounding rectangle whose edges are still aligned parallel to the axis is sometimes called the *ferret box*. But normally the term minimum bounding rectangle (MBR) denotes the rectangle that ignores this constraint and is rotated so that it is properly aligned with the enclosed object.

For both definitions of the box, m is an invariant of the position and of the scale of the object. In addition, when using the MBR, m is also invariant w.r.t. rotation. But the ferret box is easier to compute.

A natural generalization of a bounding box is the convex hull (see Figure 2.20). Accordingly, the *degree of convexity* is defined as

$$m = \frac{\text{area of the object}}{\text{area of the convex hull}} \in [0,1]. \tag{2.68}$$

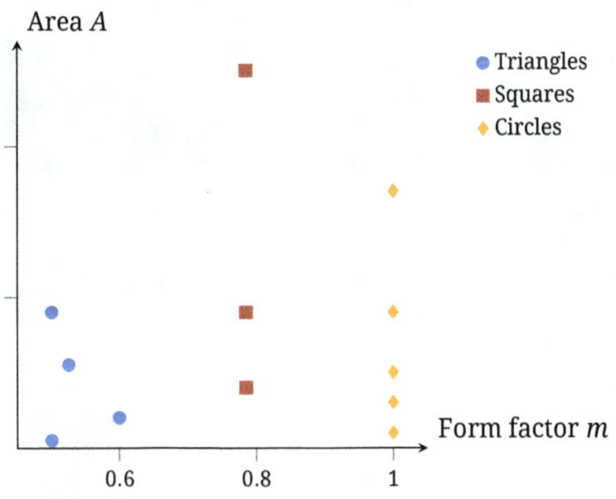

Fig. 2.21: Degree of compactness (form factor) for exemplary shapes. The form factor of different triangles can vary while it is constant for all squares and circles, respectively.

Tab. 2.2: Topology of the letters of the German alphabet.

Feature	**Letter**																			
	A	Ä	B	C	D	E	F	...	O	Ö	P	...	U	Ü	V	W	X	Y	Z
B	1	3	1	1	1	1	1	...	1	3	1	...	1	3	1	1	1	1	1
L	1	1	2	0	1	0	0	...	1	1	1	...	0	0	0	0	0	0	0
E	0	2	-1	1	0	1	1	...	0	2	0	...	1	3	1	1	1	1	1

Again, m is invariant w.r.t. translation, scaling and rotation. The *degree of compactness* or *form factor* relates the perimeter to the area and is defined as

$$m = \frac{4\pi\text{Area}}{\text{perimeter}^2} \in [0,1]. \tag{2.69}$$

The coefficient 4π has been chosen so that $0 \leq m \leq 1$, because the circle has the smallest perimeter of all areas of the same size in Euclidean geometry (see Figure 2.21).

Topological features

Topological features depart from the tangible geometry and describe an object such that the features become invariant with respect to rubber-sheeting transformations. Suitable features include, for example, the number of connected components (B) or the genus, i.e., the number of holes (L). The Euler number is defined as

$$E = B - L. \tag{2.70}$$

Table 2.2 lists the number of connected components, the genus, and the Euler number of the letters.

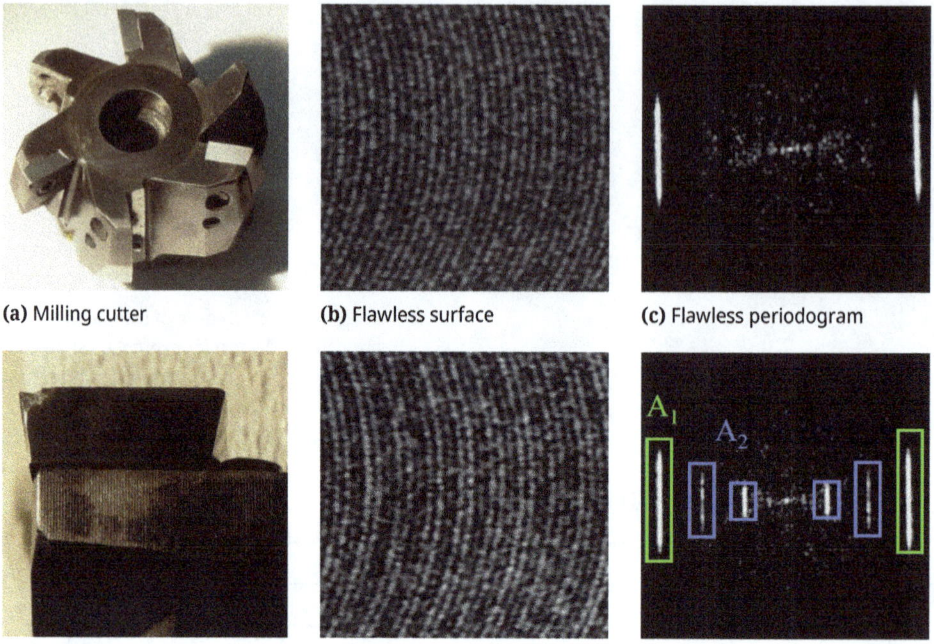

(a) Milling cutter (b) Flawless surface (c) Flawless periodogram

(d) Broken cog (e) Defective surface (f) Defective periodogram

Fig. 2.22: Classification of faulty milling cutters.

Periodogram (frequency spectrum)

In some cases, the classification task is hard to carry out in the time domain (or, in the case of images, in the spatial domain) but becomes easy in the frequency domain. This fact will be illustrated by the following example. This example is interesting for an auxiliary reason. Most of the examples so far had the defect that the pattern and feature spaces were identical because the pattern could be directly used as a feature due to the simplicity of the example. The following example distinguishes between a preceding pattern, a pattern transformation, and a derived feature.

Assume a surface is manufactured by a face-milling cutter (see Figure 2.22a). If one of the cogs is broken (see Figure 2.22d), the milling cutter will damage the surface. A direct inspection of the resulting surfaces reveals, in the case of a fault, a periodic repetition of two neighboring stronger grooves (see Figure 2.22e). A fault-free surface is more uniform, with equally strong grooves (see Figure 2.22b). Although the human eye is able to perceive the difference, its formalization in a suitable computer representation is difficult. Once again, let the gray-scale image be

$$g\colon \begin{cases} \mathbb{R}^2 \to \mathbb{R} \\ (x,y) \mapsto g(x,y) \end{cases} \tag{2.71}$$

and let $\mathcal{F}$ denote the operator of Fourier transformation. Figures 2.22c and 2.22f illustrates the Fourier transforms, i.e.,

$$G(f_x, f_y) = \mathcal{F}\{g\}(f_x, f_y) = \iint_{-\infty}^{\infty} g(x, y) \mathrm{e}^{-2\pi \mathrm{j}(x f_x + y f_y)} \, \mathrm{d}x \, \mathrm{d}y. \tag{2.72}$$

These figures clearly show a difference. The left periodogram has only one symmetric peak at the fundamental frequency, but the right periodogram has some additional peaks at the subharmonic frequencies. Hence, a reasonable feature is derived by comparing the intensities of these frequencies,

$$m = \frac{\iint_{A_2} |\mathcal{F}\{g\}(f_x, f_y)|^2 \, \mathrm{d}f_x \, \mathrm{d}f_y}{\iint_{A_1} |\mathcal{F}\{g\}(f_x, f_y)|^2 \, \mathrm{d}f_x \, \mathrm{d}f_y}. \tag{2.73}$$

In the flawless case, this ratio becomes small, because in the numerator the subharmonics vanish. In contrast, the ratio becomes large in the faulty case. For more details on this topic, see Beyerer et al. [2016].

2.6.2 Model-driven features

The principal idea behind model-driven features is to consider a parametric mathematical model which is able to generate the patterns. Different parameters of the model will yield different patterns. In particular, each class can be characterized by some subset of the parameter space and can be represented by one or more prototypical examples from that class. The variation in classes is caused by small perturbations of the parameters of these prototypes, i.e., the variation can be modeled as noise in the parameter space. Such a model is often referred to as a "generating process". Here, however, it is used for classification by estimating the parameters that generate a given pattern and using these parameters as the feature vector $\mathbf{m}$.

The model-driven approach has several advantages. First of all, it allows contributing expert knowledge. Usually this leads to feature vectors (or parameter vectors) that are highly specific to the problem and therefore illustrative. This normally coincides with short parameter vectors and hence a low dimension. Finally, the system can be comfortably evaluated, because new patterns can be artificially generated.

Autoregressive signal models

Of course, in order to take advantage of the model-driven approach, one initially needs a model. A rather generic model is the autoregressive signal model (AR model). The idea is that the state of the system at a point n only depends on finitely many previous states from a causal neighborhood. The number of model parameters is the number of those states. The (one-dimensional) system equation can be written as

$$\underline{g}_n = \sum_{i=1}^{K} a_i \underline{g}_{n-i} + \underline{e}_n, \tag{2.74}$$

where $\underline{g}_n$ denotes the state of the system at n and $\underline{e}_n$ denotes the disturbance, while the a_i are the parameters of the model.

In the context of the analysis of time series, the causal neighborhood is naturally in the past, but the AR model can also be applied to structured images (textures) if the pixels are enumerated appropriately. A pragmatic approach is to define all pixels below and to the left of a given pixel as the neighborhood of that pixel.

The two-dimensional AR model can be stated as follows: Let g_{mn} be a two-dimensional weakly stationary process with zero mean and let $\underline{e}_{mn}$ be a white noise process[1]. Let further $\mathcal{U}$ with $(0,0) \notin \mathcal{U}$ denote a finite and causal neighborhood of the origin (0,0) and let $a_{kl} \in \mathbb{R}$ denote the AR coefficients. The two-dimensional AR model is then given by

$$\underline{g}_{mn} = \sum_{(k,l)\in\mathcal{U}} a_{kl} \underline{g}_{m-k,n-l} + \underline{e}_{mn}. \tag{2.75}$$

The number $|\mathcal{U}|$ of elements in the neighborhood is called the *order* of the AR model. As $\mathcal{U}$ only refers to "past" states and there is a defined starting point (the origin), a recursive evaluation of the system model is possible. However, first one needs to find the AR parameters for the given image. To simplify notation, we write the system equation as

$$\underline{g}_{mn} = \mathbf{a}^{\mathsf{T}} \underline{\gamma}_{mn} + \underline{e}_{mn}, \tag{2.76}$$

where $\mathbf{a} = (\ldots, a_{kl}, \ldots)^{\mathsf{T}}$ with $(k, l) \in \mathcal{U}$ are the AR parameters and $\underline{\gamma}_{mn}$ is the vector of the pixel values of g in the neighborhood $\mathcal{U}$ around the point (m, n),

$$\underline{\gamma}_{mn} = (\ldots, \underline{g}_{m-k,n-l}, \ldots)^{\mathsf{T}} \qquad (k, l) \in \mathcal{U}. \tag{2.77}$$

The unknown parameters are the coefficient vector $\mathbf{a}$ and the variance of the noise σ^2. Hence the feature vector is given by $\mathbf{m} = \left(\sigma^2, \mathbf{a}^{\mathsf{T}}\right)^{\mathsf{T}}$. The objective of the optimization is to minimize the variance of the noise, which here can be interpreted as the prediction error of the AR model:

$$\begin{aligned}
\sigma^2 &= \operatorname{Var}\{\underline{e}_{mn}\} = \operatorname{E}\{\underline{e}_{mn}^2\} = \operatorname{E}\left\{(\underline{g}_m - \mathbf{a}^{\mathsf{T}}\underline{\gamma}_{mn})^2\right\} \\
&= \operatorname{E}\left\{\underline{g}_{mn}^2 - 2\mathbf{a}^{\mathsf{T}}\underline{\gamma}_{mn}\underline{g}_{mn} + \mathbf{a}^{\mathsf{T}}\underline{\gamma}_{mn}\underline{\gamma}_{mn}^{\mathsf{T}}\mathbf{a}\right\} \\
&= \operatorname{E}\left\{\underline{g}_{mn}^2\right\} - 2\mathbf{a}^{\mathsf{T}}\operatorname{E}\left\{\underline{\gamma}_{mn}\underline{g}_{mn}\right\} + \mathbf{a}^{\mathsf{T}}\operatorname{E}\left\{\underline{\gamma}_{mn}\underline{\gamma}_{mn}^{\mathsf{T}}\right\}\mathbf{a} \to \text{minimize.}
\end{aligned} \tag{2.78}$$

Do not be confused by the fact that the left side of the equation is a constant (σ^2), but the remainder of the equation seems to depend on the position (m, n): since all involved

---

1 See Appendix C for an explanation of the terms "weakly stationary", "white noise", etc.

processes are at least weakly stationary (see Appendix C), the value of the last line does not actually depend on (m, n).

To calculate the expectations of $\underline{\gamma}_{mn}$ and $\underline{g}_{mn}$, we assume that the process is locally ergodic (see, again, Appendix C) in the neighborhood $\mathcal{U}'_{mn}$. This means that the expectation can be estimated by an average over a neighborhood within the same realization. With this in mind, the necessary condition for finding the optimal $\mathbf{a}$ is that the gradient of Equation (2.78) with respect to $\mathbf{a}$ vanishes:

$$\begin{aligned}
&& \nabla_{\mathbf{a}}\sigma^2 &= -2\,\mathrm{E}\{\underline{\gamma}_{mn}\underline{g}_{mn}\} + 2\,\mathrm{E}\{\underline{\gamma}_{mn}\underline{\gamma}_{mn}^{\mathsf{T}}\}\,\mathbf{a} \overset{!}{=} \mathbf{0} \\
&\Leftrightarrow & \mathbf{a} &= \left(\mathrm{E}\{\underline{\gamma}_{mn}\underline{\gamma}_{mn}^{\mathsf{T}}\}\right)^{-1}\mathrm{E}\{\underline{\gamma}_{mn}\underline{g}_{mn}\} \\
&\text{(ergodicity)} & &\approx \left(\left|\mathcal{U}'_{mn}\right|^{-1}\sum_{(m,n)\in\mathcal{U}'_{mn}}\gamma_{mn}\gamma_{mn}^{\mathsf{T}}\right)^{-1}\cdot\left|\mathcal{U}'_{mn}\right|^{-1}\sum_{(m,n)\in\mathcal{U}'_{mn}}\gamma_{mn}g_{mn} \\
&\Rightarrow & \hat{\mathbf{a}} &= \left(\sum_{(m,n)\in\mathcal{U}'_{mn}}\gamma_{mn}\gamma_{mn}^{\mathsf{T}}\right)^{-1}\cdot\sum_{(m,n)\in\mathcal{U}'_{mn}}\gamma_{mn}g_{mn}
\end{aligned} \tag{2.79}$$

with $|\mathcal{U}'_{mn}|$ being the total number of elements in the neighborhood.

In order to obtain an estimator $\widehat{\sigma^2}$ of σ^2, the estimator $\hat{\mathbf{a}}$ is put into Equation (2.78) and ergodicity is exploited once again. To simplify the notation, the following ad hoc abbreviations are introduced:

$$G = \sum_{(m,n)\in\mathcal{U}'_{mn}} g_{mn}^2 \tag{2.80}$$

$$\mathbf{H} = \sum_{(m,n)\in\mathcal{U}'_{mn}} \gamma_{mn}g_{mn} \tag{2.81}$$

$$\Gamma = \sum_{(m,n)\in\mathcal{U}'_{mn}} \gamma_{mn}\gamma_{mn}^{\mathsf{T}}. \tag{2.82}$$

Using these, the estimator of Equation (2.79) can be written as $\hat{\mathbf{a}} = \Gamma^{-1}\mathbf{H}$. Further, Γ is symmetric, i.e., $\Gamma^{\mathsf{T}} = \Gamma$ and the values of the expectation in Equation (2.78) are

$$\mathrm{E}\{\underline{g}_{mn}^2\} = \left|\mathcal{U}'_{mn}\right|^{-1} G \tag{2.83}$$

$$\mathrm{E}\{\underline{\gamma}_{mn}\underline{g}_{mn}\} = \left|\mathcal{U}'_{mn}\right|^{-1} \mathbf{H} \tag{2.84}$$

$$\mathrm{E}\{\underline{\gamma}_{mn}\underline{\gamma}_{mn}^{\mathsf{T}}\} = \left|\mathcal{U}'_{mn}\right|^{-1} \Gamma. \tag{2.85}$$

With this notation, the estimator $\widehat{\sigma^2}$ becomes

$$\begin{aligned}
\widehat{\sigma^2} &= \mathrm{E}\{\underline{g}_{mn}^2\} - 2\hat{\mathbf{a}}^{\mathsf{T}}\,\mathrm{E}\{\underline{\gamma}_{mn}\underline{g}_{mn}\} + \hat{\mathbf{a}}^{\mathsf{T}}\,\mathrm{E}\{\underline{\gamma}_{mn}\underline{\gamma}_{mn}^{\mathsf{T}}\}\,\hat{\mathbf{a}} \\
&= \frac{1}{|\mathcal{U}'_{mn}|}\left(G - 2\left(\Gamma^{-1}\mathbf{H}\right)^{\mathsf{T}}\mathbf{H} + \left(\Gamma^{-1}\mathbf{H}\right)^{\mathsf{T}}\Gamma\left(\Gamma^{-1}\mathbf{H}\right)\right)
\end{aligned}$$

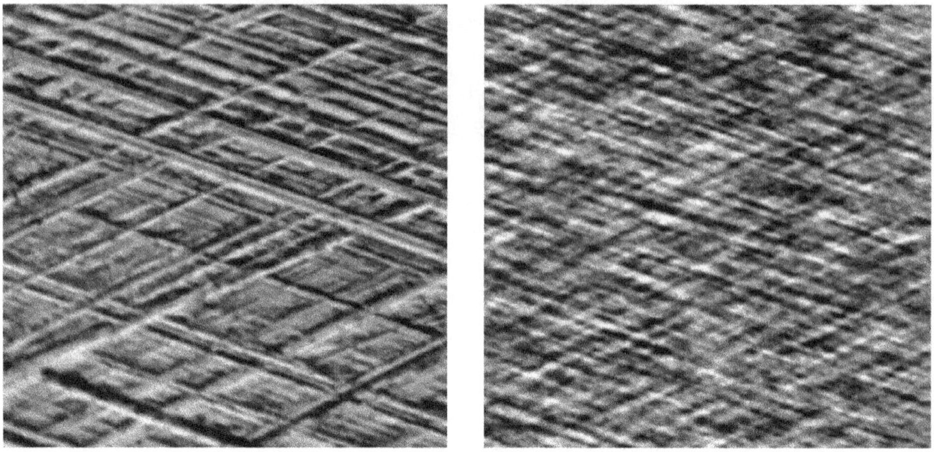

(a) Original honing texture

(b) Artificially generated texture (AR model of order 84)

Fig. 2.23: Synthetic honing textures using an AR model as an example of model-driven features.

$$= \frac{1}{|\mathcal{U}'_{mn}|}\left(G - 2\mathbf{H}^{\mathsf{T}}\boldsymbol{\Gamma}^{-1}\mathbf{H} + \mathbf{H}^{\mathsf{T}}\boldsymbol{\Gamma}^{-1}\boldsymbol{\Gamma}\boldsymbol{\Gamma}^{-1}\mathbf{H}\right)$$
$$= \frac{1}{|\mathcal{U}'_{mn}|}\left(G - \mathbf{H}^{\mathsf{T}}\boldsymbol{\Gamma}^{-1}\mathbf{H}\right). \tag{2.86}$$

Figure 2.23 presents an example for a honing texture. Figure 2.23a depicts the original image and Figure 2.23b presents an artificially generated texture based on an estimated AR model of order 84. The AR model is a rather generic model that does not make use of any context specific knowledge. The question of what order the AR model should have cannot be answered in general. Usually one has to resort to some trial-and-error approach until the achieved result is acceptable. Generally this leads to an unnecessary high dimension of the parameter space. Furthermore, the AR coefficients do not have an easily interpretable meaning in terms of the modeled pattern.

The next section deals with an adjusted, purpose-specific model for the same example, i.e., honing textures, as a counterpart to the general AR model.

A physically justified model for a honing texture

Honing is a machining process that scrubs the construction material with some abrasive grain material. The grain material, for example particles of ceramic or diamond, are superimposed on the so called “hone head”. The honing tool follows a predefined path while being pressed against the construction material. For example, honing is used for finishing the surface of cylinder bores of combustion engines. In this special case, the honing tool is rotated around its longitudinal axis and performs an oscillating stroke movement (see Figure 2.24a).

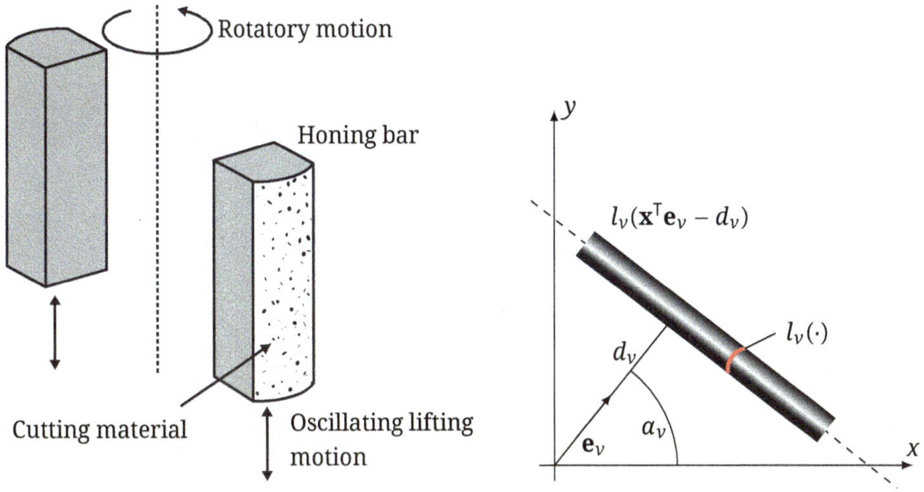

(a) Generation of a honing texture by simultaneous rotation and stroke movements

(b) Parametric model of a single honing groove

Fig. 2.24: Physical formation process and parametric model of a honing texture.

The honing texture is spawned by the additive superposition of groove profiles $g_\nu : \mathbb{R} \to \mathbb{R}$ for $\nu \in \mathbb{N}$. A groove profile specifies the gray-scale value orthogonal to the principal direction of the groove. This means it has large values around the origin and vanishes at infinity. In addition, a single groove is specified by its direction $\mathbf{e}_\nu$ and its distance from the origin d_ν (see Figure 2.24b):

$$\mathbf{x} \mapsto g_\nu\left(\mathbf{x}^\mathsf{T}\mathbf{e}_\nu - d_\nu\right). \tag{2.87}$$

Hence, the whole texture is given by

$$t(\mathbf{x}) = \sum_{\nu=-\infty}^{\infty} g_\nu\left(\mathbf{x}^\mathsf{T}\mathbf{e}_\nu - d_\nu\right). \tag{2.88}$$

As the supports of different grooves are not disjoint, their gray-scale values are added in the overlapping regions. This is an error of the model, but the error is negligible and this assumption significantly simplifies the calculation.

With $\alpha_\nu \in [0, \pi)$, the directional vector of the groove can be written as $\mathbf{e}_\nu = (\cos\alpha_\nu, \sin\alpha_\nu)^\mathsf{T}$. Then the parameters of the groove model are the angle α_ν, the distance d_ν, and the groove profile function g_ν. Due to the movement of the honing tool, the grooves normally have one out of two principal directions, i.e., a simplified parametric stochastic model is

$$p(\alpha_\nu) = \frac{1}{2}\delta(\alpha_\nu - \beta_1) + \frac{1}{2}\delta(\alpha_\nu - \beta_2) \tag{2.89}$$

with two parameters β_1 and β_2 and with $\delta(\alpha)$ denoting the Dirac distribution that is nonzero only at $\alpha = 0$.

(a) Original honing texture

(b) Artificially generated texture (physically motivated model with 14 parameters)

Fig. 2.25: Synthetic honing texture as an example of model-driven features using a physically motivated model. Compare this result to the AR model in Figure 2.23.

Likewise, the density of the grooves depends primarily on the density and distribution of the abrasive grain material on the honing tool. The distances d_v of the grooves from the origin are chosen such that they are uniformly distributed. The number q of grooves in an interval of size L is assumed to be Poisson distributed

$$P(q) = \mathrm{e}^{-\lambda L} \frac{(\lambda L)^q}{q!} \qquad q = 0, \ldots, \infty. \tag{2.90}$$

Lastly, the groove profile function $g_v(\cdot)$ is assumed to be from a parametric family of functions that is totally defined by its expectation $(\mathrm{E}\{\underline{g}_v\})(\cdot)$.

In summary, in order to learn the model, the parameters β_1, β_2, λ, and $(\mathrm{E}\{\underline{g}_v\})(\cdot)$ need to be estimated. Figure 2.25 illustrates the results. In comparison to the general AR model (see Figure 2.23), the artificially generated surface resembles the original surface much better albeit the number of parameters is much smaller. For more details on this approach, see Beyerer [1994].

2.6.3 Construction of invariant features

A general problem of choosing features is to choose those that are invariant with respect to variations within the same class. These variations can be one out of two types: Firstly, the observable patterns vary because they belong to different objects within the same class. Secondly, the same object can induce different patterns due to disturbances (see Figure 2.26). This section sheds light on the question of how to construct mappings from varying patterns onto invariant features in a systematic way.

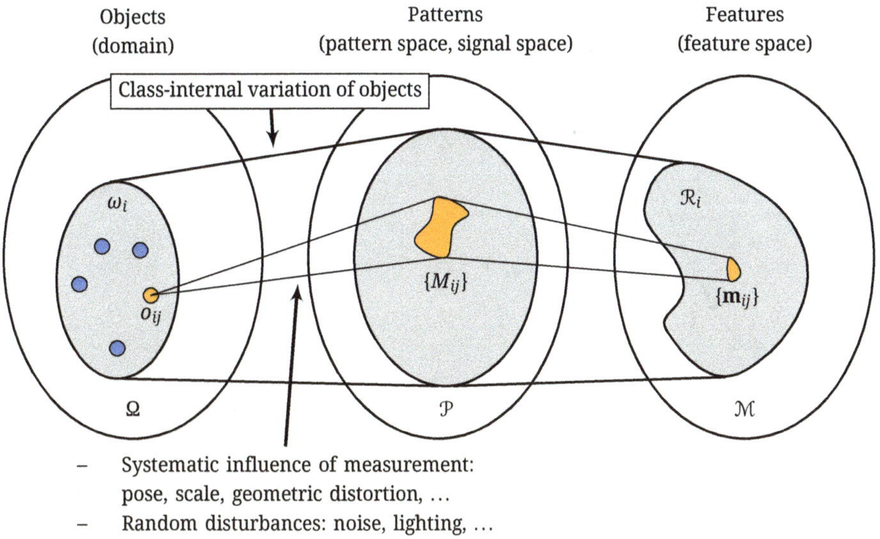

Fig. 2.26: Variations of the objects and variation of patterns due to the measurement leads to variation of the features.

For this reason this section is strongly related to Section 2.4.8. The reader is advised to recall the important concepts and definitions from that section. Section 2.4.8 tried to enforce the robustness of the distance measurement in the feature against variations of the features. The new contribution of this section is to make the feature itself invariant with respect to variations of the pattern.

Recall the situation from Section 2.4.8 and Figure 2.12. The feature space $\mathcal{M}$ is a smooth manifold and the disturbance is modeled as a Lie transformation group Π that acts on the feature space. The group action is denoted by $A: \Pi \times \mathcal{M} \rightarrow \mathcal{M}$. For an arbitrary but fixed feature $\mathbf{m}_i$, the orbit $\Pi \mathbf{m}_i$ is given by $\{A(\mathbf{p}, \mathbf{m}_i) \mid \mathbf{p} \in \Pi\}$. Refer to Appendix B for the definitions of these terms and Figure 2.27, where the orbit of a feature is illustrated.

The objective is to find a new feature $\widetilde{\mathbf{m}}$ and a suitable feature transformation $\mathbf{m}_i \mapsto \widetilde{\mathbf{m}}(\mathbf{m}_i)$ such that $\widetilde{\mathbf{m}}(\mathbf{m}_i)$ is constant on each orbit $\{A(\mathbf{p}, \mathbf{m}_i) \mid \mathbf{p} \in \Pi\}$.

To begin with, consider one toy example to illustrate the computational complexity of a brute force solution. Consider a two-dimensional point $\mathbf{x} = (x_1, x_2)^\top \in \mathbb{R}^2$. There are several options for what a suitable Lie transformation group Π could be like:

1. The translation group $\Pi = \tau = \mathbb{R}^2$ acting by $\mathbf{x}' = \mathbf{x} + \mathbf{a}$ for $\mathbf{a} \in \tau$. As the name suggests, the points are just linearly moved around in the plane. The number of degrees of freedom of this group, i.e., the dimension of the Lie transformation group, equals two.
2. The congruence group

$$\Pi = C = \left\{ (\mathbf{R}, \mathbf{a}) \,\middle|\, \mathbf{R} = \begin{pmatrix} \cos\alpha & \sin\alpha \\ -\sin\alpha & \cos\alpha \end{pmatrix}, \alpha \in \mathbb{R}, \mathbf{a} \in \mathbb{R}^2 \right\} \tag{2.91}$$

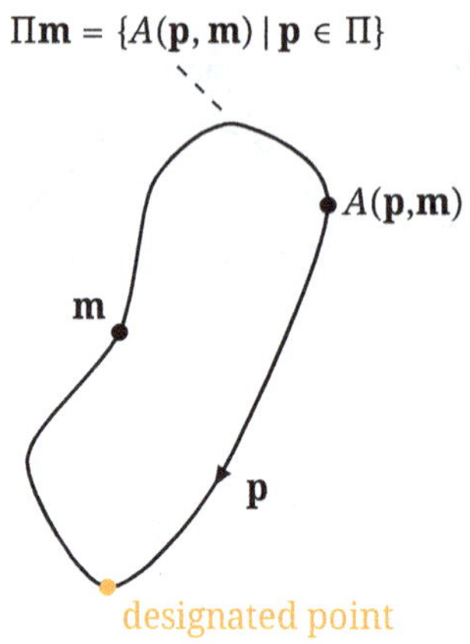

Fig. 2.27: Scheme of a feature $\mathbf{m}$ and its orbit $\Pi\mathbf{m}$.

additionally comprises rotations and has three degrees of freedom. It acts by $\mathbf{x}' = \mathbf{R}\mathbf{x} + \mathbf{a}$.

3. The similarity group

$$\Pi = S = \left\{ (\mathbf{T}, \mathbf{a}) \,\middle|\, \mathbf{T} = k \begin{pmatrix} \cos\alpha & \sin\alpha \\ -\sin\alpha & \cos\alpha \end{pmatrix}, k, \alpha \in \mathbb{R}, \mathbf{a} \in \mathbb{R}^2 \right\} \tag{2.92}$$

also allows scaling. The dimension of the Lie transformation group is four. As a congruence group, it acts by $\mathbf{x}' = \mathbf{T}\mathbf{x} + \mathbf{a}$.

4. The affine group

$$\Pi = A = \left\{ (\mathbf{P}, \mathbf{a}) \,\middle|\, \det \mathbf{P} \neq 0, \mathbf{a} \in \mathbb{R}^2 \right\} \tag{2.93}$$

includes translations, rotations, scalings, and shearing, and has six degrees of freedom. It also acts by $\mathbf{x}' = \mathbf{P}\mathbf{x} + \mathbf{a}$.

Assume that each class is represented by one feature vector $\mathbf{m}_j$ for $j = 1, \ldots, c$ and $\mathbf{m}$ is the previously unseen feature vector that should be classified. One approach for classification would be to choose $\hat{\omega} = \omega_i$ with

$$(i, \mathbf{p}^*) = \underset{\substack{j \in \{1,\ldots,c\} \\ \mathbf{p} \in \Pi}}{\arg\min} \left\| \mathbf{m} - A(\mathbf{p}, \mathbf{m}_j) \right\| . \tag{2.94}$$

This means that one has to calculate all the transformations of all the classes and choose the class that comes closest to the provided unseen feature vector. Unfortunately, the complexity grows exponentially with the dimension of the Lie transformation group. In order to obtain a feeling for this implication, consider the six-dimensional affine group from above and assume that each dimension is discretized by 10^3 steps. This results in the computation of 10^{18} values per class. A machine able to perform 10^9 of these computations per second would need 31.7 years to classify just one sample. These numbers clearly show that a brute force approach is not an option.

Given a feature $\mathbf{m}$ and its orbit $\Pi\mathbf{m} = \{A(\mathbf{p},\mathbf{m}) \mid \mathbf{p} \in \Pi\}$ defined by the group action $A(\mathbf{p},\mathbf{m})$ as illustrated in Figure 2.27, there are three approaches to systematically constructing invariant features:

1. The integral method:

$$\widetilde{\mathbf{m}} = \int_{\Pi} f\left(A(\mathbf{p},\mathbf{m})\right) \mathrm{d}\mathbf{p}. \tag{2.95}$$

2. The differential method:

$$\widetilde{\mathbf{m}} \quad \text{with} \quad \frac{\partial \widetilde{\mathbf{m}}\left(A(\mathbf{p},\mathbf{m})\right)}{\partial \mathbf{p}} \stackrel{!}{=} 0. \tag{2.96}$$

3. Normalization: Trace the feature vector back to a designated point of the orbit.

The first (respectively, second) method integrates (respectively, differentiates) with respect to the Lie transformation group. Before we are going to explain how this can be done, we will explain the general idea of both methods.

The integral method

The integral method can be seen as some kind of averaging. The new feature vector $\widetilde{\mathbf{m}}$ is invariant under group actions on $\mathbf{m}$ if the integral is calculated on the whole orbit of $\mathbf{m}$ anyway. I.e., the orbits are always the same but with different starting points. The function f maps a group element to some range such that f is integrable. Normally, one chooses the real or complex numbers as the range. The function f is a design parameter and must be carefully chosen for the specific problem. For example, if the feature vector itself is a vector of real or complex number, f might be chosen to be a polynomial. Of course, any kind of averaging implies a loss of information. But this is intentional, because an invariant feature is supposed to hide the difference between the transformed patterns of the same object. On the other hand, one must ensure that not too much information is lost, because the integral must still return different values for different classes. For example, the choice $f \equiv 0$ forces the integral to be always zero. Without a doubt, this is a perfect—but obviously too restrictive—invariant. There is no choice of f that is generally applicable to just any feature.

The differential method

The differential method copes with the problem of the first method of how to find a good f. The differential method does not need an additional design parameter, but tackles the problem directly. The goal is to find a new feature vector $\widetilde{\mathbf{m}}$ that is a function of the old feature vector $\mathbf{m}$ by design. The requirement that $\widetilde{\mathbf{m}}$ must not vary if the group element varies, is exactly what is expressed by the condition that the derivative with respect to the group element is zero. This leads to a partial differential equation whose solution is the wanted function

$$\mathbf{m} \mapsto \widetilde{\mathbf{m}}(\mathbf{m}). \tag{2.97}$$

In general, the differential equation is nonlinear and not easy to solve.

Actually, the integral and differential methods are two sides of the same story and are normally equally easy or difficult. The integral method requires a good educated guess for the correct function f and then directly provides a straightforward expression of the mapping $\mathbf{m} \mapsto \widetilde{\mathbf{m}}(\mathbf{m})$. The differential method does not need such a function f appearing from nowhere, but it is questionable whether there is a solution to the resulting differential equation.

Before the last method, "normalization", is discussed, the question of how one integrates or differentiates with respect to a group remains to be answered. This question cannot be satisfactorily answered in this book, because a rigorous answer requires mathematical concepts such as "simply connected Riemannian manifolds", "geodesic completeness", "maximal normal neighborhood", and "cut locus". Instead, we will present a very simple example so that the reader gets a glimpse that it is indeed possible.

An example of the integral method

Reconsider the example from Section 2.4.8. The feature vector is a point in the two-dimensional Euclidean plane and the Lie transformation group (see Appendix B) is the set of the rotation matrices of the plane

$$\mathcal{G} = \mathrm{SO}(2) = \left\{ \begin{pmatrix} \cos\alpha & \sin\alpha \\ -\sin\alpha & \cos\alpha \end{pmatrix} \middle| \alpha \in \mathbb{R} \right\}. \tag{2.98}$$

The group acts on the points by the usual multiplication of a vector by a matrix and the orbits are circles centered at the origin. The integral approach leads to

$$\begin{aligned}
\widetilde{\mathbf{m}} &= \int_{\mathcal{G}} f\left(A(g, \mathbf{m})\right) \mathrm{d}g \\
&= \int_{-\pi}^{\pi} f\left(\begin{pmatrix} \cos\alpha & \sin\alpha \\ -\sin\alpha & \cos\alpha \end{pmatrix} \begin{pmatrix} m_1 \\ m_2 \end{pmatrix}\right) \mathrm{d}\alpha \\
&= \int_{-\pi}^{\pi} f\left(m_1 \cos\alpha + m_2 \sin\alpha, m_2 \cos\alpha - m_1 \sin\alpha\right) \mathrm{d}\alpha \\
&\quad \left(\text{educated guess: } f(u, v) = \frac{1}{2\pi}(u^2 + v^2)\right) \\
&= \frac{1}{2\pi} \int_{-\pi}^{\pi} \left((m_1 \cos\alpha + m_2 \sin\alpha)^2 + (m_2 \cos\alpha - m_1 \sin\alpha)^2\right) \mathrm{d}\alpha \\
&= \frac{1}{2\pi} \int_{-\pi}^{\pi} \left(m_1^2 + m_2^2\right) \underbrace{\left(\cos^2\alpha + \sin^2\alpha\right)}_{=1} \mathrm{d}\alpha = m_1^2 + m_2^2.
\end{aligned} \tag{2.99}$$

This result is correct, because up to a missing root, $m_1^2 + m_2^2$ is the (squared) distance of the point from the origin. This is an invariance with respect to rotations around the origin, because it equals the radius of the orbit. Although we lose information about the precise location of the feature **m**, we do not lose too much information, because the distance from the origin still suffices to distinguish different orbits. The drawback is that we had to guess the suitable function f.

The general idea of calculating an integral with respect to a Lie transformation group is to express a group element $g \in \mathcal{G}$ by its so-called normal coordinates. In this case the normal coordinate representation is $g(\alpha) = \left(\begin{smallmatrix} \cos\alpha & \sin\alpha \\ -\sin\alpha & \cos\alpha \end{smallmatrix}\right)$. The domain of the normal coordinates is some isometry of the $\mathbb{R}^l$ and hence the integral can be pulled back to an already known integral over the real numbers. But finding the normal coordinate representation of a Lie group is not always as easy as this example might pretend.

An example of the differential method

The same course of action is applied to the differential approach. If one uses the normal coordinate representation, the differential approach of the toy example in Equation (2.98) simplifies to

$$\begin{aligned}
0 &\overset{!}{=} \frac{\partial \widetilde{\mathbf{m}}(A(g, \mathbf{m}))}{\partial g} \\
&= \frac{\partial}{\partial \alpha} \widetilde{\mathbf{m}} \left(\begin{pmatrix} \cos\alpha & \sin\alpha \\ -\sin\alpha & \cos\alpha \end{pmatrix} \begin{pmatrix} m_1 \\ m_2 \end{pmatrix} \right) \\
&\text{(assumption: } \widetilde{m}: \mathbb{R}^2 \to \mathbb{R}, \left(\begin{smallmatrix} u \\ v \end{smallmatrix}\right) \mapsto \widetilde{m}\left(\begin{smallmatrix} u \\ v \end{smallmatrix}\right)) \\
&= (m_2 \cos\alpha - m_1 \sin\alpha) \frac{\partial}{\partial u} \widetilde{m} \begin{pmatrix} m_1 \cos\alpha + m_2 \sin\alpha \\ m_2 \cos\alpha - m_1 \sin\alpha \end{pmatrix} \\
&\quad - (m_1 \cos\alpha + m_2 \sin\alpha) \frac{\partial}{\partial v} \widetilde{m} \begin{pmatrix} m_1 \cos\alpha + m_2 \sin\alpha \\ m_2 \cos\alpha - m_1 \sin\alpha \end{pmatrix} \\
&\text{(substitution: } \xi = m_2 \cos\alpha - m_1 \sin\alpha, \chi = (m_1 \cos\alpha + m_2 \sin\alpha)) \\
&= \xi \frac{\partial}{\partial u} \widetilde{m} \begin{pmatrix} \xi \\ \chi \end{pmatrix} - \chi \frac{\partial}{\partial v} \widetilde{m} \begin{pmatrix} \xi \\ \chi \end{pmatrix}.
\end{aligned} \tag{2.100}$$

Close inspection of the last line reveals that

$$\widetilde{m} \begin{pmatrix} u \\ v \end{pmatrix} = u^2 + v^2 \tag{2.101}$$

is one solution of the partial differential equation, because $\frac{\partial}{\partial u} \widetilde{m}\left(\begin{smallmatrix} u \\ v \end{smallmatrix}\right) = 2u$ and $\frac{\partial}{\partial v} \widetilde{m}\left(\begin{smallmatrix} u \\ v \end{smallmatrix}\right) = 2v$, and therefore $\xi \cdot 2\chi - \chi \cdot 2\xi = 0$ follows.

By definition, $\widetilde{m}$ is invariant under group actions and we obtain

$$\widetilde{m} \begin{pmatrix} m_1 \\ m_2 \end{pmatrix} = \widetilde{\mathbf{m}} \left(\begin{pmatrix} \cos\alpha & \sin\alpha \\ -\sin\alpha & \cos\alpha \end{pmatrix} \begin{pmatrix} m_1 \\ m_2 \end{pmatrix} \right) = m_1^2 + m_2^2 \tag{2.102}$$

for any α. This is the same result as obtained by the integral approach. Instead of guessing some function f, the problem is rather to find a formula for the solution.

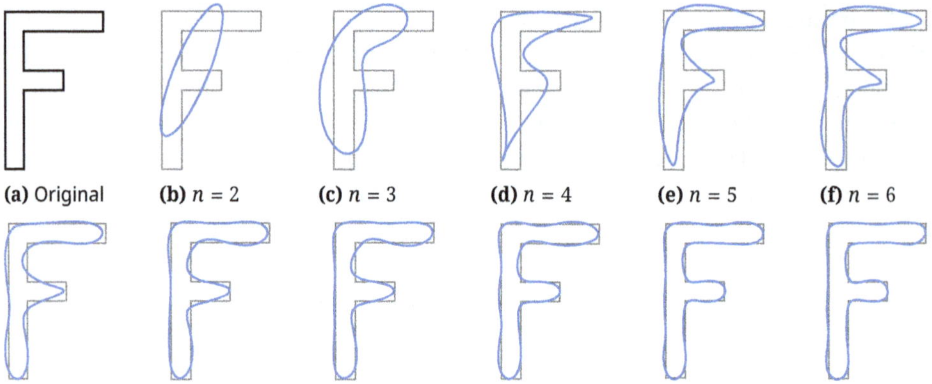

(a) Original (b) $n = 2$ (c) $n = 3$ (d) $n = 4$ (e) $n = 5$ (f) $n = 6$

(g) $n = 7$ (h) $n = 8$ (i) $n = 9$ (j) $n = 10$ (k) $n = 11$ (l) $n = 12$

Fig. 2.28: Synthesis of a two-dimensional contour with n Fourier coefficients $Z_{-(n-1)}, \ldots, Z_0, \ldots, Z_{n-1} \in \mathbb{C}$.

Normalization by example

In contrast to the first two methods, normalization does not provide a predetermined course of action. The general idea is to pull back each feature vector to a designated point on the orbit (see Figure 2.27). In other words, each orbit is characterized by a single canonical representative. Hence, if $\mathbf{m}$ is the feature vector, one needs to find the corresponding group element $g \in \mathcal{G}$ such that $A(g, \mathbf{m})$ maps to the representative of the orbit of g. Obviously, g depends on $\mathbf{m}$ and the question of how g can be calculated remains an open question for the general case. In this section, an example of two-dimensional contours is presented.

A two-dimensional contour (see Figure 2.28) can be considered as a continuous closed curve in the Euclidean plane given by

$$\mathbf{z}(l) = \begin{pmatrix} x(l) \\ y(l) \end{pmatrix} \quad \text{with} \quad x\colon [0, L] \to \mathbb{R} \quad \text{and} \quad y\colon [0, L] \to \mathbb{R} \tag{2.103}$$

with boundary condition $x(0) = x(L)$ and $y(0) = y(L)$. As x and y are continuous functions with period L, they are especially suited to be expanded in Fourier series,

$$x(l) = \sum_{n=-\infty}^{\infty} X_n \mathrm{e}^{\frac{2\pi l}{L}\mathrm{j}n} \qquad \text{with} \qquad X_n = \frac{1}{L}\int_0^L x(l)\mathrm{e}^{-\frac{2\pi l}{L}\mathrm{j}n}\,\mathrm{d}l, \tag{2.104}$$

$$y(l) = \sum_{n=-\infty}^{\infty} Y_n \mathrm{e}^{\frac{2\pi l}{L}\mathrm{j}n} \qquad \text{with} \qquad Y_n = \frac{1}{L}\int_0^L y(l)\mathrm{e}^{-\frac{2\pi l}{L}\mathrm{j}n}\,\mathrm{d}l. \tag{2.105}$$

Note that X_n and Y_n are complex values, but the additional restriction $X_n^* = X_{-n}$ and $Y_n^* = Y_{-n}$ holds, so that the imaginary parts are canceled pairwise, because x and y are real functions. Hence, $\mathbf{z}(l)$ can be written as

$$\mathbf{z}(l) = \sum_{n=-\infty}^{\infty} Z_n \mathrm{e}^{\frac{2\pi l}{L}\mathrm{j}n} \qquad \text{with} \qquad Z_n = \begin{pmatrix} X_n \\ Y_n \end{pmatrix}. \tag{2.106}$$

The Fourier series can be written in a more compact form if $\mathbf{z}(l)$ is not a two-dimensional real vector, but considered as a complex function, i.e.,

$$z(l) = x(l) + \mathrm{j}y(l). \tag{2.107}$$

The Fourier coefficients are equally added

$$Z_n = X_n + \mathrm{j}Y_n \quad \text{or equivalently} \quad Z_n = \frac{1}{L}\int_0^L z(l)\mathrm{e}^{-\frac{2\pi l}{L}\mathrm{j}n}\,\mathrm{d}l. \tag{2.108}$$

The property $X_n^* = X_{-n}$ and $Y_n^* = Y_{-n}$ does not carry over to the coefficients Z_n, because $z(l)$ is a true complex function and therefore the imaginary parts do not cancel in general. On the one hand, one has

$$Z_n^* = X_n^* + (\mathrm{j}Y_n)^* = X_n^* - \mathrm{j}Y_n^* \tag{2.109}$$

and on the other hand, one has

$$Z_{-n} = X_{-n} + \mathrm{j}Y_{-n} = X_n^* + \mathrm{j}Y_n^*. \tag{2.110}$$

From now on, we look at $z(l)$ as a true complex function and it is assumed that the coefficients $Z_0, Z_1, Z_{-1}, Z_2, Z_{-2}$ are not restricted to be pairwise complex conjugates.

At this point, one has a feature vector

$$\mathbf{m} = (Z_0, Z_1, Z_{-1}, \ldots, Z_n, Z_{-n})^\mathsf{T} \tag{2.111}$$

which is a truncated Fourier series that approximately describes a closed contour. Now let us examine how normalization is used to make this feature vector invariant by pulling it back to a canonical representative.

If the contour $z(l)$ is translated by a fixed (complex) value $a \in \mathbb{C}$ to $z'(l) = z(l) + a$, then the new feature vector $\mathbf{m}'$ becomes

$$\mathbf{m}' = \mathbf{m} + (a, 0, \ldots, 0)^\mathsf{T} = (Z_0 + a, Z_1, Z_{-1}, \ldots, Z_n, Z_{-n})^\mathsf{T}. \tag{2.112}$$

A translation only affects the first coefficient Z_0. Actually, this coefficient is nothing else than the center of mass of the contour. Hence, omitting this coefficient (or implicitly setting it to zero) describes the same contour with its center of mass moved to the origin. A translation invariant feature vector is therefore

$$\widetilde{\mathbf{m}} = (Z_1, Z_{-1}, \ldots, Z_n, Z_{-n})^\mathsf{T}. \tag{2.113}$$

Scaling invariance is also easy to obtain. If the contour $z(l)$ is scaled by a real, positive value $a \in \mathbb{R}^{>0}$ to $z'(l) = a\,z(l)$, all coefficients are scaled by the same value

$$\mathbf{m}' = a\,\mathbf{m} = (aZ_1, aZ_{-1}, \ldots, aZ_n, aZ_{-n})^\mathsf{T}. \tag{2.114}$$

Hence, dividing all coefficients by the absolute value of the first element yields a scaling invariant feature vector

$$\widetilde{\mathbf{m}} = \left(\frac{Z_1}{|Z_1|}, \frac{Z_{-1}}{|Z_1|}, \ldots, \frac{Z_n}{|Z_1|}, \frac{Z_{-n}}{|Z_1|} \right)^{\mathsf{T}}. \tag{2.115}$$

This step resulted in a feature vector whose first component has an absolute value of one, but an arbitrary direction, on the complex unit circle. A rotation (but not scale) invariant feature vector is obtained if all coefficients are multiplied in such a way that the coefficient Z_1 points in the direction of the real axis, i.e., if the first coefficient becomes a positive real number. This is true because all coefficients are multiplied by the same value $\mathrm{e}^{\mathrm{j}\alpha}$ if the contour $z(l)$ is multiplied by $\mathrm{e}^{\mathrm{j}\alpha}$. This is a rotation in the complex plane. Let

$$\varphi_1 = \operatorname{Arg} Z_1 \in (-\pi, \pi] \tag{2.116}$$

denote the argument of the first coefficient. This means that

$$Z_1 = |Z_1| \, \mathrm{e}^{\mathrm{j}\varphi_1}. \tag{2.117}$$

Hence

$$\begin{aligned} \widetilde{\mathbf{m}} &= \left(Z_1 \mathrm{e}^{-\mathrm{j}\varphi_1}, Z_{-1} \mathrm{e}^{-\mathrm{j}\varphi_1}, \ldots, Z_n \mathrm{e}^{-\mathrm{j}\varphi_1}, Z_{-n} \mathrm{e}^{-\mathrm{j}\varphi_1} \right)^{\mathsf{T}} \\ &= \left(|Z_1|, Z_{-1} \mathrm{e}^{-\mathrm{j}\varphi_1}, \ldots, Z_n \mathrm{e}^{-\mathrm{j}\varphi_1}, Z_{-n} \mathrm{e}^{-\mathrm{j}\varphi_1} \right)^{\mathsf{T}} \end{aligned} \tag{2.118}$$

is a rotation (but not scale) invariant feature vector. In other words, the orientation of the contour is encoded in the phases of the coefficients.

If the last two steps are combined, one obtains a scale and rotation invariant feature vector. Of course, this actually means that all coefficients are divided by Z_1. As the first coefficient becomes one, it can implicitly be omitted.

In summary, let

$$\mathbf{m} = (Z_0, Z_1, Z_{-1}, Z_2, Z_{-2}, \ldots, Z_n, Z_{-n})^{\mathsf{T}} \tag{2.119}$$

be the feature vector of the Fourier series approximation of a contour with n coefficients. Then

$$\widetilde{\mathbf{m}} = \left(\widetilde{Z}_{-1}, \widetilde{Z}_2, \widetilde{Z}_{-2}, \ldots, \widetilde{Z}_n, \widetilde{Z}_{-n} \right)^{\mathsf{T}} = \left(\frac{Z_{-1}}{Z_1}, \frac{Z_2}{Z_1}, \frac{Z_{-2}}{Z_1}, \ldots, \frac{Z_n}{Z_1}, \frac{Z_{-n}}{Z_1} \right)^{\mathsf{T}} \tag{2.120}$$

is an example of a translation, scale and rotation invariant feature vector of the contour. The values $\widetilde{Z}_0 = 0$ and $\widetilde{Z}_1 = 1$ are implicitly omitted. More generally, because $z'(l) = a\mathrm{e}^{\mathrm{j}\alpha} z(l)$, with $a \in \mathbb{R}^{>0}$ and $\alpha \in \mathbb{R}$, leads to $Z'_k = Z_k a \mathrm{e}^{\mathrm{j}\alpha}$, each ratio $\widetilde{m} = \frac{Z_n}{Z_m}$, with $m,n \neq 0$, is a feature that is invariant w.r.t. translation, scaling, and rotation.

Besides translation, scale, and rotation invariance, another desirable property for features of a periodic function as a contour is the invariance with regard to the starting point. A change of the starting point of the contour can be expressed as $z'(l) = z(l + \alpha)$.

Tab. 2.3: Contour transformations and approaches for constructing invariant features (normalization).

	Translation	Scaling	Rotation	Start point variation
Effect on contour	$z'(l) = z(l) + a$	$z'(l) = az(l)$	$z'(l) = z(l)\mathrm{e}^{\mathrm{j}a}$	$z'(l) = z(l + a)$
Effect on coefficients	$Z'_n = Z_n + a\delta_0^n$	$Z'_n = aZ_n$	$Z'_n = Z_n\mathrm{e}^{\mathrm{j}a}$	$Z'_n = Z_n\mathrm{e}^{\mathrm{j}2\pi na/L}$
Normalization	Ignore Z_0	Ratios $\frac{Z_n}{Z_m}$	Ratios $\frac{Z_n}{Z_m}$, Absolute val. $\lvert Z_n\rvert$	Absolute val. $\lvert Z_n\rvert$, Phase diff. $n\varphi_m - m\varphi_n$

Using Equation (2.108), one can derive a formula that states how the coefficients Z_n change, when the starting point is translated with $a \in \mathbb{R}$:

$$Z'_n = \frac{1}{L}\int_0^L z(l + a)\mathrm{e}^{-\frac{2\pi l}{L}\mathrm{j}n}\,\mathrm{d}l \overset{k:=l+a}{=} \frac{1}{L}\int_a^{L+a} z(k)\mathrm{e}^{-\frac{2\pi(k-a)}{L}\mathrm{j}n}\,\mathrm{d}k \tag{2.121}$$

$$= \frac{1}{L}\int_0^L z(k)\mathrm{e}^{-\frac{2\pi k}{L}\mathrm{j}n}\,\mathrm{d}k\,\mathrm{e}^{\frac{2\pi a}{L}\mathrm{j}n} = Z_n\mathrm{e}^{\frac{2\pi a}{L}\mathrm{j}n}. \tag{2.122}$$

As with rotation, changing the starting point affects only the phases of the coefficients but not the magnitudes. Thus, a start point invariant feature can be obtained by regarding only the magnitudes of the coefficients $|Z_n|$. Building upon Equation (2.120), the feature

$$\widetilde{\mathbf{m}} = \left(\widetilde{Z}_{-1}, \widetilde{Z}_2, \widetilde{Z}_{-2}, \ldots, \widetilde{Z}_n, \widetilde{Z}_{-n}\right)^{\mathrm{T}} \tag{2.123}$$

$$= \left(\left|\frac{Z_{-1}}{Z_1}\right|, \left|\frac{Z_2}{Z_1}\right|, \left|\frac{Z_{-2}}{Z_1}\right|, \ldots, \left|\frac{Z_n}{Z_1}\right|, \left|\frac{Z_{-n}}{Z_1}\right|\right)^{\mathrm{T}} \tag{2.124}$$

is invariant to translation, scaling, rotation, and change of starting point.

The four discussed transformations, their effects on the contour and Fourier coefficients, as well as normalization approaches for constructing invariant features are summarized in Table 2.3.

2.7 Dimensionality reduction of the feature space

Generally, a high dimension of the feature space is unfavorable, for reasons that will be explained in Section 6.1. The last section of this chapter will treat the question of how a high dimension of the feature space can be reduced.

The presentation starts with the concepts of principal component analysis (PCA) and independent component analysis (ICA). Both methods derive new features by combining the original features and projecting the result to a subspace of smaller dimension. They share the objective of approximately representing the collection of all samples $\mathcal{D}$ with the desired (lower) dimensional space, so that when one would reconstruct the samples in the original feature space, the mean square error between the original features and the reconstructed features is minimized. In other words, these methods do not take the class affiliation into consideration but regard the whole collection of samples $\mathcal{D}$ at once.

The third method this section will treat is the multiple discriminant analysis (MDA). This method initially focuses on optimal class separation, but apart from that, it operates similarly to the other two methods. That is, this method also calculates combinations of the original features and projects them to a subspace. But the way the combinations and projections are calculated differs.

As fourth method, the *t*-distributed stochastic neighbor embedding will be examined for the use case of dimensionality reduction. Although it is typically used for visualization purposes only, it preserves the local proximity of high-dimensional data points and is therefore suitable for the task of dimensionality reduction.

The fifth method described in this section is called autoencoder (AE), an artificial neural network for data compression. The basic idea is to transform the high-dimensional features to low-dimensional features that are transformed back to features of the original dimension, while aiming to reproduce the original features without error. The intermediate features can then be used as low-dimensional representation of the high-dimensional input features.

All these methods suffer from two drawbacks. First, they only work for features on an interval scale at least, because subtraction and scalar multiplication must be defined. Second, the descriptive meaning of the original features might get lost. Instead of concrete features these methods might generally return opaque transformations.

For these reasons, the last method presents a systematic way of selecting a subset of the original feature vector such that the smaller set of features is still good enough. Throwing away some components is the same as a projection but it keeps the axis "as is". The advantage of this method is that it works for arbitrary kinds of features and retains the meaning of the original features. The disadvantage is that it is less powerful, because it does not make use of combinations of the features.

2.7.1 Principal component analysis

The idea of principal component analysis (PCA) is to find a lower dimensional subspace such that the data is optimally represented in terms of the mean square error. This subsection proceeds as follows. First, the method is presented for the case where the subspace is chosen to be zero-, one-, or two-dimensional, because these cases can be easily depicted and they descriptively convey the underlying idea. Then the general case is presented for an arbitrary number of dimensions. At the end, the method of principal component analysis is generalized to *kernelized principal component analysis*, which uses in addition a nonlinear transformation in order to improve the representation.

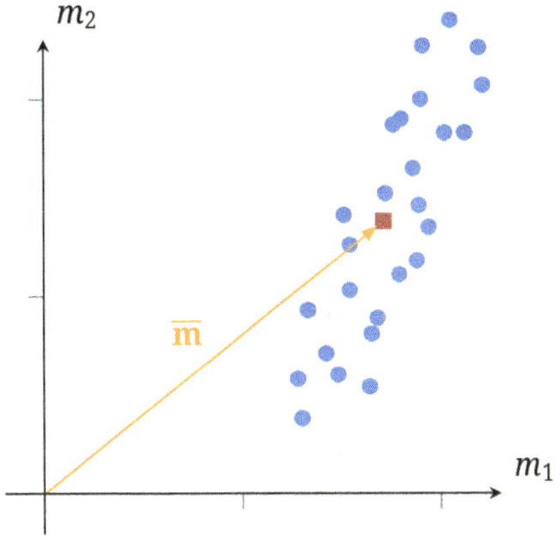

Fig. 2.29: Principal component analysis, first step: Finding the point with minimal reconstruction error.

First steps and general deduction

Let $\mathcal{D} = \{\mathbf{m}_1, \ldots, \mathbf{m}_N\}$ be a set of feature vectors. A zero-dimensional space is a point and the projection onto this space is that point itself. Hence one tries to find that point $\mathbf{m}_0$ such that the sum of all squared distances between it and the points $\mathbf{m}_1, \ldots, \mathbf{m}_N$ is minimal. This means the objective is to find the minimum of

$$J_0(\mathbf{m}) = \sum_{k=1}^{N} \|\mathbf{m} - \mathbf{m}_k\|^2. \tag{2.125}$$

This can be rewritten as

$$\begin{aligned} J_0(\mathbf{m}) &= \sum_{k=1}^{N} \|\mathbf{m} - \mathbf{m}_k\|^2 = \sum_{k=1}^{N} \|(\mathbf{m} - \overline{\mathbf{m}}) - (\mathbf{m}_k - \overline{\mathbf{m}})\|^2 \\ &= \sum_{k=1}^{N} \|\mathbf{m} - \overline{\mathbf{m}}\|^2 - 2\sum_{k=1}^{N} (\mathbf{m} - \overline{\mathbf{m}})^\top (\mathbf{m}_k - \overline{\mathbf{m}}) + \sum_{k=1}^{N} \|\mathbf{m}_k - \overline{\mathbf{m}}\|^2 \\ &= \sum_{k=1}^{N} \|\mathbf{m} - \overline{\mathbf{m}}\|^2 - 2(\mathbf{m} - \overline{\mathbf{m}})^\top \underbrace{\sum_{k=1}^{N} (\mathbf{m}_k - \overline{\mathbf{m}})}_{=0} + \sum_{k=1}^{N} \|\mathbf{m}_k - \overline{\mathbf{m}}\|^2 \\ &= \underbrace{\sum_{k=1}^{N} \|\mathbf{m} - \overline{\mathbf{m}}\|^2}_{\text{to be minimized}} + \underbrace{\sum_{k=1}^{N} \|\mathbf{m}_k - \overline{\mathbf{m}}\|^2}_{\text{fixed, independent of } \mathbf{m}}. \end{aligned} \tag{2.126}$$

This shows that the point with the least squared distance to the $\mathbf{m}_1, \ldots, \mathbf{m}_N$ is the center of mass of these points (see Figure 2.29),

$$\mathbf{m}_0 = \arg\min_{\mathbf{m}} J_0(\mathbf{m}) = \overline{\mathbf{m}}. \tag{2.127}$$

Iteratively, one can now seek for the one-dimensional line that best represents the points. That line is given by

$$\mathbf{m}_{\mathbf{e}}(a) = \overline{\mathbf{m}} + a\mathbf{e} \quad \text{with} \quad a \in \mathbb{R}, \mathbf{e} \in \mathbb{R}^d \text{ and } \|\mathbf{e}\| = 1 \tag{2.128}$$

where $\mathbf{e}$ denotes the normalized directional vector and a the scalar parameter of the line. Let $\breve{\mathbf{m}}_k = \overline{\mathbf{m}} + a_k\mathbf{e}$ denote the orthogonal projection of the feature $\mathbf{m}_k$ onto the line. The optimization functional is

$$J_1(a_1,\dots,a_N,\mathbf{e}) = \sum_{k=1}^{N} \|\breve{\mathbf{m}}_k - \mathbf{m}_k\|^2 = \sum_{k=1}^{N} \|\overline{\mathbf{m}} + a_k\mathbf{e} - \mathbf{m}_k\|^2. \tag{2.129}$$

The scalars $a_1, \dots, a_N$ vary if the directional $\mathbf{e}$ vector varies. Hence, we minimize the functional in two steps. Firstly, $a_1, \dots, a_N$ are chosen such that J_1 is minimized for an arbitrary, but fixed, $\mathbf{e}$. The result is a set of scalars, depending on $\mathbf{e}$, i.e., $a_1(\mathbf{e}), \dots, a_N(\mathbf{e})$. Secondly, $\mathbf{e}$ is chosen to minimize J_1. Once again we rewrite J_1 as (note that $\|\mathbf{e}\| = 1$)

$$\begin{aligned}
J_1(a_1,\dots,a_N,\mathbf{e}) &= \sum_{k=1}^{N} \|\overline{\mathbf{m}} + a_k\mathbf{e} - \mathbf{m}_k\|^2 \\
&= \sum_{k=1}^{N} \|a_k\mathbf{e} - (\mathbf{m}_k - \overline{\mathbf{m}})\|^2 \\
&= \sum_{k=1}^{N} \|a_k\mathbf{e}\|^2 - 2\sum_{k=1}^{N} a_k\mathbf{e}^{\mathsf{T}}(\mathbf{m}_k - \overline{\mathbf{m}}) + \sum_{k=1}^{N} \|\mathbf{m}_k - \overline{\mathbf{m}}\|^2 \\
&= \sum_{k=1}^{N} a_k^2 - 2\sum_{k=1}^{N} a_k\mathbf{e}^{\mathsf{T}}(\mathbf{m}_k - \overline{\mathbf{m}}) + \sum_{k=1}^{N} \|\mathbf{m}_k - \overline{\mathbf{m}}\|^2.
\end{aligned} \tag{2.130}$$

As the minimum is an inner point, it suffices to find the point where the first derivatives are zero:

$$\frac{\partial}{\partial a_k} J_1(a_1,\dots,a_N,\mathbf{e}) = 2a_k - 2\mathbf{e}^{\mathsf{T}}(\mathbf{m}_k - \overline{\mathbf{m}}) \stackrel{!}{=} 0 \Leftrightarrow a_k = \mathbf{e}^{\mathsf{T}}(\mathbf{m}_k - \overline{\mathbf{m}}). \tag{2.131}$$

Putting this solution into the last line of Equation (2.130) yields

$$\begin{aligned}
J_1(\mathbf{e}) &= \sum_{k=1}^{N} a_k^2 - 2\sum_{k=1}^{N} a_k^2 + \sum_{k=1}^{N} \|\mathbf{m}_k - \overline{\mathbf{m}}\|^2 \\
&= -\sum_{k=1}^{N} a_k^2 + \sum_{k=1}^{N} \|\mathbf{m}_k - \overline{\mathbf{m}}\|^2 \\
&= -\sum_{k=1}^{N} \mathbf{e}^{\mathsf{T}}(\mathbf{m}_k - \overline{\mathbf{m}})(\mathbf{m}_k - \overline{\mathbf{m}})^{\mathsf{T}}\mathbf{e} + \sum_{k=1}^{N} \|\mathbf{m}_k - \overline{\mathbf{m}}\|^2 \\
&= -\mathbf{e}^{\mathsf{T}}\mathbf{S}\mathbf{e} + \underbrace{\sum_{k=1}^{N} \|\mathbf{m}_k - \overline{\mathbf{m}}\|^2}_{\text{fix, independent of } \mathbf{e}} \quad \text{with } \mathbf{S} := \sum_{k=1}^{N} (\mathbf{m}_k - \overline{\mathbf{m}})(\mathbf{m}_k - \overline{\mathbf{m}})^{\mathsf{T}}.
\end{aligned} \tag{2.132}$$

The matrix

$$\mathbf{S} := \sum_{k=1}^{N} (\mathbf{m}_k - \overline{\mathbf{m}})(\mathbf{m}_k - \overline{\mathbf{m}})^{\mathsf{T}} \in \mathbb{R}^{d\times d} \tag{2.133}$$

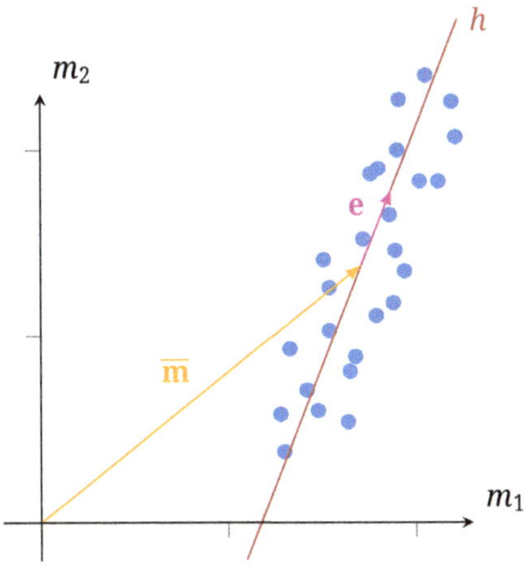

Fig. 2.30: Principal component analysis, second step: finding the line with minimal reconstruction error.

is called the *scatter matrix* and the term

$$\mathbf{e}^\mathsf{T}\mathbf{S}\mathbf{e} \qquad \text{with } \|\mathbf{e}\| = \mathbf{e}^\mathsf{T}\mathbf{e} = 1 \tag{2.134}$$

must be maximized in order to minimize $J_1(\mathbf{e})$. The method of Lagrange multipliers, with multiplier $\lambda(\mathbf{e}^\mathsf{T}\mathbf{e} - 1)$, yields

$$\begin{aligned} &\frac{\partial}{\partial \mathbf{e}}\left(\mathbf{e}^\mathsf{T}\mathbf{S}\mathbf{e} - \lambda(\mathbf{e}^\mathsf{T}\mathbf{e} - 1)\right) \stackrel{!}{=} 0 \\ \Leftrightarrow\ & 2\mathbf{S}\mathbf{e} - 2\lambda\mathbf{e} = 0 \\ \Leftrightarrow\ & \mathbf{S}\mathbf{e} = \lambda\mathbf{e} \\ \Rightarrow\ & \mathbf{e}^\mathsf{T}\mathbf{S}\mathbf{e} = \lambda\mathbf{e}^\mathsf{T}\mathbf{e} = \lambda. \end{aligned} \tag{2.135}$$

The line before the last line shows that the sought value of λ is an eigenvalue of the matrix $\mathbf{S}$. Since $\mathbf{S}$ is symmetric by construction (see Equation (2.133)), it is diagonalizable and such an eigenvalue must exist. The last line reveals that the greatest eigenvalue must be picked to maximize $\mathbf{e}^\mathsf{T}\mathbf{S}\mathbf{e}$.

In summary, the best line has a base point at the center of mass and the same direction as the eigenvector with the largest eigenvalue of the scatter matrix (see Figure 2.30).

In order to complete the usual notation, let the column-wise concatenation of the zero mean feature vectors

$$\mathbf{M} := \left(\mathbf{m}_1 - \overline{\mathbf{m}}, \quad \dots, \quad \mathbf{m}_N - \overline{\mathbf{m}}\right) \in \mathbb{R}^{d\times N} \tag{2.136}$$

denote the so-called *data matrix*. Then the scatter matrix can be written as

$$\mathbf{S} = \sum_{k=1}^{N}(\mathbf{m}_k - \overline{\mathbf{m}})(\mathbf{m}_k - \overline{\mathbf{m}})^\mathsf{T} = \mathbf{M}\mathbf{M}^\mathsf{T}. \tag{2.137}$$

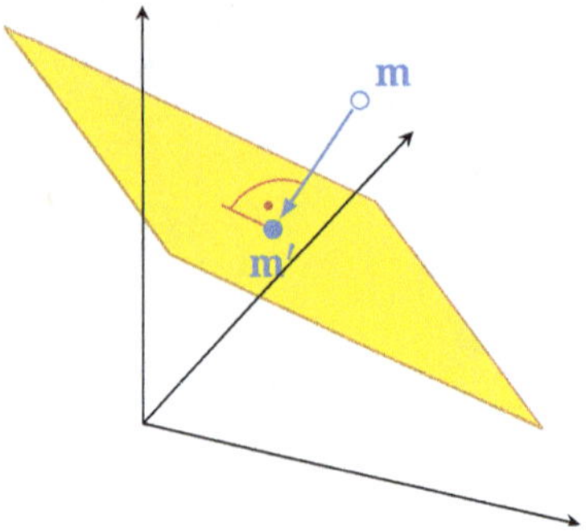

Fig. 2.31: Principal component analysis, general case: finding the d'-dimensional subspace with minimal reconstruction error.

We now turn to the general case. Again, $\check{\mathbf{m}}_k$ will denote the projection of $\mathbf{m}_k$ to a d'-dimensional affine subspace given by

$$\overline{\mathbf{m}} + \sum_{i=1}^{d'} a_i \mathbf{e}_i \tag{2.138}$$

with $\{\mathbf{e}_1, \ldots, \mathbf{e}_{d'}\}$ constituting an orthonormal basis. Then the objective function is

$$\begin{aligned} J_{d'}(a_{11}, \ldots, a_{Nd'}, \mathbf{e}_1, \ldots, \mathbf{e}_{d'}) &= \sum_{k=1}^{N} \|\check{\mathbf{m}}_k - \mathbf{m}_k\|^2 \\ &= \sum_{k=1}^{N} \left\| \left(\overline{\mathbf{m}} + \sum_{i=1}^{d'} a_{ki} \mathbf{e}_i \right) - \mathbf{m}_k \right\|^2 \end{aligned} \tag{2.139}$$

for $d' \leq d$. A generalized variant of the same course of action as above leads to the following result: the optimal affine subspace with dimension d' has a base point at its average value $\overline{\mathbf{m}}$ and is spanned by the d' eigenvectors of the d' largest eigenvalues of the scatter matrix $\mathbf{S}$ (see Figure 2.31).

Hence, the usual procedure to calculate the d'-dimensional principal component analysis consists of the following steps:

1. Calculate the average

$$\overline{\mathbf{m}} = \sum_{i=1}^{N} \mathbf{m}_i \in \mathbb{R}^d, \tag{2.140}$$

the data matrix

$$\mathbf{M} = \left(\mathbf{m}_1 - \overline{\mathbf{m}}, \quad \ldots, \quad \mathbf{m}_N - \overline{\mathbf{m}} \right) \in \mathbb{R}^{d \times N}, \tag{2.141}$$

and the scatter matrix

$$\mathbf{S} = \mathbf{M}\mathbf{M}^\top \in \mathbb{R}^{d \times d} \tag{2.142}$$

of all feature vectors $\mathbf{m}_1, \ldots, \mathbf{m}_N$.

2. Calculate the normalized eigenvectors $\mathbf{e}_1, \ldots, \mathbf{e}_d$ of $\mathbf{S}$ and sort them such that the corresponding eigenvalues $\lambda_1, \ldots, \lambda_d$ are decreasing, i.e., $\lambda_1 > \lambda_2 > \ldots > \lambda_d$. (N.b.: There are d eigenvectors, because $\mathbf{S}$ is symmetric and therefore diagonalizable.)

3. Construct a matrix

$$\mathbf{A} := (\mathbf{e}_1, \dots, \mathbf{e}_{d'}) \in \mathbb{R}^{d \times d'} \tag{2.143}$$

with the first d' eigenvectors as its columns.

4. Transform each feature vector $\mathbf{m}_i$ into a new feature vector

$$\mathbf{m}'_i = \mathbf{A}^\mathsf{T}(\mathbf{m}_i - \overline{\mathbf{m}}) \qquad \text{for } i = 1, \dots, N \tag{2.144}$$

of smaller dimension d'.

Relation to the Karhunen–Loève transformation

Before the implications of the PCA are discussed in detail in the upcoming paragraphs, this section will shed some light on the relation between the PCA and the Karhunen–Loève transformation. The latter regards the same concepts from the perspective of a stochastic process. Generally, the Karhunen–Loève transformation tries to find an optimal representation by orthogonal functions for a stochastic process $\underline{m}(t) \in \mathbb{R}$ and some time index t from a compact set of $\mathbb{R}$. In this context, the coefficients of the system of orthogonal functions are considered to be random variables. Of course, if the support of the random process is chosen to be a bounded set of natural numbers, i.e., $t \in \{1, \dots, d\}$, then the stochastic process can be written as a vector $(\underline{m}(1), \dots, \underline{m}(d))^\mathsf{T}$ and one is in the same situation as for principal component analysis.

Let $\underline{\mathbf{m}}$ be a random vector and let

$$\boldsymbol{\mu} = \mathrm{E}\{\underline{\mathbf{m}}\}, \tag{2.145}$$

$$\boldsymbol{\Sigma} = \mathrm{Cov}\{\underline{\mathbf{m}}\} \tag{2.146}$$

denote the expectation and the covariance matrix respectively. As $\boldsymbol{\Sigma}$ is symmetric and positive definite, it can be decomposed into its eigenvectors. Let the $\boldsymbol{\epsilon}_i$ be the normalized eigenvectors and $\kappa_i > 0$ be the eigenvalues in decreasing order. Define the matrices

$$\mathbf{E} = (\boldsymbol{\epsilon}_1, \dots, \boldsymbol{\epsilon}_d), \tag{2.147}$$

$$\boldsymbol{\Lambda} = \begin{pmatrix} \kappa_1 & 0 & \dots & 0 \\ 0 & \ddots & \ddots & \vdots \\ \vdots & \ddots & \ddots & 0 \\ 0 & \dots & 0 & \kappa_d \end{pmatrix} \tag{2.148}$$

where $\mathbf{E}$ is a column-wise concatenation of the eigenvectors. Then the covariance matrix can be rewritten as

$$\boldsymbol{\Sigma} = \sum_{i=1}^{d} \kappa_i \boldsymbol{\epsilon}_i \boldsymbol{\epsilon}_i^\mathsf{T} = \mathbf{E}\boldsymbol{\Lambda}\mathbf{E}^\mathsf{T}. \tag{2.149}$$

Because the eigenvectors constitute a orthonormal basis, $\mathbf{E}$ is orthogonal and $\mathbf{E}^\mathsf{T} = \mathbf{E}^{-1}$. Therefore Equation (2.149) yields

$$\boldsymbol{\Lambda} = \mathbf{E}^\mathsf{T}\boldsymbol{\Sigma}\mathbf{E}. \tag{2.150}$$

The Karhunen–Loève transformation of $\underline{\mathbf{m}}$ is defined as

$$\underline{\widetilde{\mathbf{m}}} = \mathbf{E}^{\mathsf{T}}\left(\underline{\mathbf{m}} - \mu\right). \tag{2.151}$$

The transformed random variable has zero mean

$$\mathrm{E}\{\underline{\widetilde{\mathbf{m}}}\} = \mathbf{0} \tag{2.152}$$

and the covariance matrix is

$$\begin{aligned}\mathrm{Cov}\{\underline{\widetilde{\mathbf{m}}}\} &= \mathrm{E}\left\{\mathbf{E}^{\mathsf{T}}\left(\underline{\mathbf{m}} - \mu\right)\left(\underline{\mathbf{m}} - \mu\right)^{\mathsf{T}}\mathbf{E}\right\} = \mathbf{E}^{\mathsf{T}}\,\mathrm{E}\left\{\left(\underline{\mathbf{m}} - \mu\right)\left(\underline{\mathbf{m}} - \mu\right)^{\mathsf{T}}\right\}\mathbf{E}\\ &= \mathbf{E}^{\mathsf{T}}\boldsymbol{\Sigma}\mathbf{E} = \boldsymbol{\Lambda}.\end{aligned} \tag{2.153}$$

This equation shows that the variance of a single component is

$$\kappa_i = \mathrm{Var}\{\underline{\widetilde{m}}_i\} \tag{2.154}$$

and that all components are pairwise uncorrelated:

$$\mathrm{Cov}\left\{\underline{\widetilde{m}}_i, \underline{\widetilde{m}}_j\right\} = 0 \qquad \text{for } i \neq j. \tag{2.155}$$

That being said, we now return to principal component analysis. Instead of a random vector $\underline{\mathbf{m}}$, one has a set of feature vectors $\mathbf{m}_k$ that are nothing else than realizations of $\underline{\mathbf{m}}$ and the empirical mean $\overline{\mathbf{m}}$ is an unbiased estimator of the expectation vector

$$\hat{\mu} = \overline{\mathbf{m}}. \tag{2.156}$$

Except for a correction factor, something similar holds for the scatter matrix. An unbiased estimator for the covariance matrix is

$$\hat{\boldsymbol{\Sigma}} = \frac{1}{N-1}\mathbf{S}. \tag{2.157}$$

The component-wise variance of the transformed feature can be unbiasedly estimated by the scaled eigenvalues of the scatter matrix

$$\hat{\kappa}_i = \frac{1}{N-1}\lambda_i. \tag{2.158}$$

This situation is depicted in Figure 2.32.

Some characteristics of principal component analysis

We now try to calculate the approximation error that arises if the features are projected to a space of lower dimension. The eigenvectors $\{\mathbf{e}_1, \ldots, \mathbf{e}_d\}$ constitute an orthonormal basis and the corresponding coefficients are

$$m_i' = \mathbf{e}_i^{\mathsf{T}}\left(\mathbf{m} - \overline{\mathbf{m}}\right) \tag{2.159}$$

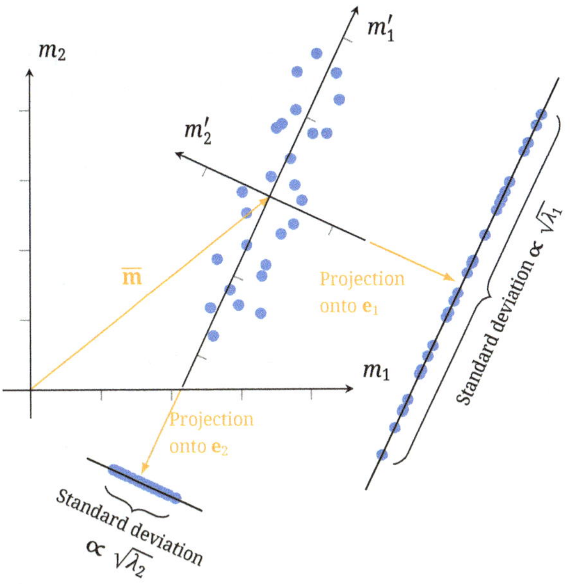

Fig. 2.32: The variance of the dataset is encoded in the principal components so that the variance along a component is proportional to the corresponding eigenvalue.

and the reconstruction of the ith component in the original feature space is given by

$$\mathbf{e}_i m_i' = \underbrace{\mathbf{e}_i \mathbf{e}_i^\mathsf{T}}_{\in \mathbb{R}^{d\times d}} (\mathbf{m} - \overline{\mathbf{m}})\,. \tag{2.160}$$

Hence $\mathbf{m}_{[1]} = \left(\mathbf{I} - \mathbf{e}_1\mathbf{e}_1^\mathsf{T}\right)(\mathbf{m} - \overline{\mathbf{m}})$ is the feature vector with the first component removed. More generally, we denote the transformed vector $\mathbf{m}$ without the entries corresponding to the first i eigenvectors $\mathbf{e}_i$ by:

$$\mathbf{m}_{[i]} = \left(\mathbf{I} - \mathbf{e}_1\mathbf{e}_1^\mathsf{T} \cdots - \mathbf{e}_i\mathbf{e}_i^\mathsf{T}\right)(\mathbf{m} - \overline{\mathbf{m}})\,. \tag{2.161}$$

Because distinct eigenvectors are orthogonal, the sequence $\mathbf{m}_{[1]}, \mathbf{m}_{[2]}, \mathbf{m}_{[3]}, \ldots$ can be calculated recursively:

$$m_1' = \mathbf{e}_1^\mathsf{T}(\mathbf{m} - \overline{\mathbf{m}}) \qquad \mathbf{m}_{[1]} = \left(\mathbf{I} - \mathbf{e}_1\mathbf{e}_1^\mathsf{T}\right)(\mathbf{m} - \overline{\mathbf{m}}) \tag{2.162}$$

$$m_2' = \mathbf{e}_2^\mathsf{T}\mathbf{m}_{[1]} \qquad \mathbf{m}_{[2]} = \left(\mathbf{I} - \mathbf{e}_2\mathbf{e}_2^\mathsf{T}\right)\mathbf{m}_{[1]} \tag{2.163}$$

$$m_3' = \mathbf{e}_3^\mathsf{T}\mathbf{m}_{[2]} \qquad \mathbf{m}_{[3]} = \left(\mathbf{I} - \mathbf{e}_3\mathbf{e}_3^\mathsf{T}\right)\mathbf{m}_{[2]} \tag{2.164}$$

$$\vdots$$

The Equations (2.162) to (2.164) and so on can be thought of as follows: At first, the direction of maximum variance is determined and the variation of the data w.r.t. this direction is removed. Then, within the data modified in this way, again the direction of maximum variance is determined, and so on. Therefore, in a greedy manner, the maximum variance directions are identified recursively and the pertaining components of the data are consecutively subtracted.

For a single zero-mean feature vector $\mathbf{m}_k - \overline{\mathbf{m}}$ with $k = 1, \ldots, N$, the projection squared error onto the d'-dimensional subspace is

$$\left\| (\mathbf{I} - \sum_{i=1}^{d'} \mathbf{e}_i \mathbf{e}_i^\top)(\mathbf{m}_k - \overline{\mathbf{m}}) \right\|^2 = \left\| (\sum_{i=d'+1}^{d} \mathbf{e}_i \mathbf{e}_i^\top)(\mathbf{m}_k - \overline{\mathbf{m}}) \right\|^2 \tag{2.165}$$

and the total squared error for all feature vectors is the sum of the remaining squared eigenvalues:

$$\sum_{k=1}^{N} \left\| (\mathbf{I} - \sum_{i=1}^{d'} \mathbf{e}_i \mathbf{e}_i^\top)(\mathbf{m}_k - \overline{\mathbf{m}}) \right\|^2 = \sum_{i=d'+1}^{d} \lambda_i^2 . \tag{2.166}$$

By construction, the principal component analysis yields the best (w.r.t. mean square error) d'-dimensional approximation. Furthermore, we already know that the eigenvalues are proportional to the standard deviation of the data with respect to the corresponding direction. Now, assume some other arbitrary d'-dimensional projection and calculate the standard deviation of the dimensions being thrown away. Then these deviations will be greater than the term above. In this sense, the sum $\sum_{i=d'+1}^{d} \lambda_i^2$ is minimal or loses as little information as possible.

Abusing some concepts and notation, we can clarify what is meant by "loss of information". In virtue of the coerced normalization, one can regard the sequence of eigenvalues $\lambda_1, \ldots, \lambda_d$ as a probability distribution

$$\chi_i := \frac{\lambda_i}{\sum_{j=1}^{d} \lambda_j} . \tag{2.167}$$

From an information theoretic point of view, the entropy of this distribution is

$$H(\chi_1, \ldots, \chi_d) = -\sum_{i=1}^{d} \chi_i \ln \chi_i . \tag{2.168}$$

For any other linear transformation of the feature space, the corresponding entropy of the variances in each direction is larger. In other words, the principal component analysis yields that linear transformation for which the entropy of the "variances" becomes minimal. Hence, the variances are as unequally distributed as possible. In a lax interpretation, one could say that the first dimension bears as much information about the data as possible, the second dimension bears most of the remaining information, the third dimension bears most of the information without the first two dimensions, and so on.

The following list recapitulates the essential characteristics of a principal component analysis.

- The components of the transformed feature vectors are pairwise uncorrelated.

- The variances of the components of the transformed feature are maximally unequally distributed for all linear transformations. (The variances have minimal entropy.)
- The PCA yields the best d'-dimensional approximation in terms of the squared deviation.
- The PCA does not aim for the optimal separability of the classes, but tries to provide the best representation of all the data $\mathcal{D}$ as a whole. Nonetheless, experience shows that the PCA yields feature spaces of good quality with low dimensions.
- The descriptive meaning of the original features is lost.

Applications and examples: Eigenfaces

A well-known application of principal component analysis within the field of pattern recognition is identity recognition: given an image of a face, what is the name of that person? With the "eigenfaces" approach (now superseded by more sophisticated methods), faces are represented as the deviations from a mean face. The mean face as well as the "directions" of the deviation are calculated using PCA.

Let $g(x, y)$ denote the gray-scale image of a face with $(x, y) \in \{1, \dots, n\}^2$. Note that all images are required to be of the same size, but there is no technical reason to require them to be square. However, this restriction simplifies the following discussion. In addition, all images should show the face in the same pose and be aligned with a common reference frame (e.g., eye centers on the same height) for this technique to work well. The pixels are arranged into a vector $\mathbf{m} \in \mathbb{R}^d$ with dimension $d = n^2$. Note that here the pattern itself is used as the feature vector. As above, let

$$\mathbf{M} = \left(\mathbf{m}_1 - \overline{\mathbf{m}}, \quad \dots, \quad \mathbf{m}_N - \overline{\mathbf{m}}\right) \in \mathbb{R}^{d \times N} \tag{2.169}$$

denote the data matrix and $\mathbf{S} = \mathbf{M}\mathbf{M}^\mathsf{T} \in \mathbb{R}^{d \times d}$ denote the scatter matrix.

Usually, the next step would be to calculate the eigenvectors and eigenvalues of $\mathbf{S}$. In practice, however, this is infeasible due to the size of $\mathbf{S}$ and the resulting computational complexity of the eigen-decomposition. Consider, for example, small facial images measuring 32×32 pixels, i.e., $n = 32$ (in real applications the images will be larger). Then the "feature vectors" $\mathbf{m}$ will be of dimension $d = n^2 = 32^2$ and the scatter matrix $\mathbf{S}$ will have $d^2 = n^4 = 1{,}048{,}576$ entries.

The costly eigen-decomposition can be avoided by exploiting the structure of the problem: the dimensionality of the space induced by the training sample is smaller than the dimensionality of the feature space. In other words, the number N of features in the training sample is much smaller than d. This is an odd situation: in most cases, the number of samples is much larger than d. As we will see in Chapter 4, $N \gg d$ is (often) actually required in order to successfully estimate the decision boundaries in the feature space. Note, however, that at the moment we do not wish to derive a classifier, rather, we wish to find a compact representation of the facial images that *can be used* with a classifier.

Fig. 2.33: Mean face computed from the YALE faces dataset of Georghiades et al. [2001].

Nevertheless, as here $N < d$, consider instead the matrix

$$\mathbf{K} := \mathbf{M}^\mathsf{T}\mathbf{M} \in \mathbb{R}^{N\times N}. \tag{2.170}$$

By construction, this matrix is symmetric and therefore diagonalizable. Let $\boldsymbol{\eta}_i$ denote the eigenvectors of $\mathbf{K}$ for $i = 1, \ldots, N$. Then the first N ($N < d = n^2$) greatest eigenvectors $\mathbf{e}_i$ of $\mathbf{M}$ can be calculated from the $\boldsymbol{\eta}_i$ of $\mathbf{K}$ using

$$\mathbf{e}_i = \frac{\mathbf{M}\boldsymbol{\eta}_i}{\|\mathbf{M}\boldsymbol{\eta}_i\|} \qquad \text{for } i = 1, \ldots, N. \tag{2.171}$$

Since the eigenvectors are computed from images, they can themselves be converted into images. Figures 2.33 and 2.34 show the mean vector $\overline{\mathbf{m}}$ and the eigenvectors $\mathbf{e}_i$, $i = 1, \ldots, 10$ of the extended YALE face dataset B (Georghiades et al. [2001]) interpreted as gray-scale images. This dataset contains pictures of the faces of 39 subjects. The images were recorded under different lighting conditions and cropped and rotated so that the faces of two different images are aligned. One can clearly see that the eigenvectors represent major modes of change: lighting, pose, and facial structure. The eigenvalues corresponding to the eigenvectors are shown in Figure 2.35. As expected from the previous discussion, the eigenvalues are very unequally distributed; most of the variation in the dataset is represented by the first two components. The third, fourth, etc. eigenvalues are of much smaller magnitude, which means that the associated components explain finer, but less common, details.

For classification, a d'-dimensional feature vector (where $d' \ll d$) according to $\mathbf{m}' = (\mathbf{e}_1, \cdots, \mathbf{e}_{d'})^\mathsf{T}(\mathbf{m} - \overline{\mathbf{m}})$ is used. Note that this approach is not restricted to facial image recognition. It can also be used with 3D facial data from depth sensors, or indeed any other type of data.

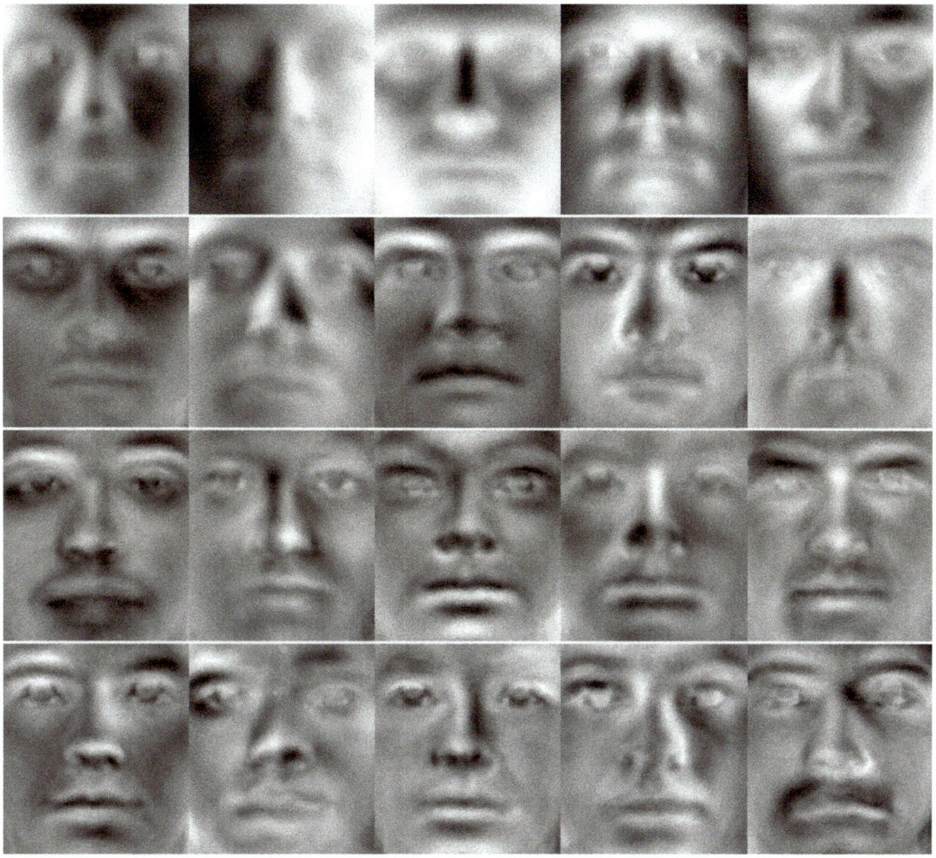

Fig. 2.34: First 20 eigenfaces computed from the YALE faces dataset of Georghiades et al. [2001]. The first components clearly correspond to different lighting conditions, while the other components correspond to changes in pose and facial structure.

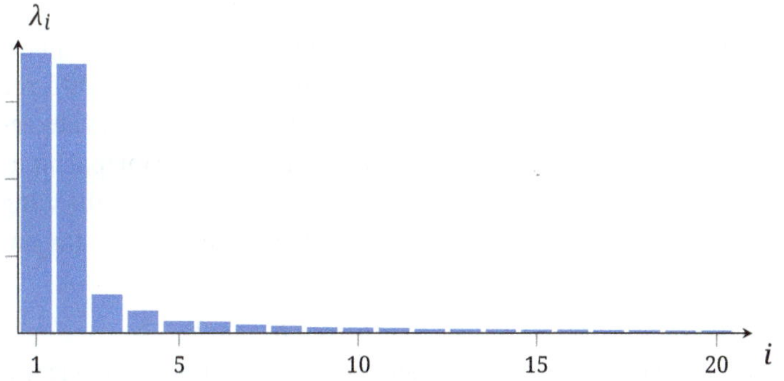

Fig. 2.35: First 20 eigenvalues corresponding to the eigenfaces in Figure 2.34. Note that most of the variation is captured with just the first two components which correspond to lighting directions.

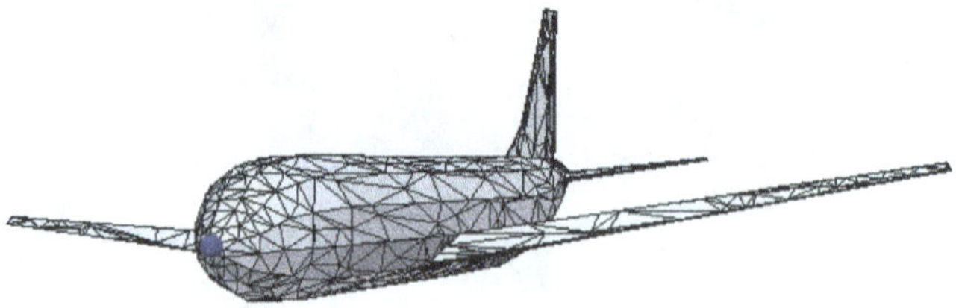

Fig. 2.36: Wireframe model of an airplane. Image source: Fraunhofer IOSB Karlsruhe and Laubenheimer [2004].

Applications and examples: Classification of airplane models
In fact, a similar approach has been used to classify airplane types in aerial images. As with eigenfaces, the two-dimensional images could be fed directly into PCA to derive a feature descriptor. However, environmental conditions such as lighting, partial occlusion, etc. add non-relevant variation, which in turn results in much larger feature vectors than expected. To circumvent this problem, a parametric 3D wireframe model, like the one shown in Figure 2.36, can be adjusted to match the image. The parameters of this model are then used as a feature vector (Laubenheimer [2004]).

More formally, the wireframe models consist of N nodes with three coordinates $(x_i, y_i, z_i)^\mathsf{T}, i = 1, \dots, N$ and associations between these coordinates. In this example, there are $N = 2000$ nodes for every wireframe model in the database. The coordinates are collected into a vector $\mathbf{g} = (x_1, y_1, z_1, \dots, x_N, y_N, z_N)^\mathsf{T}$, where the order of the points is arbitrary, but consistent among the different models. For example, the first node could always be on the nose of the airplane, while the 20th point could always be on the tip of the left wing.

A simple parametric model chooses a small number of representative airplanes as a basis to represent unknown models. If, for example, the basis consists of wireframe models of the three common Airbus models and three common Boeing models, a new wireframe model is parametrized by

$$\begin{aligned}\mathbf{g}_{\text{new}} &= \alpha_1 \mathbf{g}_{\text{Airbus A320}} + \alpha_2 \mathbf{g}_{\text{Airbus A300}} + \alpha_3 \mathbf{g}_{\text{Airbus A340}} \\ &\quad + \alpha_4 \mathbf{g}_{\text{Boeing 747}} + \alpha_5 \mathbf{g}_{\text{Boeing 707}} + \alpha_6 \mathbf{g}_{\text{Boeing 737}} .\end{aligned} \tag{2.172}$$

This model, however, can only represent planes that are somewhat similar to the models chosen as the basis. Of course, one can simply provide a larger selection of different planes, but then the size of the feature vector will increase, which typically has adverse effects on classification performance. A better model should capture the principal modes of change in an airplane: the size, the position of the wings, the orientation of the wings, etc. Such a model can be derived using PCA. The base models are collected into the data matrix,

$$\mathbf{M} = \left(\mathbf{g}_{\text{Airbus A320}}, \mathbf{g}_{\text{Boeing 747}}, \mathbf{g}_{\text{Airbus A300}}, \mathbf{g}_{\text{Boeing 707}}, \dots\right), \tag{2.173}$$

and PCA is used to extract the mean model $\overline{\mathbf{g}}$ and eigenvectors $\mathbf{e}_i$. A new model is then represented by

$$\mathbf{g}_{\text{new}} = \overline{\mathbf{g}} + m_1 \mathbf{e}_1 + m_2 \mathbf{e}_2 + m_3 \mathbf{e}_3 + \dots \tag{2.174}$$

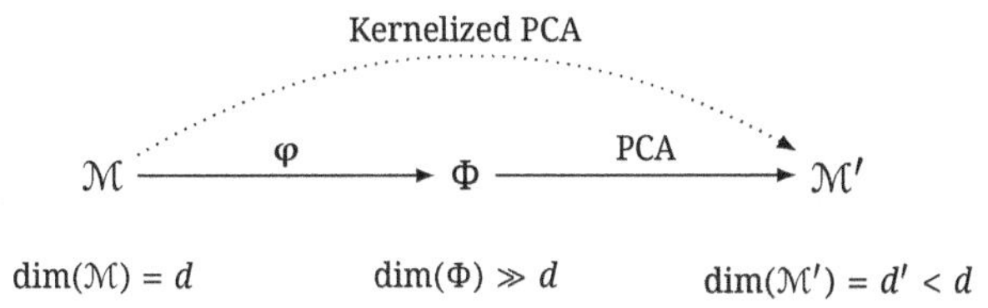

Fig. 2.37: Concept of kernelized PCA.

As it turns out, the first eigenvector $\mathbf{e}_1$ mainly accounts for the size of the airplane body, the second and third eigenvectors $\mathbf{e}_2$ and $\mathbf{e}_3$ mainly encode the position and orientation of the wings, and the other eigenvectors encode minor details. PCA has found the most relevant modes of change purely from the data provided, without any guidance from a human expert. With this model, most of the inherent variation in the airplane models can be expressed using only the first three or four eigenvectors. The resulting feature descriptor $\mathbf{m} = (m_1,m_2,m_3,m_4)^\mathsf{T}$ is very compact, yet sufficient for the task of describing different types of airplanes. More information about this approach can be found in Laubenheimer [2004].

2.7.2 Kernelized principal component analysis

Standard principal component analysis aims to find the best orthogonal transformation of the feature space such that the projection onto the d' first dimensions (or principal components) is the best approximation among all other orthogonal transformations.

But sometimes it might yield better results if PCA is not applied to the original feature space, but to an intermediate higher dimensional space. In other words, there is a nonlinear function $\varphi : \mathcal{M} \to \Phi$ that maps from the feature space $\mathcal{M}$ with dimension d to a new Hilbert space Φ with higher—possibly infinite—dimension. Then the principal component analysis is applied to this intermediate space in order to obtain the feature space $\mathcal{M}'$ with reduced dimension d'. This idea is depicted in Figure 2.37. Only if φ is chosen to be truly nonlinear does this approach provide any benefits in comparison with standard principal component analysis. Otherwise, the results do not differ.

The reason behind the name *kernelized PCA*, and why it is covered by a whole section on its own, is that there is a clever calculation trick. This trick is referred to as the *kernel trick* and will be revisited and explained in greater detail in Section 7.7.

The naive approach to realize Figure 2.37 would be to explicitly choose Φ and φ and perform a principal component analysis in the high-dimensional space Φ. In this case, one needs to compute the inner products of vectors explicitly mapped by φ with possibly prohibitive computational costs. As the final goal is to ease the complexity by dimension reduction, this seems like a step in the wrong direction. The trick is to rewrite all the formulas so that the map φ only occurs in pairs within the inner product of Φ.

This means that terms like

$$\langle \varphi(\cdot), \varphi(\cdot) \rangle : \mathcal{M} \times \mathcal{M} \to \mathbb{R} \tag{2.175}$$

are the only places where φ appears. Then all these terms can be replaced by a so called *kernel function*

$$k \colon \mathcal{M} \times \mathcal{M} \to \mathbb{R} \tag{2.176}$$

that absorbs two mappings and the inner product into one simply evaluable function so that φ never needs to be calculated explicitly.

This being said, the upcoming course of action in this section is already clear. One starts with the "regular" PCA on Φ and tries to rewrite all formulas so that all φ vanish and k remains.

Standard PCA centers the data first, i.e., in the first step $\mathbf{m}_k - \overline{\mathbf{m}}$ is calculated. Without explicit knowledge about φ it is neither possible nor computationally feasible to calculate $\varphi(\mathbf{m}_k) - \overline{\varphi(\mathbf{m})}$ in the same way. Hence, one assumes that φ already generates zero-mean data $\varphi(\mathbf{m}_1), \ldots, \varphi(\mathbf{m}_N)$ with $\sum_{k=1}^{N} \varphi(\mathbf{m}_k) = 0$. Of course, this assumption is rather far-fetched, and at the end this condition will be dropped again, but provisionally this is assumed to be true.

The following derivation mainly follows that of Schölkopf et al., which can be found in Schölkopf et al. [1997].

Let $\mathbf{m}_1, \ldots, \mathbf{m}_N \in \mathcal{M}$ denote the original features. Then $\varphi(\mathbf{m}_1), \ldots, \varphi(\mathbf{m}_N) \in \Phi$ are the non-linearly transformed, high-dimensional features. Furthermore,

$$\mathbf{D} = (\varphi(\mathbf{m}_1), \ldots, \varphi(\mathbf{m}_N)) \tag{2.177}$$

is the data matrix (see Equation (2.136)) and

$$\mathbf{C} = \frac{1}{N} \sum_{k=1}^{N} \varphi(\mathbf{m}_k)\varphi(\mathbf{m}_k)^{\mathsf{T}} = \frac{1}{N}\mathbf{D}\mathbf{D}^{\mathsf{T}} \tag{2.178}$$

is the scatter matrix (see Equation (2.137)). Two aspects are important to be noted: First, the scatter matrix is additionally normalized; Second, these terms assume that $\frac{1}{N} \sum_{k=1}^{N} \varphi(\mathbf{m}_k) = \mathbf{0}$.

Following the usual procedure, the eigenvalue equation

$$\lambda \mathbf{v} = \mathbf{C}\mathbf{v} \qquad \text{for } \mathbf{v} \in \Phi, \lambda \in \mathbb{R} \tag{2.179}$$

must be solved next. As always, $\mathbf{C}$ is diagonalizable and all eigenvalues are non-negative $\lambda \geq 0$, because of the special form $\mathbf{C} = \mathbf{D}\mathbf{D}^{\mathsf{T}}$. As Equation (2.179) will not be explicitly solved, the definition of $\mathbf{C}$ is put into Equation (2.179):

$$\lambda \mathbf{v} = \mathbf{C}\mathbf{v} = \left(\frac{1}{N} \sum_{k=1}^{N} \varphi(\mathbf{m}_k)\varphi(\mathbf{m}_k)^{\mathsf{T}} \right) \mathbf{v} = \frac{1}{N} \sum_{k=1}^{N} \underbrace{\left(\varphi(\mathbf{m}_k)^{\mathsf{T}}\mathbf{v}\right)}_{(*)\in\mathbb{R}} \varphi(\mathbf{m}_k) \tag{2.180}$$

$$\overset{\lambda \neq 0}{\Rightarrow} \mathbf{v} = \sum_{k=1}^{N} \alpha_k \varphi(\mathbf{m}_k) = \mathbf{D}\boldsymbol{\alpha} \qquad \text{for } \boldsymbol{\alpha} \in \mathbb{R}^N. \tag{2.181}$$

Equation (2.180) reveals an important observation. The eigenvalue zero ($\lambda = 0$) can occur for two reasons: either the right side is a non-trivial linear combination, i.e., at least two coefficients ($*$) are nonzero, or all coefficients equal zero. In the latter case, the eigenvector $\mathbf{v}$ is perpendicular to every $\varphi(\mathbf{m}_k)$. But if the eigenvalue is nonzero ($\lambda \neq 0$), then the right side must be a nonzero linear combination. Hence, $\mathbf{v}$ is a non-trivial linear combination of the $\varphi(\mathbf{m}_k)$. To summarize, any eigenvector solution of Equation (2.179) that corresponds to a nonzero eigenvalue is in the span of $\{\varphi(\mathbf{m}_1), \ldots, \varphi(\mathbf{m}_N)\}$.

Hence, in case $\lambda \neq 0$, one can rewrite $\mathbf{v}$ as a linear combination of the $\varphi(\mathbf{m}_1), \ldots, \varphi(\mathbf{m}_N)$ (see Equation (2.181)) or as the product of the data matrix Equation (2.177) applied to a vector.

These observations are very important, because they are the key that allows reducing the complexity. Although Φ might be infinite dimensional, all the action takes place in a subspace that has dimension N at most. There are at most N eigenvectors with a nonzero eigenvalue and these eigenvectors are in the span $\{\varphi(\mathbf{m}_1), \ldots, \varphi(\mathbf{m}_N)\}$; all other (possibly infinitely many) eigenvalues are zero and their eigenvectors are orthogonal to that subspace. Intuitively, this is not a surprise. If there are only N feature vectors (data points), these points can span at most a $N - 1$ dimensional subspace. (Two points are always on one line, three points are always on one plane, and so on.) This does not change, if the points are mapped into a space with higher dimension first. As one is only interested in principal components with nonzero variance (no other components bear any information at all), one can presume $\lambda > 0$ from now on.

The eigenvector Equation (2.180) corresponds to a system of linear equations with possibly infinitely many rows. Because we know that all interesting solutions with $\lambda > 0$ are in the span $\{\varphi(\mathbf{m}_1), \ldots, \varphi(\mathbf{m}_N)\}$, it suffices to consider the projection onto this space. This means one can multiply Equation (2.180) from the left by $\mathbf{D}^\top$ (see Equation (2.177)) without losing any interesting solution.

In conclusion, Equation (2.180) with Equation (2.181) and left multiplication with $\mathbf{D}^\top$ leads to

$$\lambda\mathbf{v} = \mathbf{C}\mathbf{v} \Rightarrow \lambda\mathbf{D}\alpha = \mathbf{C}\mathbf{D}\alpha \Rightarrow \lambda\mathbf{D}^\top\mathbf{D}\alpha = \mathbf{D}^\top\mathbf{C}\mathbf{D}\alpha. \tag{2.182}$$

In order to go on, define the kernel matrix

$$\mathbf{K} := \mathbf{D}^\top\mathbf{D} \in \mathbb{R}^{N\times N} \tag{2.183}$$

for short and obtain

$$\lambda\mathbf{D}^\top\mathbf{D}\alpha = \mathbf{D}^\top\mathbf{C}\mathbf{D}\alpha \Rightarrow \lambda\mathbf{K}\alpha = \mathbf{D}^\top\mathbf{D}\mathbf{D}^\top\mathbf{D}\alpha \Rightarrow \lambda\mathbf{K}\alpha = \mathbf{K}^2\alpha. \tag{2.184}$$

Now compare the definition of the kernel matrix $\mathbf{K}$ (Equation (2.183)) with the definition of the data matrix $\mathbf{C}$ (Equation (2.177)) and note that this is the same trick that has already been used to reduce the complexity of the eigenface problem (Equation (2.170)). Furthermore, one can see that

$$K_{ij} = \varphi(\mathbf{m}_i)^\top\varphi(\mathbf{m}_j) \tag{2.185}$$

holds. This means each matrix entry $\mathbf{K}_{ij}$ is the inner product of the corresponding feature vectors in the high-dimensional space. For later use, we introduce the kernel function

$$k\colon \begin{cases} \mathcal{M} \times \mathcal{M} \to \mathbb{R} \\ (\mathbf{m}_i, \mathbf{m}_j) \mapsto \varphi(\mathbf{m}_i)^\mathsf{T} \varphi(\mathbf{m}_j) \end{cases} \tag{2.186}$$

and set

$$K_{ij} = k(\mathbf{m}_i, \mathbf{m}_j). \tag{2.187}$$

Again, because $\mathbf{K}$ is symmetric and positive-definite, and because only nonzero solutions are of interest, one factor $\mathbf{K}$ can be canceled out in Equation (2.184). There remains

$$\lambda \boldsymbol{\alpha} = \mathbf{K} \boldsymbol{\alpha} \tag{2.188}$$

for $\mathbf{K} \in \mathbb{R}^{N \times N}$ and $\boldsymbol{\alpha} \in \mathbb{R}^N$. This means it suffices to solve an N-dimensional eigenvector problem in order to obtain the eigenvalues λ. The usual PCA requires that the eigenvectors that are used for the projection matrix be normalized. This means the high-dimensional eigenvectors $\mathbf{v} \in \Phi$ needs to be normalized (see Equation (2.179)), but Equation (2.188) determines the eigenvectors $\boldsymbol{\alpha} \in \mathbb{R}^N$. A normalization condition can be derived from Equation (2.181):

$$\begin{aligned} 1 \overset{!}{=} \|\mathbf{v}\|^2 = \|\mathbf{D}\boldsymbol{\alpha}\|^2 = \boldsymbol{\alpha}^\mathsf{T}\mathbf{D}^\mathsf{T}\mathbf{D}\boldsymbol{\alpha} = \boldsymbol{\alpha}^\mathsf{T}(\mathbf{K}\boldsymbol{\alpha}) = \boldsymbol{\alpha}^\mathsf{T}(\lambda\boldsymbol{\alpha}) = \lambda\|\boldsymbol{\alpha}\|^2 \\ \Rightarrow \|\boldsymbol{\alpha}\|^2 = \frac{1}{\lambda}. \end{aligned} \tag{2.189}$$

Let $\lambda_1, \ldots, \lambda_{d'}$ denote the d' greatest eigenvalues of Equation (2.188) and $\boldsymbol{\alpha}_1, \ldots, \boldsymbol{\alpha}_{d'} \in \mathbb{R}^N$ the corresponding eigenvectors with normalization condition $\|\boldsymbol{\alpha}_i\|^2 = \lambda_i^{-1}$ for $i = 1, \ldots, d'$. Let $\mathbf{m} \in \mathcal{M}$ be another feature vector whose first d' principal components are to be calculated.

Let

$$\mathbf{A} := (\mathbf{v}_1, \ldots, \mathbf{v}_{d'}) \tag{2.190}$$

be the projection matrix from the high-dimensional space Φ into $\mathcal{M}'$ and write

$$\tilde{\mathbf{A}} := (\boldsymbol{\alpha}_1, \ldots, \boldsymbol{\alpha}_{d'})\,. \tag{2.191}$$

Then the projection matrix can be written as

$$\mathbf{A} = \mathbf{D}\tilde{\mathbf{A}}. \tag{2.192}$$

Again, with the usual PCA this would require computing $\mathbf{m}' = \mathbf{A}^\mathsf{T}\varphi(\mathbf{m})$. This can be rewritten as

$$\begin{aligned} \mathbf{m}' = \mathbf{A}^\mathsf{T}\varphi(\mathbf{m}) = \left(\mathbf{D}\tilde{\mathbf{A}}\right)^\mathsf{T}\varphi(\mathbf{m}) = \tilde{\mathbf{A}}^\mathsf{T}\left(\mathbf{D}^\mathsf{T}\varphi(\mathbf{m})\right) \\ = \tilde{\mathbf{A}}^\mathsf{T}\begin{pmatrix} \varphi(\mathbf{m}_1)^\mathsf{T}\varphi(\mathbf{m}) \\ \vdots \\ \varphi(\mathbf{m}_N)^\mathsf{T}\varphi(\mathbf{m}) \end{pmatrix} = \tilde{\mathbf{A}}^\mathsf{T}\begin{pmatrix} k(\mathbf{m}_1, \mathbf{m}) \\ \vdots \\ k(\mathbf{m}_N, \mathbf{m}) \end{pmatrix}. \end{aligned} \tag{2.193}$$

The last step finishes the derivation of the kernelized PCA. In summary, Equation (2.188) must be solved with $\mathbf{K}$ being defined as in Equation (2.187) under the normalization condition from Equation (2.189). The eigenvectors found must be organized into a projection matrix $\widetilde{\mathbf{A}}$ and the last equation above returns the principal components $\mathbf{m}'$ for any feature vector $\mathbf{m}$. All steps require evaluating the kernel function k at most. The transformation φ is never needed explicitly.

Attention: Although $k\colon \mathcal{M} \times \mathcal{M} \to \mathbb{R}$ seems to come out of the blue, it is still assumed that it corresponds to some inner product on some unknown vector space Φ of unknown dimension and that there is a (nonlinear) mapping φ from $\mathcal{M}$ into Φ such that $k(\cdot,\cdot) = \langle \varphi(\cdot), \varphi(\cdot)\rangle$. Moreover, it is assumed that $\sum_{k=1}^{N} \varphi(\mathbf{m}_k) = \mathbf{0}$. This means that the transformed dataset has zero mean.

Two last questions still need to be answered:

- Which functions are allowed for the kernel $k\colon \mathcal{M} \times \mathcal{M} \to \mathbb{R}$ without a map φ being explicitly given?
- How can the condition $\sum_{k=1}^{N} \varphi(\mathbf{m}_k) = \mathbf{0}$ be relaxed if φ is not given?

First, some very common examples will be given:

- Polynomial kernel: $k(\mathbf{m}, \mathbf{s}) = (\mathbf{m}^{\mathsf{T}}\mathbf{s} + c)^n$ for fixed $c \in \mathbb{R}$, $n \in \mathbb{N}$.
- Radial kernel: $k(\mathbf{m}, \mathbf{s}) = \exp\left(-\frac{\|\mathbf{m}-\mathbf{s}\|^2}{2\sigma^2}\right)$ for fixed $\sigma^2 \in \mathbb{R}^{>0}$.
- Sigmoid kernel: $k(\mathbf{m}, \mathbf{s}) = \tanh\left(\kappa \mathbf{m}^{\mathsf{T}}\mathbf{s} + \theta\right)$ for fixed $\kappa, \theta \in \mathbb{R}$.

In general, Mercer's condition (see Theorem 7.4) can be used to check whether a function $k\colon \mathcal{M} \times \mathcal{M} \to \mathbb{R}$ is permissible.

We now tackle the problem of data with nonzero mean. Recall that the data matrix is defined as

$$\mathbf{D} = (\varphi(\mathbf{m}_1), \ldots, \varphi(\mathbf{m}_N)) \tag{2.194}$$

and the kernel matrix as

$$\mathbf{K} = \mathbf{D}^{\mathsf{T}}\mathbf{D} \in \mathbb{R}^{N\times N}. \tag{2.195}$$

Let further

$$\widetilde{\mathbf{D}} = \left(\varphi(\mathbf{m}_1) - \overline{\varphi(\mathbf{m})}, \ldots, \varphi(\mathbf{m}_N) - \overline{\varphi(\mathbf{m})}\right) \tag{2.196}$$

with $\overline{\varphi(\mathbf{m})} = \frac{1}{N}\sum_{k=1}^{N} \varphi(\mathbf{m}_k)$ be the centered data matrix and

$$\widetilde{\mathbf{K}} = \widetilde{\mathbf{D}}^{\mathsf{T}}\widetilde{\mathbf{D}} \in \mathbb{R}^{N\times N} \tag{2.197}$$

the analogue of $\mathbf{K}$. The aim is to rewrite $\widetilde{\mathbf{K}}$ in terms of $\mathbf{K}$, in such a way that no explicit evaluation of φ is necessary. Actually, this is straightforward. Let $\mathbf{U}$ denote the $(N \times N)$-matrix of ones. Then the centered data matrix can be rewritten as

$$\widetilde{\mathbf{D}} = \mathbf{D} - \frac{1}{N}\mathbf{D}\mathbf{U} \tag{2.198}$$

and putting this into the definition of the centered kernel matrix yields

$$\begin{aligned}\tilde{\mathbf{K}} &= \left(\mathbf{D} - \frac{1}{N}\mathbf{D}\mathbf{U}\right)^{\mathsf{T}}\left(\mathbf{D} - \frac{1}{N}\mathbf{D}\mathbf{U}\right)\\ &= \mathbf{D}^{\mathsf{T}}\mathbf{D} - \frac{1}{N}\mathbf{D}^{\mathsf{T}}\mathbf{D}\mathbf{U} - \frac{1}{N}\mathbf{U}\mathbf{D}^{\mathsf{T}}\mathbf{D} + \frac{1}{N^2}\mathbf{U}\mathbf{D}^{\mathsf{T}}\mathbf{D}\mathbf{U}\\ &= \mathbf{K} - \frac{1}{N}\mathbf{K}\mathbf{U} - \frac{1}{N}\mathbf{U}\mathbf{K} + \frac{1}{N^2}\mathbf{U}\mathbf{K}\mathbf{U}.\end{aligned} \tag{2.199}$$

The eigenvector equation need to be solved with $\tilde{\mathbf{K}}$ instead of $\mathbf{K}$, but as one can see, no explicit evaluation of φ is required.

Moreover, the projection must be redefined. As in Equation (2.193), let $\mathbf{m}$ be the vector under consideration and $\tilde{\mathbf{A}}$ be the projection matrix. Write $\mathbf{u} \in \mathbb{R}^N$ for the vector of ones. A similar calculation as above leads to

$$\mathbf{m}' = \tilde{\mathbf{A}}^{\mathsf{T}}\left(\begin{pmatrix} k(\mathbf{m}_1, \mathbf{m}) \\ \vdots \\ k(\mathbf{m}_N, \mathbf{m}) \end{pmatrix} - \frac{1}{N}\mathbf{K}\mathbf{u}\right) \tag{2.200}$$

as the new projection formula.

To finish this section, the following list give a ready to use sequence of instructions for the kernelized PCA, as the section before did for the usual PCA. Let $\mathbf{m}_1, \ldots, \mathbf{m}_N \in \mathcal{M} = \mathbb{R}^d$ be a training set and $\mathbf{m} \in \mathcal{M}$ an additional feature vector. $k\colon \mathcal{M} \times \mathcal{M} \to \mathbb{R}$ denotes a permissible kernel function. Fix some $d' < \min\{d, N\}$.

1. Calculate the matrices

$$\mathbf{K} \in \mathbb{R}^{N\times N} \qquad \text{with } \mathbf{K}_{ij} = k(\mathbf{m}_i, \mathbf{m}_j) \tag{2.201}$$

and

$$\tilde{\mathbf{K}} = \mathbf{K} - \frac{1}{N}\mathbf{K}\mathbf{U} - \frac{1}{N}\mathbf{U}\mathbf{K} + \frac{1}{N^2}\mathbf{U}\mathbf{K}\mathbf{U} \quad \text{with } \mathbf{U} := \begin{pmatrix} 1 & \cdots & 1 \\ \vdots & & \vdots \\ 1 & \cdots & 1 \end{pmatrix} \in \mathbb{R}^{N\times N}. \tag{2.202}$$

2. Solve the eigenvector equation

$$\lambda\boldsymbol{\alpha} = \tilde{\mathbf{K}}\boldsymbol{\alpha} \qquad \text{with } \lambda \in \mathbb{R}, \boldsymbol{\alpha} \in \mathbb{R}^N \tag{2.203}$$

under the normalization condition $\|\boldsymbol{\alpha}\|^2 = \lambda^{-1}$. Let $\boldsymbol{\alpha}_1, \ldots, \boldsymbol{\alpha}_N \in \mathbb{R}^N$ denote the solutions such that the corresponding eigenvalues $\lambda_1, \ldots, \lambda_N$ are decreasing. (N.b.: There are N eigenvectors, because $\tilde{\mathbf{K}}$ is symmetric and therefore diagonalizable.)

3. Construct a matrix

$$\tilde{\mathbf{A}} = (\boldsymbol{\alpha}_1, \ldots, \boldsymbol{\alpha}_{d'}) \in \mathbb{R}^{N\times d'} \tag{2.204}$$

with the first d' eigenvectors as columns.

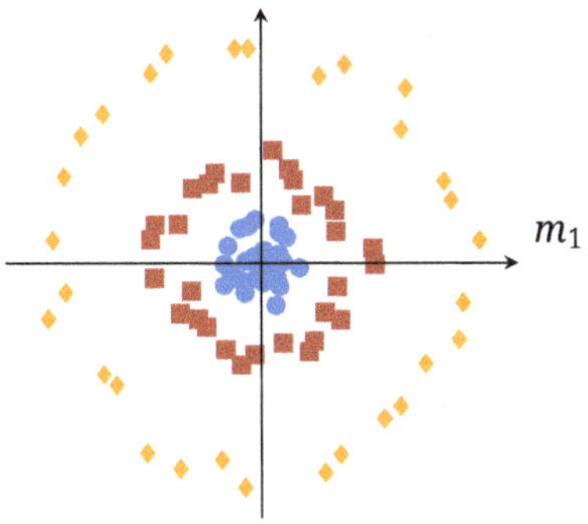

(a) Original feature space

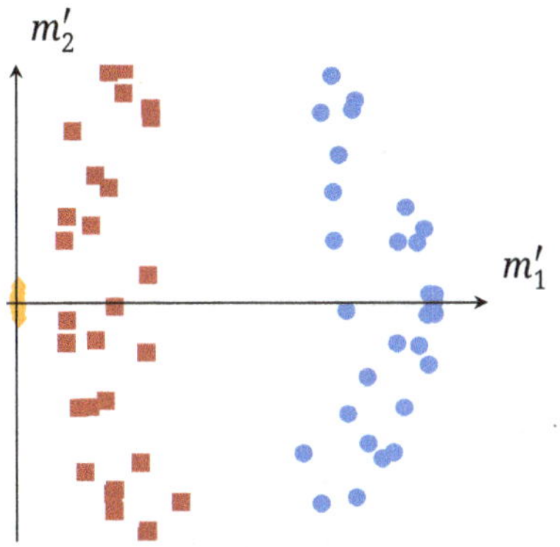

(b) Transformed feature space

Fig. 2.38: Kernelized PCA with radial kernel function $k(\mathbf{m}, \mathbf{s}) = \exp\left(-\frac{1}{2}\|\mathbf{m} - \mathbf{s}\|^2\right)$.

4. Transform the feature vector $\mathbf{m} \in \mathbb{R}^d$ into the feature vector

$$\mathbf{m}' = \widetilde{\mathbf{A}}^{\mathsf{T}} \left(\begin{pmatrix} k(\mathbf{m}_1, \mathbf{m}) \\ \vdots \\ k(\mathbf{m}_N, \mathbf{m}) \end{pmatrix} - \frac{1}{N}\mathbf{K}\mathbf{u} \right) \qquad \text{with } \mathbf{u} := (1, \ldots, 1)^{\mathsf{T}} \in \mathbb{R}^N \tag{2.205}$$

of smaller dimension d'.

Figure 2.38 gives an example of a kernelized PCA. In this example, the final dimension d' is not chosen to be smaller but is equal to the original dimension $d = d' = 2$.

2.7.3 Independent component analysis

Independent component analysis (ICA) is very similar to principal component analysis but additionally aims at finding a linear transformation such that the components become stochastically independent (Hyvärinen et al. [2004]). Recall that the principal component analysis interpreted as a special case of the Karhunen–Loève transformation yields uncorrelated components (see Equations (2.152), (2.153) and (2.155)). Surely, if independency

$$p_{\underline{\mathbf{m}}''_1, \ldots, \underline{\mathbf{m}}''_{d''}}(\mathbf{m}''_1, \ldots, \mathbf{m}''_{d''}) = p_{\underline{\mathbf{m}}''_1}(\mathbf{m}''_1) \cdots p_{\underline{\mathbf{m}}''_{d''}}(\mathbf{m}''_{d''}) \tag{2.206}$$

holds for the transformed vector $\mathbf{m}'' = \begin{pmatrix} \mathbf{m}''_1 & \ldots & \mathbf{m}''_{d''} \end{pmatrix}$, then being uncorrelated,

$$\operatorname{Cov}\{\underline{\mathbf{m}}''_i, \underline{\mathbf{m}}''_j\} = 0 \qquad \text{for } i \neq j, \tag{2.207}$$

follows as well. Thus, one could say that the independent component analysis goes one step further than the PCA. Actually, ICA can be seen as a two-step process. First, the

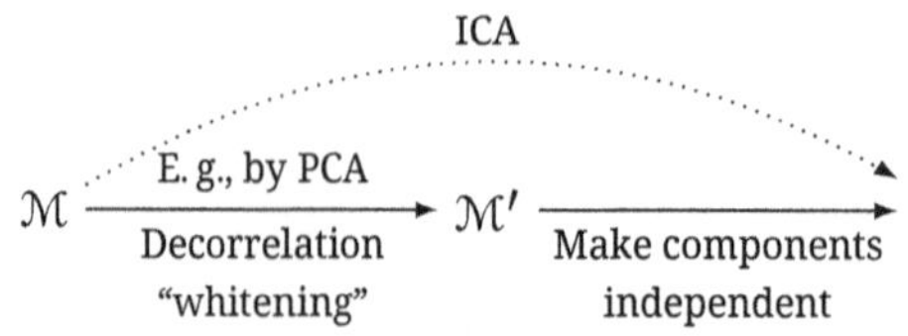

Fig. 2.39: Concept of independent component analysis.

components are decorrelated. Second, the components are orthogonally transformed so that they become independent (see Figure 2.39).

Before we dive more deeply into ICA, we will review some fundamentals from probability theory.

Definition 2.10 (Correlation coefficient, uncorrelated). Let $\underline{a}$, $\underline{b}$ be two random variables with common density $p_{\underline{a},\underline{b}}$. The *correlation coefficient* $\rho_{\underline{a},\underline{b}} \in [-1,1]$ is defined as

$$\rho_{\underline{a},\underline{b}} = \frac{\mathrm{Cov}\{\underline{a},\underline{b}\}}{\sqrt{\mathrm{Var}\{\underline{a}\}\,\mathrm{Var}\{\underline{b}\}}} = \frac{\mathrm{E}\{(\underline{a}-\mathrm{E}\{\underline{a}\})\,(\underline{b}-\mathrm{E}\{\underline{b}\})\}}{\sqrt{\mathrm{E}\left\{(\underline{a}-\mathrm{E}\{\underline{a}\})^2\right\}\mathrm{E}\left\{(\underline{b}-\mathrm{E}\{\underline{b}\})^2\right\}}}. \tag{2.208}$$

The random variables $\underline{a}$, $\underline{b}$ are called *uncorrelated* iff either of the following two equivalent conditions is met:

$$\mathrm{Cov}\{\underline{a},\underline{b}\} = 0 \qquad \Leftrightarrow \qquad \rho_{\underline{a}\underline{b}} = 0. \tag{2.209}$$

Definition 2.11 (Independence). Let $p(a)$ and $p(b)$ denote the marginal densities of the random variables $\underline{a}$ and $\underline{b}$ respectively. Moreover, let $p(a \mid b)$ and $p(b \mid a)$ be their conditional densities. $\underline{a}$ and $\underline{b}$ are called *independent* if any one (and therefore all) of the following equivalent conditions holds:

$$p(a,b) = p(a) \cdot p(b) \tag{2.210}$$

$$\Leftrightarrow p(a \mid b) = p(a) \tag{2.211}$$

$$\Leftrightarrow p(b \mid a) = p(b). \tag{2.212}$$

Theorem 2.12 (Induced uncorrelatedness). If two random variables are independent, then they are also uncorrelated. The converse is not necessarily true.

Example: Let $\underline{a} \sim \mathcal{N}(0,1)$ and $\underline{b} := \underline{a}^2$. Certainly, $\underline{a}$ and $\underline{b}$ are not stochastically independent, since if $\underline{b}$ is observed, one knows the magnitude $|\underline{a}|$ of $\underline{a}$. If $\underline{a}$ is observed, one knows $\underline{b}$. Nevertheless, $\underline{a}$ and $\underline{b}$ are uncorrelated, because $\mathrm{Cov}\{\underline{a},\underline{b}\} \propto \mathrm{E}\left\{\underline{a}^3\right\} = 0$ since the third central moment of a Gaussian distribution is zero.

As already mentioned, the ICA is composed of two steps (see Figure 2.40). Instead of directly looking for a matrix $\mathbf{U}$ such that

$$\underline{\mathbf{m}}'' = \mathbf{U}\,(\underline{\mathbf{m}} - \mathrm{E}\{\underline{\mathbf{m}}\}) \tag{2.213}$$

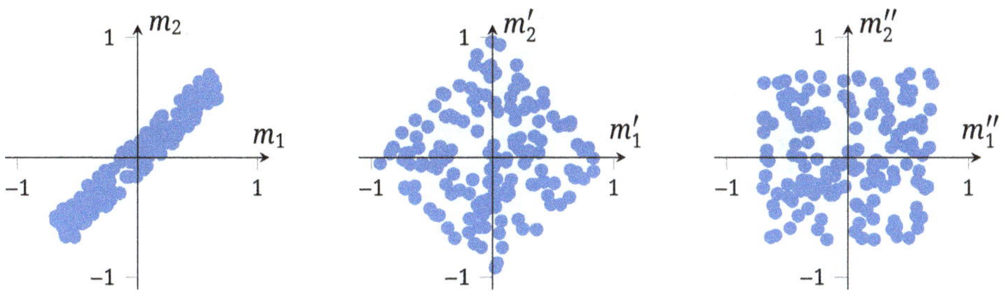

(a) Original feature space

(b) Intermediate space of decorrelated features (after whitening)

(c) Transformed space of independent features

Fig. 2.40: Effect of an independent component analysis.

yields independent components, $\mathbf{U} = \mathbf{YZ}$ is split into a product

$$\underline{\mathbf{m}}'' = \mathbf{Y}\underbrace{\mathbf{Z}\left(\underline{\mathbf{m}} - \mathrm{E}\{\underline{\mathbf{m}}\}\right)}_{\text{decorrelate}} = \mathbf{Y}\underline{\mathbf{m}}' \tag{2.214}$$

and $\mathbf{Z}$ is chosen to be the scaled transformation matrix of the Karhunen–Loève transformation

$$\mathbf{Z} = \sqrt{\mathbf{\Lambda}^{-1}}\mathbf{E}^\mathsf{T} = \begin{pmatrix} \frac{1}{\sqrt{\kappa_1}} & 0 & \cdots & 0 \\ 0 & \ddots & \ddots & \vdots \\ \vdots & \ddots & \ddots & 0 \\ 0 & \cdots & 0 & \frac{1}{\sqrt{\kappa_d}} \end{pmatrix} \mathbf{E}^\mathsf{T}. \tag{2.215}$$

See Equations (2.147) and (2.148) for the definition of $\mathbf{\Lambda}$ and $\mathbf{E}$. The scaling is necessary, because otherwise the covariance matrix of the transformed feature would equal $\mathbf{\Lambda}$ (see Equations (2.153) to (2.155)) but we want it to be the identity. In practice, if only the set of features $\mathbf{m}_1, \ldots, \mathbf{m}_N$ is known, one uses the PCA to obtain an unbiased estimator

$$\hat{\mathbf{Z}} = \begin{pmatrix} \sqrt{\frac{N-1}{\lambda_1}} & 0 & \cdots & 0 \\ 0 & \ddots & \ddots & \vdots \\ \vdots & \ddots & \ddots & 0 \\ 0 & \cdots & 0 & \sqrt{\frac{N-1}{\lambda_d}} \end{pmatrix} \mathbf{A}^\mathsf{T}. \tag{2.216}$$

See Equation (2.143) for the definition of $\mathbf{A}$.

Secondly, one needs to find the matrix $\mathbf{Y}$. As the lack of correlation is a necessary condition for independence, the lack of correlation of the intermediate feature vector $\underline{\mathbf{m}}'$ due to whitening must be retained:

$$\mathbf{I} \overset{!}{=} \mathrm{Cov}\{\underline{\mathbf{m}}''\} = \mathrm{Cov}\{\mathbf{Y}\underline{\mathbf{m}}'\} = \mathbf{Y}\underbrace{\mathrm{Cov}\{\underline{\mathbf{m}}'\}}_{=\mathbf{I}}\mathbf{Y}^\mathsf{T} = \mathbf{YY}^\mathsf{T}. \tag{2.217}$$

This observation reveals that **Y** has to be an orthogonal matrix. The advantage is that **Y** has only $\frac{d(d-1)}{2}$ degrees of freedom. Hence, the remaining task is to find an orthogonal matrix such that $\underline{\mathbf{m}}'' = \mathbf{Y}\underline{\mathbf{m}}'$ has independent components.

Unfortunately, this goal is not always achievable, nor is such a matrix unique. The idea is to find an objective function that measures the "magnitude of independence" with respect to **Y**. Actually, there a two major approaches to define something like a "magnitude of independence"; they lead to different algorithms for ICA. The first approach leads to the non-Gaussian family of ICA algorithms, which, as the name suggests, is inspired by the central limit theorem. This approach is not part of the subject matter of this textbook. The second approach is inspired by Shannon's information theory and uses the concept of *mutual information* to measure the independence of random variables. This textbook follows this second approach.

Definition 2.13 (Differential entropy). Let $\underline{a}$ and $\underline{b}$ be two absolutely continuous random variables with density $p(a,b)$. Furthermore, let $p(a)$ and $p(b)$ denote their marginal densities and $p(a \mid b)$ and $p(b \mid a)$ denote their conditional densities.

1. The *differential entropy* of the random variable $\underline{a}$ (respectively $\underline{b}$) is defined as

$$h(\underline{a}) = -\int p(a) \ln(p(a))\, \mathrm{d}a = -\mathrm{E}\{\ln(p(\underline{a}))\}, \tag{2.218}$$

$$h(\underline{b}) = -\int p(b) \ln(p(b))\, \mathrm{d}b = -\mathrm{E}\{\ln(p(\underline{b}))\}. \tag{2.219}$$

2. The *conditional differential entropy* of the random variables $\underline{a}$ and $\underline{b}$ is defined as

$$h(\underline{a} \mid \underline{b}) = -\iint p(a,b) \ln(p(a \mid b))\, \mathrm{d}a\, \mathrm{d}b = -\mathrm{E}\{\ln(p(\underline{a} \mid \underline{b}))\}, \tag{2.220}$$

$$h(\underline{b} \mid \underline{a}) = -\iint p(a,b) \ln(p(b \mid a))\, \mathrm{d}a\, \mathrm{d}b = -\mathrm{E}\{\ln(p(\underline{b} \mid \underline{a}))\}. \tag{2.221}$$

The original concept of entropy is only defined for the discrete case. Differential entropy, sometimes referred to as continuous entropy was Shannon's attempt to generalize the discrete case. Obtaining the formula of the differential entropy is syntactically straightforward: $h(\cdot)$ uses densities and integrals where $H(\cdot)$ uses discrete probabilities and sums (the result must be handled with care). However, it turned out that the differential entropy is an entirely different concept as the repeated discretization of a random variable does not make the discrete entropy $H(\cdot)$ converge to its continuous counterpart $h(\cdot)$. Also, the differential entropy is not always positive and is not invariant under continuous coordinate transformations. Both are important points for a "real" entropy and a measure of information uncertainty and are fulfilled by the original, discrete entropy. Nevertheless, for the purposes of this book, these problems can be ignored.

Definition 2.14 (Mutual information (transinformation)). The setting is the same as in Definition 2.13. The *mutual information* between the random variables $\underline{a}$ and $\underline{b}$ is defined as

$$I(\underline{a}, \underline{b}) = h(\underline{a}) - h(\underline{a} \mid \underline{b}) = h(\underline{b}) - h(\underline{b} \mid \underline{a}) = I(\underline{b}, \underline{a}). \tag{2.222}$$

The mutual information I is a symmetric function and the aforementioned problems of the differential entropy have no consequences here. From a descriptive point of view, the mutual information measures how much an observation of a realization of the random variable $\underline{a}$ reveals about possible outcomes of the random variable $\underline{b}$ and vice versa. In addition, let $p(a)p(b)$ denote the product density of the marginal densities. Then

$$\begin{aligned} I(\underline{a}, \underline{b}) &= h(\underline{a}) - h(\underline{a} \mid \underline{b}) \\ &= -\int p(a) \ln(p(a))\, da + \iint p(a,b) \ln(p(a \mid b))\, da\, db \\ &= -\iint p(a,b) \ln(p(a))\, da\, db + \iint p(a,b) \ln(p(a \mid b))\, da\, db \\ &= \iint p(a,b) \ln \frac{p(a \mid b)}{p(a)}\, da\, db = \iint_{\operatorname{supp} p(a,b)} p(a,b) \ln \frac{p(a,b)}{p(a)p(b)}\, da\, db \\ &= D(p(a,b) \| p(a)p(b)). \end{aligned} \tag{2.223}$$

This shows that the mutual information between two random variables is the Kullback–Leibler divergence (see 2.4.6) between the joint distribution and the product of the marginal distributions. This divergence becomes zero if the random variables are independent.

That yields the objective function for the second step of the ICA. Let the random vector after decorrelation as in Equation (2.214) be denoted by $\underline{\mathbf{m}}' = \begin{pmatrix} \underline{m}'_1 & \dots & \underline{m}'_d \end{pmatrix}$. Then

$$J(\mathbf{Y}) = D\left(p(\mathbf{m}'') \| p(m''_1) \cdots p(m''_d)\right) \quad \text{with } \underline{\mathbf{m}}'' = \mathbf{Y}\underline{\mathbf{m}}' \text{ and } \mathbf{Y}^{\mathsf{T}}\mathbf{Y} = \mathbf{I} \tag{2.224}$$

needs to be minimized, i.e.,

$$\mathbf{Y}^* = \arg\min_{\mathbf{Y}} J(\mathbf{Y}) \tag{2.225}$$

is the transformation matrix sought. In practice, this optimization problem can only be solved numerically and $J(\mathbf{Y}^*) = 0$ cannot be guaranteed. Still, the resulting transform will make the features approximately independent.

2.7.4 Multiple discriminant analysis

The goal of principal component analysis (and therefore of independent component analysis, too) is to find a projection to a lower dimensional space with respect to the entirety of all points $\mathcal{D}$. There is no guarantee that the first components are suitable for making a separation between the classes, because PCA is totally ignorant of these classes. This situation is depicted in Figure 2.41. In both cases, the position of the entirety of the points is the same and therefore the principal components analysis yields the same projection. But in Figure 2.41b, both classes exploit the whole range of the first principal component. Hence, a projection onto this component (i.e., a reduction to $d' = 1$) is not

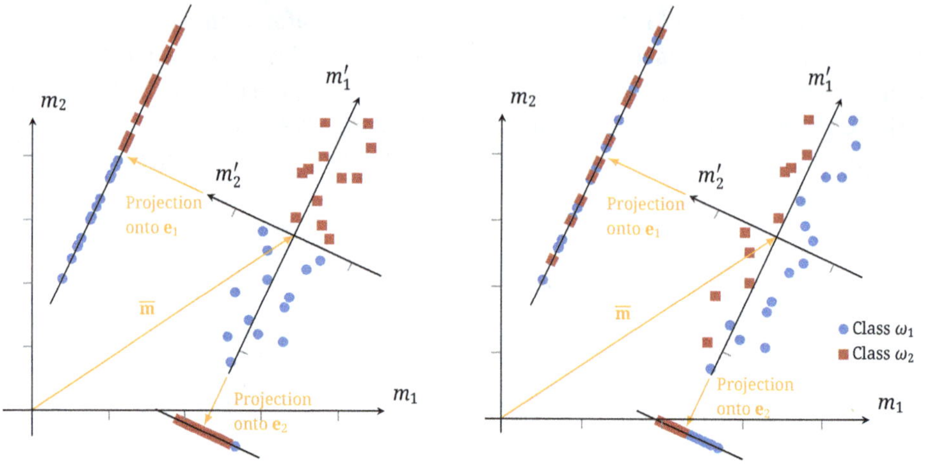

(a) First component suffices to separate the classes.

(b) Both components are necessary to separate the classes.

Fig. 2.41: The case for multiple discriminant analysis: PCA does not take class information into account. In particular, it does not aim for optimal class separability.

able to separate the classes. At least a second component is necessary. Actually the second component alone would suffice. Although the first component bears more information about the whole set, there is no information that is productive with respect to the actual problem. In contrast, in Figure 2.41a, a reduction to the first component still suffices to distinguish between the classes.

This shortcoming is tackled by multiple discriminant analysis (MDA). As the name suggests, MDA considers different classes and aims for an optimal separation right from the beginning. If the problem has c classes, then MDA finds the best projection onto a $(c-1)$-dimensional subspace.

The case of two classes

To begin, we will consider the simplest case of $c = 2$ classes and a projection onto a line, i.e., $d' = c - 1 = 1$. An objective function will be derived to find the best line in this case. This toy example will justify the later definition of the objective function in the higher dimensional case.

Let $\mathcal{D} = \{\mathbf{m}_1, \ldots, \mathbf{m}_N\}$ be the set of data. As already stated, $c = 2$ is assumed and

$$\mathcal{D}_1 = \{\mathbf{m}_i | \omega(\mathbf{m}_i) = \omega_1\} \qquad \mathcal{D}_2 = \{\mathbf{m}_i | \omega(\mathbf{m}_i) = \omega_2\} \tag{2.226}$$

will denote the partition of the dataset. Additionally, $|\mathcal{D}_1| = N_1$ and $|\mathcal{D}_2| = N_2$ denote the cardinalities of these sets. The goal is to find a vector $\mathbf{w} \in \mathbb{R}^d$ such that

$$m' = \mathbf{w}^{\mathsf{T}}\mathbf{m} \tag{2.227}$$

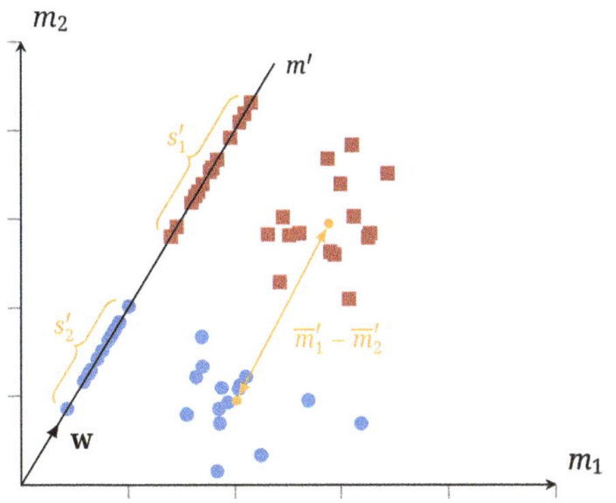

Fig. 2.42: Quantities in the two-class case of multiple discriminant analysis.

yields a projection that optimally separates both classes. Figure 2.42 illustrates the situation.

A good choice of **w** would be one that on the one hand arranges that the projected mean points of both classes are spread out and on the other hand that the standard deviation of each class is concentrated at the same time. This means one should optimize the ratio between these two quantities. To that end, one defines the mean of the projected classes by

$$\overline{m}_i' = \frac{1}{N_i} \sum_{m' \in \mathcal{D}_i'} m' = \frac{1}{N_i} \sum_{\mathbf{m} \in \mathcal{D}_i} \mathbf{w}^\mathsf{T}\mathbf{m} = \mathbf{w}^\mathsf{T} \left(\frac{1}{N_i} \sum_{\mathbf{m} \in \mathcal{D}_i} \mathbf{m} \right) = \mathbf{w}^\mathsf{T}\overline{\mathbf{m}}_i \qquad \text{for } i = 1, 2 \quad (2.228)$$

and the squared standard deviation of the projected classes by

$$\begin{aligned} s_i'^2 &= \frac{1}{N_i} \sum_{m' \in \mathcal{D}_i'} \left(m' - \overline{m}_i'\right)^2 = \frac{1}{N_i} \sum_{\mathbf{m} \in \mathcal{D}_i} \left(\mathbf{w}^\mathsf{T}\mathbf{m} - \mathbf{w}^\mathsf{T}\overline{\mathbf{m}}_i\right)^2 \\ &= \mathbf{w}^\mathsf{T} \frac{1}{N_i} \left(\sum_{\mathbf{m} \in \mathcal{D}_i} (\mathbf{m} - \overline{\mathbf{m}}_i)(\mathbf{m} - \overline{\mathbf{m}}_i)^\mathsf{T} \right) \mathbf{w} \qquad \text{for } i = 1, 2. \end{aligned} \quad (2.229)$$

The objective function will be the ratio

$$J(\mathbf{w}) = \frac{\left|\overline{m}_1' - \overline{m}_2'\right|^2}{s_1'^2 + s_2'^2} \quad (2.230)$$

and needs to be maximized. This form of the objective function is called the *Fisher linear discriminant*. This functional is not in an optimal form to solve the maximization problem, because the dependence on **w** is not explicit, nor is this form suited to be generalized to higher dimensions. So the next step is to rewrite Equation (2.230) into the product of a matrix and a vector.

A close look at Equation (2.229) reveals that the middle factor is again the scatter matrix (here normalized by N_i):

$$\mathbf{S}_i := \frac{1}{N_i} \sum_{\mathbf{m} \in \mathcal{D}_i} (\mathbf{m} - \overline{\mathbf{m}}_i)(\mathbf{m} - \overline{\mathbf{m}}_i)^\mathsf{T} \qquad \text{for } i = 1, 2 \quad (2.231)$$

and one writes the sum of both scatter matrices as

$$\mathbf{S}_\mathrm{W} := \mathbf{S}_1 + \mathbf{S}_2. \tag{2.232}$$

Then the denominator of Equation (2.230) can be rewritten as

$$s_1'^2 + s_2'^2 = \mathbf{w}^\mathsf{T}\mathbf{S}_1\mathbf{w} + \mathbf{w}^\mathsf{T}\mathbf{S}_2\mathbf{w} = \mathbf{w}^\mathsf{T}\mathbf{S}_\mathrm{W}\mathbf{w}. \tag{2.233}$$

The suffix W in $\mathbf{S}_\mathrm{W}$ stands for "within". Similarly, with

$$\mathbf{S}_\mathrm{B} := (\overline{\mathbf{m}}_1 - \overline{\mathbf{m}}_2)(\overline{\mathbf{m}}_1 - \overline{\mathbf{m}}_2)^\mathsf{T} \tag{2.234}$$

the numerator of Equation (2.230) becomes

$$\begin{aligned}\left|\overline{m}_1' - \overline{m}_2'\right|^2 &= \left|\mathbf{w}^\mathsf{T}(\overline{\mathbf{m}}_1 - \overline{\mathbf{m}}_2)\right|^2 \\ &= \mathbf{w}^\mathsf{T}(\overline{\mathbf{m}}_1 - \overline{\mathbf{m}}_2)(\overline{\mathbf{m}}_1 - \overline{\mathbf{m}}_2)^\mathsf{T}\mathbf{w} \\ &= \mathbf{w}^\mathsf{T}\mathbf{S}_\mathrm{B}\mathbf{w}.\end{aligned} \tag{2.235}$$

The suffix B in $\mathbf{S}_\mathrm{B}$ stands for "between". Then the objective function takes the form

$$J(\mathbf{w}) = \frac{\mathbf{w}^\mathsf{T}\mathbf{S}_\mathrm{B}\mathbf{w}}{\mathbf{w}^\mathsf{T}\mathbf{S}_\mathrm{W}\mathbf{w}}. \tag{2.236}$$

This form is called the *Rayleigh coefficient* or *Rayleigh quotient*. The aim is to maximize the ratio of the deviation between the classes compared to the deviation within the classes.

The Rayleigh quotient is invariant under scaling $\mathbf{w}$, hence it suffices to maximize the numerator $\mathbf{w}^\mathsf{T}\mathbf{S}_\mathrm{B}\mathbf{w}$ for all $\mathbf{w}$ such that the denominator $\mathbf{w}^\mathsf{T}\mathbf{S}_\mathrm{W}\mathbf{w}$ equals 1,

$$\max_{\mathbf{w}\in\mathbb{R}^d} \frac{\mathbf{w}^\mathsf{T}\mathbf{S}_\mathrm{B}\mathbf{w}}{\mathbf{w}^\mathsf{T}\mathbf{S}_\mathrm{W}\mathbf{w}} = \max_{\substack{\mathbf{w}\in\mathbb{R}^d \\ \mathbf{w}^\mathsf{T}\mathbf{S}_\mathrm{W}\mathbf{w}=1}} \mathbf{w}^\mathsf{T}\mathbf{S}_\mathrm{B}\mathbf{w}. \tag{2.237}$$

This, on being rewritten with Lagrange multipliers, leads to

$$f(\mathbf{w}, \lambda) = \mathbf{w}^\mathsf{T}\mathbf{S}_\mathrm{B}\mathbf{w} - \lambda\left(\mathbf{w}^\mathsf{T}\mathbf{S}_\mathrm{W}\mathbf{w}^\mathsf{T} - 1\right) \tag{2.238}$$

with λ as the Lagrange multiplier,

$$\begin{aligned}&\nabla_\mathbf{w} f(\mathbf{w}, \lambda) = 2\mathbf{S}_\mathrm{B}\mathbf{w} - 2\lambda\mathbf{S}_\mathrm{W}\mathbf{w} \stackrel{!}{=} 0 \\ \Leftrightarrow\ &\mathbf{S}_\mathrm{B}\mathbf{w} = \lambda\mathbf{S}_\mathrm{W}\mathbf{w}.\end{aligned} \tag{2.239}$$

The last line is called a generalized eigenvalue problem. Under the assumption that $\mathbf{S}_\mathrm{W}^{-1}$ is invertible, one obtains the standard eigenvalue problem

$$\mathbf{S}_\mathrm{W}^{-1}\mathbf{S}_\mathrm{B}\mathbf{w} = \lambda\mathbf{w}. \tag{2.240}$$

Luckily, this equation does not need to be solved directly. From the definition of $\mathbf{S}_\mathrm{B}$ one can see that

$$\mathbf{S}_\mathrm{B}\mathbf{w} = (\overline{\mathbf{m}}_1 - \overline{\mathbf{m}}_2) \underbrace{(\overline{\mathbf{m}}_1 - \overline{\mathbf{m}}_2)^\mathsf{T} \mathbf{w}}_{\text{scalar}} \tag{2.241}$$

always has the same direction as $(\overline{\mathbf{m}}_1 - \overline{\mathbf{m}}_2)$. Therefore, Equation (2.240) can be simplfied by setting $\mathbf{S}_\mathrm{B}\mathbf{w} = \lambda' (\overline{\mathbf{m}}_1 - \overline{\mathbf{m}}_2)$ for some unknown scalar λ',

$$\begin{aligned} &\mathbf{S}_\mathrm{W}^{-1}\lambda' (\overline{\mathbf{m}}_1 - \overline{\mathbf{m}}_2) = \lambda \mathbf{w} \\ \Leftrightarrow\ &\mathbf{w} = \lambda^{-1}\lambda' \mathbf{S}_\mathrm{W}^{-1} (\overline{\mathbf{m}}_1 - \overline{\mathbf{m}}_2)\,. \end{aligned} \tag{2.242}$$

By virtue of the normalization condition $\mathbf{w}^\mathsf{T}\mathbf{S}_\mathrm{W}\mathbf{w} = 1$ from Equation (2.237), the unknown factor can be cancelled:

$$\begin{aligned} \mathbf{w} &= \frac{\mathbf{w}}{\sqrt{\mathbf{w}^\mathsf{T}\mathbf{S}_\mathrm{W}\mathbf{w}}} = \frac{\mathbf{S}_\mathrm{W}^{-1} (\overline{\mathbf{m}}_1 - \overline{\mathbf{m}}_2)}{\sqrt{(\overline{\mathbf{m}}_1 - \overline{\mathbf{m}}_2)^\mathsf{T} {\mathbf{S}_\mathrm{W}^{-1}}^\mathsf{T} \mathbf{S}_\mathrm{W} \mathbf{S}_\mathrm{W}^{-1} (\overline{\mathbf{m}}_1 - \overline{\mathbf{m}}_2)}} \\ &= \frac{\mathbf{S}_\mathrm{W}^{-1} (\overline{\mathbf{m}}_1 - \overline{\mathbf{m}}_2)}{\sqrt{(\overline{\mathbf{m}}_1 - \overline{\mathbf{m}}_2)^\mathsf{T} \mathbf{S}_\mathrm{W}^{-1} (\overline{\mathbf{m}}_1 - \overline{\mathbf{m}}_2)}}\,. \end{aligned} \tag{2.243}$$

This concludes the case of two classes.

The general case

The introduction of this chapter already noted that in the general case the dimension of the subspace is $c-1$, i.e., one less than the number of classes. Instead of a single projection vector $\mathbf{w} \in \mathbb{R}^d$ to obtain the projected "vector" (a one-dimensional vector, i.e., a scalar)

$$m' = \mathbf{w}^\mathsf{T}\mathbf{m} \qquad \in \mathbb{R}, \tag{2.244}$$

one has a projection matrix $\mathbf{W} \in \mathbb{R}^{d\times(c-1)}$, so that the projection becomes

$$\mathbf{m}' = \mathbf{W}^\mathsf{T}\mathbf{m} \qquad \in \mathbb{R}^{(c-1)} \tag{2.245}$$

with $\mathbf{W} = (\mathbf{w}_1, \dots, \mathbf{w}_{c-1})$ and $\mathbf{w}_1, \dots, \mathbf{w}_{c-1}$ being the columns of $\mathbf{W}$.

In order to avoid a lot of technical details, this section only gives a sketch of the overall course of action. The reader is recommended to compare the formulas with the corresponding formulas from the two-class case and note the similarities.

As before,

$$\overline{\mathbf{m}}_i = \frac{1}{N_i} \sum_{\mathbf{m}\in\mathcal{D}_i} \mathbf{m} \qquad \text{for } i = 1, \dots, c \tag{2.246}$$

denotes the mean of each class and

$$\overline{\mathbf{m}} = \frac{1}{N} \sum_{\mathbf{m}\in\mathcal{D}} \mathbf{m} = \frac{1}{N} \sum_{i=1}^{c} N_i \overline{\mathbf{m}}_i \tag{2.247}$$

the overall mean. The scatter matrices for each class are

$$\mathbf{S}_i = \frac{1}{N_i} \sum_{\mathbf{m} \in \mathcal{D}_i} (\mathbf{m} - \overline{\mathbf{m}}_i)(\mathbf{m} - \overline{\mathbf{m}}_i)^\mathsf{T} \tag{2.248}$$

and the intra-class scattering is

$$\mathbf{S}_\mathrm{W} = \sum_{i=1}^{c} \mathbf{S}_i . \tag{2.249}$$

Likewise, the inter-class scatter matrix is

$$\mathbf{S}_\mathrm{B} = \sum_{i=1}^{c} N_i (\overline{\mathbf{m}}_i - \overline{\mathbf{m}})(\overline{\mathbf{m}}_i - \overline{\mathbf{m}})^\mathsf{T} . \tag{2.250}$$

Note that for $c = 2$, this definition differs from the previous definition.

The corresponding definitions can be set up for the projected feature $\mathbf{m}' = \mathbf{W}^\mathsf{T}\mathbf{m}$:

$$\overline{\mathbf{m}}'_i = \frac{1}{N_i} \sum_{\mathbf{m}' \in \mathcal{D}'_i} \mathbf{m}' \qquad \text{(class means)} \tag{2.251}$$

$$\overline{\mathbf{m}}' = \frac{1}{N} \sum_{\mathbf{m}' \in \mathcal{D}'} \mathbf{m}' \qquad \text{(overall mean)} \tag{2.252}$$

$$\mathbf{S}'_i = \frac{1}{N_i} \sum_{\mathbf{m}' \in \mathcal{D}'_i} \left(\mathbf{m}' - \overline{\mathbf{m}}'_i\right)\left(\mathbf{m}' - \overline{\mathbf{m}}'_i\right)^\mathsf{T} \qquad \text{(scatter matrices of each class)} \tag{2.253}$$

$$\mathbf{S}'_\mathrm{W} = \sum_{i=1}^{c} \mathbf{S}'_i \qquad \text{(intra-class scattering)} \tag{2.254}$$

$$\mathbf{S}'_\mathrm{B} = \sum_{i=1}^{c} N_i \left(\overline{\mathbf{m}}'_i - \overline{\mathbf{m}}'\right)\left(\overline{\mathbf{m}}'_i - \overline{\mathbf{m}}'\right)^\mathsf{T} \qquad \text{(inter-class scattering)} \tag{2.255}$$

Unsurprisingly, some algebraic calculation reveals the relation

$$\mathbf{S}'_\mathrm{W} = \mathbf{W}^\mathsf{T}\mathbf{S}_\mathrm{W}\mathbf{W} \tag{2.256}$$

$$\mathbf{S}'_\mathrm{B} = \mathbf{W}^\mathsf{T}\mathbf{S}_\mathrm{B}\mathbf{W} \tag{2.257}$$

between the scattering of the original features and the projected features (see Equations (2.233) and (2.235)). This leads again to the Rayleigh quotient

$$J(\mathbf{W}) = \frac{\det(\mathbf{S}'_\mathrm{B})}{\det(\mathbf{S}'_\mathrm{W})} = \frac{\det(\mathbf{W}^\mathsf{T}\mathbf{S}_\mathrm{B}\mathbf{W})}{\det(\mathbf{W}^\mathsf{T}\mathbf{S}_\mathrm{W}\mathbf{W})} \tag{2.258}$$

as the objective function. The columns $\mathbf{w}_1, \ldots, \mathbf{w}_{c-1}$ of the $\mathbf{W}$ that maximizes $J(\mathbf{W})$ are the eigenvectors of the generalized eigenproblem

$$\mathbf{S}_\mathrm{B}\mathbf{w} = \lambda\mathbf{S}_\mathrm{W}\mathbf{w} \quad \Leftrightarrow \quad (\mathbf{S}_\mathrm{B} - \lambda\mathbf{S}_\mathrm{W})\mathbf{w} = 0 \tag{2.259}$$

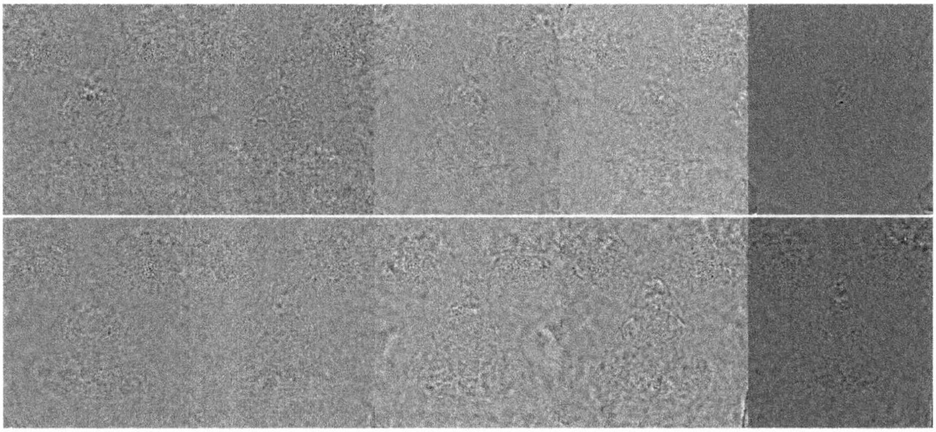

Fig. 2.43: First 10 Fisherfaces computed from the YALE faces dataset of Georghiades et al. [2001]. Unlike with the eigenfaces in Figure 2.34, there is no directly human-interpretable structure in the images.

that belong to the $(c-1)$ greatest eigenvalues $\lambda_1, \ldots, \lambda_{c-1}$. The task is to find the roots of the characteristic polynomial

$$g(\lambda) = \det(\mathbf{S}_B - \lambda \mathbf{S}_W) \tag{2.260}$$

and then solve the system of linear equations

$$(\mathbf{S}_B - \lambda \mathbf{S}_W)\mathbf{w} = 0 \tag{2.261}$$

in $\mathbf{w}$ for each $\lambda = \lambda_1, \ldots, \lambda_{c-1}$.

Example: Fisherfaces

In Section 2.7.1 it was shown how PCA can be used to represent the images of faces in an approach called *eigenfaces*. Similarly, MDA can be used to extract *Fisherfaces* from a dataset of images: the images are collected into vectors $\mathbf{m}_i$, from which the MDA matrix $\mathbf{W}$ is computed according to Equation (2.261). The columns $\mathbf{w}_i$, the Fisherfaces, of the matrix $\mathbf{W}$ can then be reorganized into images and inspected.

As an example, Fisherfaces were extracted from the extended YALE face dataset B (which was used to extract the eigenfaces, too, see Section 2.7.1). Images of the same subject were grouped into the same class, yielding 39 classes and hence 38 Fisherfaces, in all. The corresponding images of ten Fisherfaces are shown in Figure 2.43. Unlike the eigenfaces in Figure 2.34, humans have trouble interpreting the meaning of these images. If one concentrates enough, it is possible to see outlines of the eyes, the nose, and the mouth. One can also see the outline of the chin in the fifth picture from the left of the upper row, but it is difficult to imagine how Fisherfaces could be useful in determining the identity of a person. Yet, when the feature vector $\mathbf{m}' = \mathbf{W}^T\mathbf{m}$ of a given unknown image $\mathbf{m}$ is used in classification, Fisherfaces prove to be quite effective.

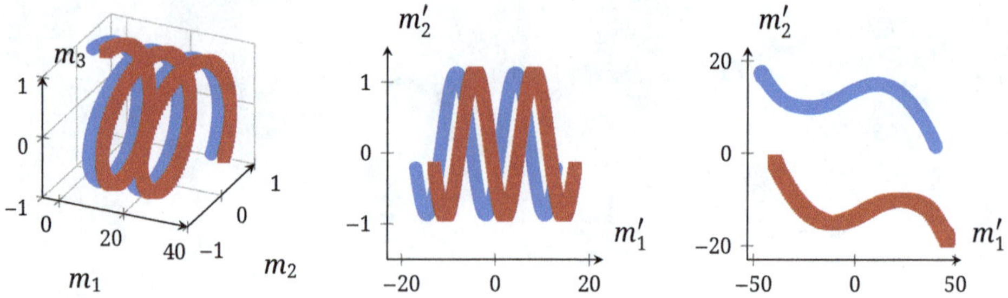

(a) 3D scatter plot of two spirals **(b)** PCA applied to (a) **(c)** *t*-SNE applied to (a)

Fig. 2.44: Visualization of PCA and *t*-SNE applied to a 3D scatter of two spirals.

In an experiment, a linear soft margin support vector machine (see Section 7.7) was trained to recognize the 39 identities in the extended YALE dataset using both eigenfaces and Fisherfaces as representations. With eigenfaces, 19 % accuracy was achieved with 76 components, 42 % accuracy with 127 components, and 65 % accuracy with 200 components. With Fisherfaces, on the other hand, the classification was 75 % accurate with only 38 components. This experiment shows that MDA is much more efficient at encoding discriminative information than PCA.

2.7.5 Dimensionality reduction with *t*-SNE

The *t*-distributed stochastic neighbor embedding introduced in Section 2.4.7 provides a way of visualizing high-dimensional data in lower dimensions. The main idea was to define a Gaussian in the original data in the high-dimensional space and map it to a heavy-tailed distribution in a low-dimensional space. By reducing the KL divergence between those two distributions, one can derive an embedding which preserves local similarities very well. Although the procedure has been developed mainly for visualization purposes, it essentially is a (non-linear) dimension reduction technique. In some scenarios, this can produce more useful results than the PCA which is demonstrated on a spiral example in Figure 2.44. In this example, *t*-SNE is able to separate the spirals very well, while PCA is not able to preserve the structure. Generalizing from the example, when dealing with high-dimensional data on non-linear manifolds, it is crucial for a clean visualization to maintain the proximity of *locally similar* points.

The general approach of creating a low-dimensional distribution P' from an original dataset bases on the iterative procedure presented in Algorithm 2.1. The process is visualized in Figure 2.45. The distances represented by colored arrows should align in both the 1D and the 2D visualization and are therefore dragged further to each other in every iteration. The adaption speed of P' can be controlled using a custom learning rate η and a momentum α (cf. Algorithm 2.1).

Algorithm 2.1: Gradient descent as iteration routine for t-SNE (van der Maaten and Hinton [2008])

Data: High-dimensional dataset $\mathcal{D} = \{\mathbf{m}_1, \mathbf{m}_2, \ldots, \mathbf{m}_N\}$, iterations T, design-parameter σ, learning rate $\eta > 0$, momentum α

Result: Low-dimensional data points $\mathcal{D}'^{(T)} = \{\mathbf{m}'_1, \mathbf{m}'_2, \ldots, \mathbf{m}'_N\}$

$P \leftarrow$ distribution based on pairwise affinities p_{ij} from Equation (2.33) and Equation (2.34)

$\mathcal{D}'^{(0)} \leftarrow$ sample from normal distribution

for $t \leftarrow 1$ **to** T **do**

  $P' \leftarrow$ low-dimensional distribution based on pairwise affinities p'_{ij} from Equation (2.35) and Equation (2.34)

  With $K = D_{\mathrm{KL}}(P \| P')$, compute gradient $\frac{\partial K}{\partial \mathcal{D}'^{(t)}}$

  $\mathcal{D}'^{(t)} \leftarrow \mathcal{D}'^{(t-1)} + \eta \frac{\partial K}{\partial \mathcal{D}'^{(t)}} + \delta_{[t \geq 2]} \alpha \left(\mathcal{D}'^{(t-1)} - \mathcal{D}'^{(t-2)} \right)$

return $\mathcal{D}'^{(T)}$

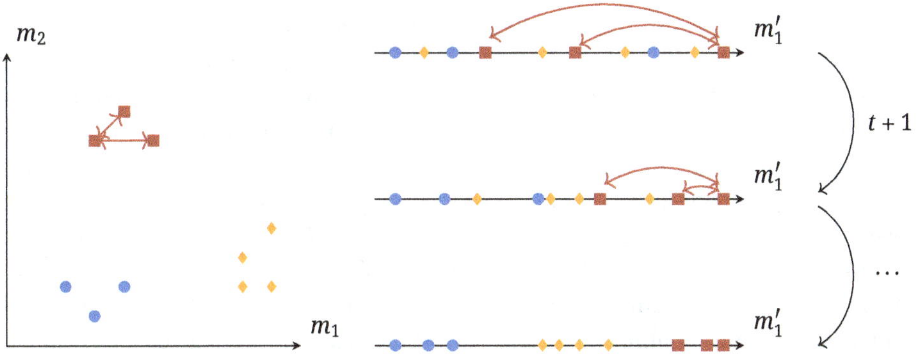

Fig. 2.45: Simplified illustration of the t-SNE algorithm, when visualizing 2D data points in 1D.

2.7.6 Autoencoder

Similar to the previously discussed methods for dimensionality reduction, the *autoencoder* aims to find a low-dimensional approximation that optimally represents the original data with respect to a distance measure. However, the function that applies the dimensionality reduction is learned by an artificial neural network (ANN) as well as the inverse function to reproduce the original data. ANNs are biologically inspired networks of artificial neurons and synapses and treated in detail in Section 7.4. For understanding the concept of the autoencoder, it is sufficient to know that ANNs can approximate arbitrary functions and thus can be applied to various tasks, also for data compression.

More precisely, let $\mathbf{m} = (m_1, m_2, \cdots, m_d)^\mathsf{T}$ be a d-dimensional feature vector and f_{enc} and f_{dec} two functions represented by an ANN termed *encoder* and *decoder*,

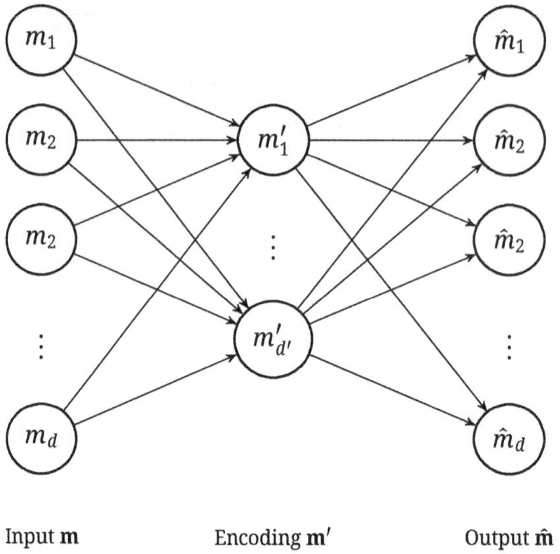

Fig. 2.46: Autoencoder with one hidden layer.

respectively. The encoder first calculates the low-dimensional feature vector $\mathbf{m}' = (m'_1, m'_2, \cdots, m'_{d'})^\mathsf{T} = f_{\text{enc}}(\mathbf{m})$ with $d' < d$. Then, the decoder computes the reconstructed feature vector $\hat{\mathbf{m}} = (\hat{m}_1, \hat{m}_2, \cdots, \hat{m}_d)^\mathsf{T} = f_{\text{dec}}(\mathbf{m}')$. Typically, a feed-forward neural network is utilized as ANN. Besides the input and the output layer, a feed-forward neural network comprises at least one hidden layer but can also have multiple hidden layers. If used as autoencoder, the layer with the smallest dimension contains the desired low-dimensional feature vector, also referred to as *encoding*. An autoencoder with one hidden layer is illustrated in Figure 2.46. Autoencoders are trained in an unsupervised fashion such that the distance D between original and reconstructed feature vectors is minimized: $D(\mathbf{m}, \hat{\mathbf{m}}) \to \min$. Like number and dimension of hidden layers, the distance function D is a degree of freedom when designing the autoencoder. For instance, the mean squared error is a suitable distance D. An application example of autoencoders for data compression is presented in Section 7.5.

2.7.7 Dimensionality reduction by feature selection

A common disadvantage of all the previous methods to reduce the dimension of the feature space is that the result is an opaque transformation matrix. The conversion leads to a new feature vector whose components are nebulous combinations of the former feature components. Hence, a (potentially existing) descriptive meaning gets lost. Moreover, the previous methods only work for features on at least an interval scale.

Feature selection means choosing a subset of features from a wider set of features that are considered to be sane for the problem at hand. In terms of the previous method, this method actually projects onto subspaces that are aligned with the same axes, i.e.,

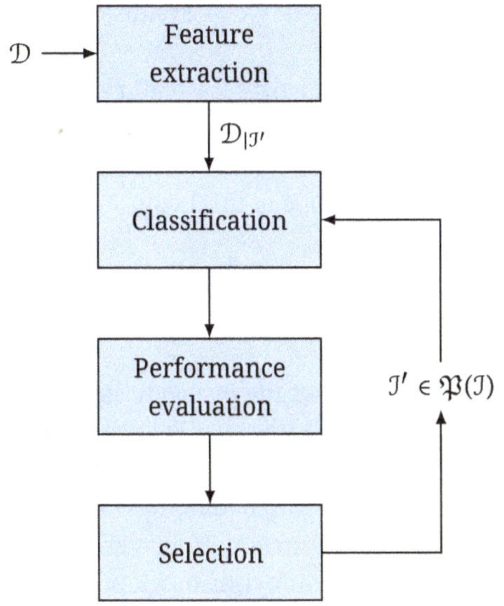

Fig. 2.47: Workflow of feature selection.

components are just left out or added but neither combined nor rotated. For this reason, this method also works for features on a lower scale.

Instead of an objective function that only depends on the data, the performance of a selection of features is directly evaluated with respect to a previously chosen classifier. For each selected set of features, the classifier is tuned on the training set, the classifier is applied to the test set, and the estimated class assignments are compared with the real classes of the data. As opposed to the other methods, the outcome of the dimension reduction thus depends on the established classifier. The workflow is depicted in Figure 2.47.

More formally, let $\mathcal{D} \in \mathcal{M}$ denote a training set with $d = \dim \mathcal{M}$ and $\mathcal{D} = \{\mathbf{m}_1, \ldots, \mathbf{m}_N\}$. Let $\mathcal{J} = \{1, \ldots, d\}$ be the index set of all dimensions and $\mathcal{J}' \in \mathfrak{P}(\mathcal{J})$ a selection of indices of the dimensions. For a feature vector $\mathbf{m}$, let $\mathbf{m}_{|\mathcal{J}'}$ denote the feature vector restricted to the selected components. Similarly, $\mathcal{D}_{|\mathcal{J}'}$ denotes the restricted set.

The task is to find the $\mathcal{J}^* \in \mathfrak{P}(\mathcal{J})$ such that the classifier has the best performance on $\mathcal{D}_{|\mathcal{J}^*}$ among all $\mathcal{J} \in \mathfrak{P}(\mathcal{J})$. To test every subset of $\mathcal{J}$, 2^d runs are necessary. Hence, this is only possible if d is small, because for each subset $\mathcal{J}'$ the classifier needs to be trained anew, tested, and evaluated on $\mathcal{D}_{|\mathcal{J}'}$. For reasonable values of d this is already prohibitive. If the desired dimension $d' < d$ is already given in advance, the number of subsets is still $\binom{d}{d'}$. Thus a brute force approach is normally impossible.

A suboptimal, but feasible approach is a greedy technique. First, the single-element set with the best feature component is selected. This requires d runs. The best component is held. Then the component out of the $d-1$ remaining ones is chosen that shows the best performance in conjunction with the already chosen first one. This procedure is repeated until the desired number of components is chosen or until joining a new component does not improve the performance. This way only $d + (d-1) \cdots + (d-d'+1)$ subsets need to be evaluated.

This approach is called a *wrapper* approach, because it *wraps* the feature selection around a classifier. Besides wrappers, there are *embedded* approaches, where a classifier implicitly performs the feature selection, and *filter* approaches, that do not depend on a classifier at all. However, these methods are outside the scope of this book.

2.7.8 Bag of words

Especially when dealing with images or video, one often has the problem that the feature extraction step assumes that all the patterns are of the same size. For example, the eigenfaces and Fischerfaces approaches in Sections 2.7.1 and 2.7.4 assume that the facial images are of the same size. One possible solution, the one pursued in these examples, is to crop and rescale the images so that they fulfill this constraint. However, doing so will inevitably remove information that is then unavailable for classification.

An alternative solution is given by the bag of visual words approach. Here, several low level features are extracted from different parts of the pattern and then combined into one higher level descriptor that characterizes the whole pattern. By construction, this descriptor always has the same dimensionality, irrespective of the size of the underlying pattern or the number of extracted low level features.

The approach has its roots in the bag of words model from natural language processing. Without going into too much detail, this model can be described as follows. Document generation is modeled as the repeated and independent drawing of words w_k that follow a probability distribution $P(w)$. The overall probability of generating the document $\boldsymbol{\tau} = (w_1, w_2, \ldots, w_K)$, that is, the sequence of K words w_k, is thus given by

$$P(\boldsymbol{\tau}) = \prod_{k=1}^{N} P(w_k). \tag{2.262}$$

In a classification setting, e.g., when the goal is to classify e-mails into "ham" $\hat{=}\ \omega_1$ or "spam" $\hat{=}\ \omega_2$, characteristic word distributions $P(w \mid \omega_1)$ for ham and $P(w \mid \omega_2)$ for spam can be estimated from a collection of documents by counting the words that occur in them. An unseen e-mail $\mathbf{p}' = (w'_1, \ldots, w'_{K'})$ with K' words can then be classified by assigning the class that maximizes the likelihood

$$L(\omega) = \prod_{k=1}^{K'} P\left(w'_k \,\middle|\, \omega\right) \quad \text{for } \omega \in \{\omega_1, \omega_2\}. \tag{2.263}$$

Alternatively, one can estimate the word distributions for every e-mail in the training set and train a classifier using these distributions as feature vectors. In other words, the $P(w)$ estimated from the document $\mathbf{p}$ (the pattern) acts as the feature vector $\mathbf{m}$, i.e., $\mathbf{m} = (P(w_1), \ldots, P(w_K))^\mathsf{T}$.

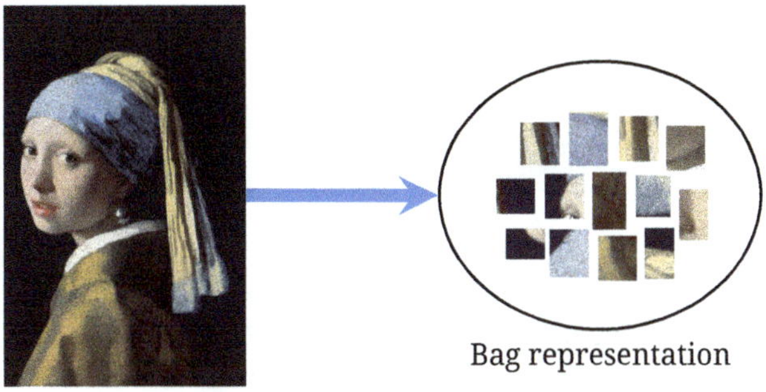

Fig. 2.48: Illustration of the underlying idea of bag of visual words: An image can be thought of as the composition of words from some visual vocabulary and can therefore be characterized by the words that appear in it.

Two things are important here. First, the order of the words in the document does not matter, nor does their surrounding text. Second, this method works irrespective of the length of the documents. Both of these are caused by the underlying idea of treating a document as an unordered collection—a bag—of words.

The same idea can be used to classify images of varying size. Here, it is assumed that the images are composed of *visual words* from some (for now) nondescript *visual vocabulary*. Similar to document classification, an image can then be characterized by observing the words that occur in the image (see Figure 2.48). Note that in general the original image cannot be reconstructed from the bag representation, because the vocabulary might not contain all of the possible visual words that appear in the image and because the position of the words in the image is not specified in the bag representation.

This approach is divided into two steps: learning the visual vocabulary from a training set, i.e., defining the visual words, and extracting a higher level descriptor from an image. Formally, let $\mathcal{D} = \{\mathbf{p}_n \mid n = 1, \ldots, N\}$ be a set of N patterns $\mathbf{p}_n \in \mathcal{P}$. These patterns should be representative for the patterns that are to be classified later on, but information about their classes is not needed. Low level features $\mathbf{m}_t(\mathbf{p}_n) \in \mathbb{R}^d$, $t = 1, \ldots, T(\mathbf{p}_n)$ are extracted from each of the N patterns $\mathbf{p}_n$. Note that the number of extracted low level features $T(\mathbf{p}_n)$ depends on $\mathbf{p}_n$. In general, a different number of low level features is extracted from each pattern, $T(\mathbf{p}_n) \neq T(\mathbf{p}_m)$ for $n \neq m$, e.g., when the features are extracted on key points as in the example below. The $\mathbf{m}_t(\mathbf{p}_n)$ are then used to partition the $\mathbb{R}^d$ into K non-overlapping tiles z_k, $k = 1, \ldots, K$ (see Figure 2.49), i.e.,

$$\biguplus_{k=1}^{K} z_k = \mathbb{R}^d. \tag{2.264}$$

The z_k form the visual vocabulary, that is, each z_k corresponds to a visual word. It is tempting to think of the low level features $\mathbf{m}_k(\mathbf{p}_n)$ as the *alphabet* from which the visual words are constructed, but this is a false analogy: unlike characters, every low level feature appears in one and only word z_k. A better analogy is to consider the $\mathbf{m}_t(\mathbf{p}_n) \in z_k$

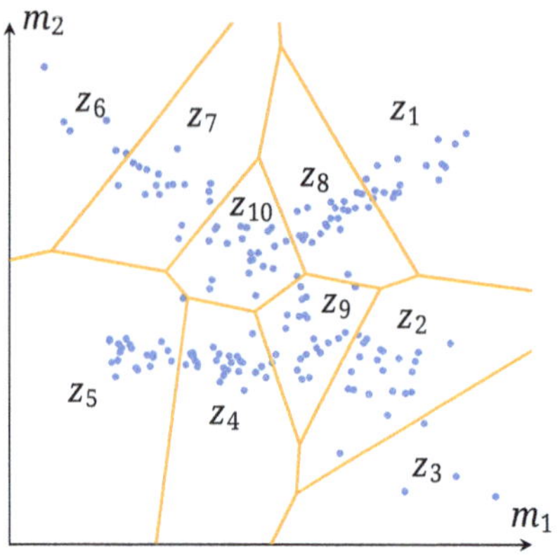

Fig. 2.49: Example of a visual vocabulary with $K = 10$ words $z_1, \dots, z_{10}$ derived from two-dimensional low level features $\mathbf{m} = (m_1, m_2)^\mathsf{T}$ using k-means clustering. Note that here the goal is not to find clusters, but rather to find a partition of the feature space.

as different spellings ("colour" and "color") or perhaps synonyms ("color" and "hue") of the same word z_k.

Once the vocabulary is determined, a K-dimensional high level descriptor $\mathbf{f}$ can be extracted from an unseen pattern $\mathbf{p}'$. Once again, low level features $\mathbf{m}_t(\mathbf{p}') \in \mathbb{R}^d$ with $t = 1, \dots, T(\mathbf{p}')$ are extracted from $\mathbf{p}'$. Note that the type of the features must be the same as the type that was used to determine the vocabulary. The high level descriptor $\mathbf{f}$ is built as a count statistic (i.e., a histogram) over the $\mathbf{m}_t(\mathbf{p}')$ w.r.t. the z_k,

$$\mathbf{f} = (f_1, f_2, \dots, f_K)^\mathsf{T} \quad \text{with}$$

$$f_k = \frac{1}{T(\mathbf{p})} \sum_{t=1}^{T(\mathbf{p})} \delta_{[\mathbf{m}_t \in z_k]}, \quad k = 1, \dots, K, \tag{2.265}$$

where $\delta_{[\cdot]}$ denotes the generalized Kronecker symbol. In other words, the entry f_k is the fraction (i.e., the frequency) of low level feature descriptors that fall into the tile z_k. Figure 2.50 shows an example descriptor derived from $T = 20$ low level features using the vocabulary from Figure 2.49.

The bag of visual words approach has some important properties:

1. As in text processing, the size of $\mathbf{f}$ does not depend on the size of the underlying pattern $\mathbf{p}$.
2. Determining the vocabulary is unsupervised, i.e., information about the patterns' classes $\omega(\mathbf{p}_n)$, $n = 1, \dots, N$ is not needed to determine the vocabulary.
3. Invariance properties of the low level features $\mathbf{m}_t$ propagate to the high level descriptors $\mathbf{p}$, e.g., if the $\mathbf{m}_t$ are invariant under rotation, scale or illumination, so will be $\mathbf{f}$.
4. Spatial (and in the case of a video, temporal) relations between image patches are discarded. On the one hand, this makes $\mathbf{f}$ robust against translation; on the other hand, this may remove discriminative information and prevent localization of objects in an image.
5. If there is a semantic interpretation of the low level features $\mathbf{m}_t$, the high level descriptor $\mathbf{f}$ can also be interpretable.

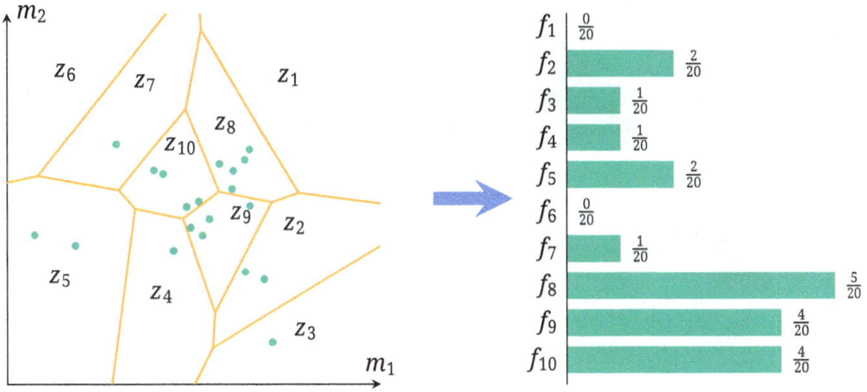

Fig. 2.50: Example of a bag of visual words descriptor constructed using the vocabulary in Figure 2.49. $T(\mathbf{p}) = 20$ low level features were extracted from the underlying pattern.

6. The dimensionality of $\mathbf{f}$ is often much larger than the dimensionality of the low level features $\mathbf{m}_t$, but the exact number of the z_k usually does not have a significant impact on the classification performance.

Lastly, bag of words can be interpreted as a *meta-feature* that subsumes multiple features extracted from an object into a single feature vector. This is similar to the Kullback–Leibler divergence from Section 2.4.6, which can be used to compare objects by estimating object-dependent probability distributions from the features. The difference is that the KL divergence compares full probability densities, while bag of words compares histograms, i.e., estimates of the probability mass in buckets of the feature space. Furthermore, bag of visual words produces a feature vector and relies on a separate, but arbitrary, classifier to derive the class assignment, while the KL divergence produces a scalar that can directly be used for classification, as in the example of Section 2.4.6.

Applications and examples: Image categorization

A highly influential paper that popularized this method in the field of computer vision was presented by Csurka et al. [2004]. In their paper, they described a bag of visual words approach for image categorization, i.e., for classifying images into one of many categories. Image categorization is difficult not only because of the large number of classes, but also because the images may show several objects at once and under various poses and lighting conditions. Furthermore, objects may be partially occluded and may be subject to non-rigid transformations, e.g., different arm positions when classifying images of humans.

To cope with these difficulties, SIFT keypoint descriptors (Lowe [2004]) were used as low level features. Without going into the details, a SIFT descriptor is a very powerful visual descriptor that characterizes the local texture around keypoints. SIFT is invariant

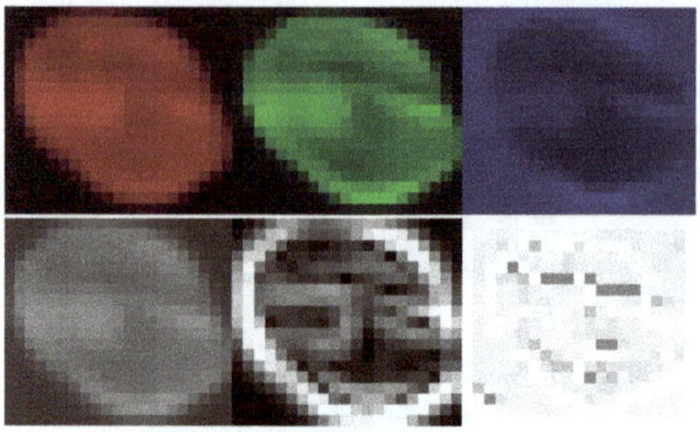

(a) Primitive features: features that require little or no computation time, e.g., color channels, gradient magnitude, texture codes.

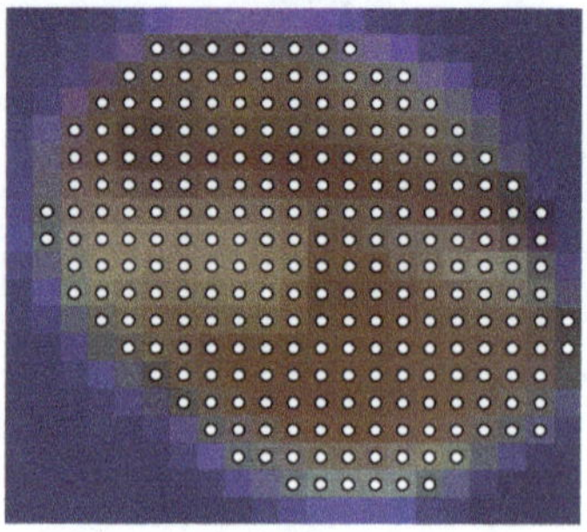

(b) Dense sampling: features are extracted from every foreground pixel.

Fig. 2.51: Modifications to use bag of words in bulk material sorting. Images from Richter et al. [2016].

under scaling and robust against rotation, non-rigid transformations, and differences in lighting. The partition of the feature space is achieved using *k*-means clustering. As in the example Figure 2.49, the goal was not to derive an optimal clustering of the features, but rather to build a descriptor that allows an accurate image categorization later on. Two classifiers, naive Bayes (see Section 3.3.3) and a kernel SVM (see Section 7.7), were evaluated. It was found that the SVM classifier produced more accurate results (Csurka et al. [2004]). Nowadays, the method has been superseded by convolutional neural networks (see Section 7.6), but the approach is still useful in other domains.

Applications and examples: Sorting of bulk materials

For example, in Richter et al. [2016], a bag of visual words approach is used for the sorting of bulk materials such as plastic pellets, minerals or foodstuffs. As can be seen in Figure 2, the environmental conditions (lighting, background, etc.) in bulk material sorters are design parameters rather than a source of nuisance and are chosen to aid the pattern recognition system. In addition, the objects under inspection are relatively simple, so that, compared to image categorization, the pattern recognition is relatively easy. The limiting factor is the processing time, since bulk material sorting is a real time task: in a typical system, the system has only a few tens of milliseconds to capture and process an image, classify the objects, and carry out the derived sorting decision.

This severely restricts the features and classifiers one can use. Using SIFT features and kernel SVM classifiers as above is not feasible. However, the specific circumstances of bulk material sorting can be exploited in a modified bag of words approach. First, since the objects are relatively simple, only primitive low level features, such as the values of the red, green and blue color channels, the gradient magnitude, or texture codes, are extracted (see Figure 2.51a). These features can encode color and texture information and require little or no computing time. However, since each individual low level feature carries little discriminative information, they are only useful when a large number of

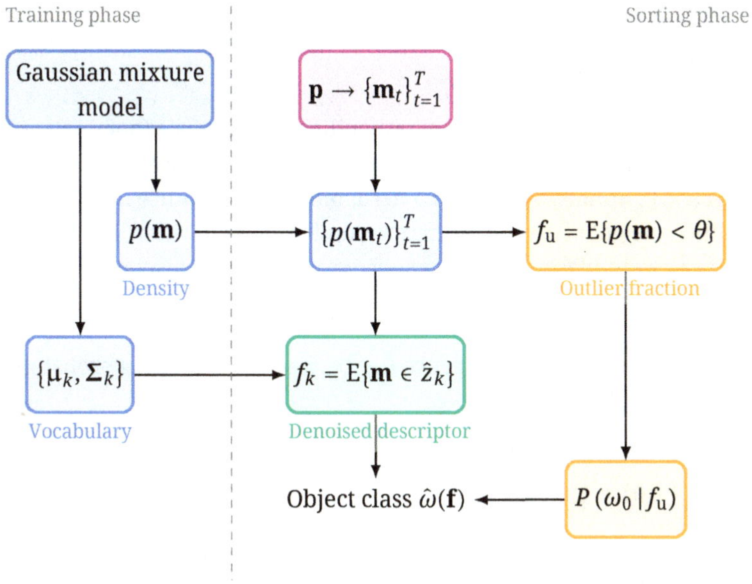

Fig. 2.52: Structure of the bag of words approach in Richter et al. [2016].

them is considered. To this end, every foreground pixel in an object image is considered a key point, that is, the primitive features are extracted on every foreground pixel of the object (see Figure 2.51b). This dense sampling comes at no additional cost, since foreground–background separation is a necessary step in the overall processing pipeline of the sorting system.

Once the primitive features are extracted from a training set, the feature space is partitioned using a Gaussian mixture model (see Section 3.3.6) instead of k-means clustering. The use of this model is twofold: first, it provides the vocabulary of visual words z_k and second, the density estimate $\hat{p}(\mathbf{m})$ is used to define the regions of outliers in the feature space. These regions are then used to de-noise the object descriptor $\mathbf{f}$ and to introduce a rejection class ω_0 to classify unknown objects. The overall structure of the approach is shown in Figure 2.52.

2.8 Exercises

(2.1) Suppose we are given the following features: m_n on a nominal scale, m_o on a ordinal scale, m_i on a interval scale, m_r on a ratio-scale, and m_a on a absolute scale. Which of these features allow which of the following transformations?

1. $f(m) = 3\,m + a$ with $a \in \mathbb{R}$
2. $f(m) = e^m$

3. $f(m) = \begin{cases} \frac{1}{m} & \text{if } m \neq 0 \\ 0 & \text{else} \end{cases}$.

(2.2) Assign the following features to their corresponding feature scale (nominal, ordinal, interval, ratio, or absolute): (school) grades, car brands, date of birth, area of the canvas of a sail, number of cows in a herd, motor temperature in °C, engine speed/revolution, height of body, clothing size, optical magnification, account balance, electrical voltage, place in a race, gender, variety of apple, display of a Geiger counter, population density, annual income in EUR, intelligence quotient (IQ).

(2.3) Compute the Kullback–Leibler divergence D_{KL} between the following probability distributions P_1 and P_2:

$$\begin{array}{llll} P_1(a) = \frac{1}{3}, & P_1(b) = 0, & P_1(c) = \frac{1}{3}, & P_1(d) = \frac{1}{3}, \\ P_2(a) = \frac{1}{3}, & P_2(b) = \frac{1}{6}, & P_2(c) = \frac{1}{6}, & P_2(d) = \frac{1}{3}. \end{array}$$

(2.4) Suppose there are three states, referred to as 1, 2, and 3, and two random variables with the probability mass functions $P_1(X)$ and $P_2(X)$ such that $P_1(1) = P_1(2) = P_1(3) = \frac{1}{3}$ and $P_2(1) = \frac{1}{6}$. How must the remaining probabilities $P_2(2) := a$ and $P_2(3) := b$ be chosen so that the Kullback–Leibler divergence $D_{\text{KL}}(P_1 \| P_2)$ is minimized?

(2.5) The traffic police invented a new, innovative test to assess a road user's ability to drive a vehicle: drivers are prompted to fire a gun at a target. The hit pattern is compared to the following reference, obtained from drunk and sober drivers:

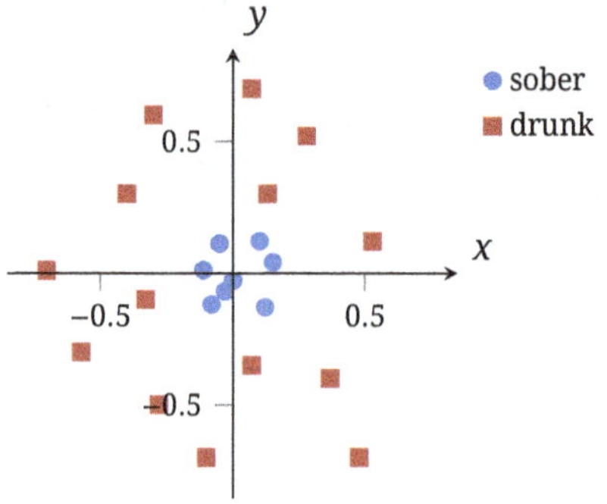

Construct a one-dimensional feature m' that can be used to assess whether a subject is driving under the influence or not. Give a formula to compute m'.

(2.6) A well-known car manufacturer manipulated the software that controls the ammonium nitrate injection to the catalytic converter so that less ammonium nitrate was used. The resulting engine fumes contain less smelly ammonia (NH_3), but more pollutant nitrogen oxides (NO_X). In a randomized test of engine fumes for these two compounds, the following scatter plot resulted:

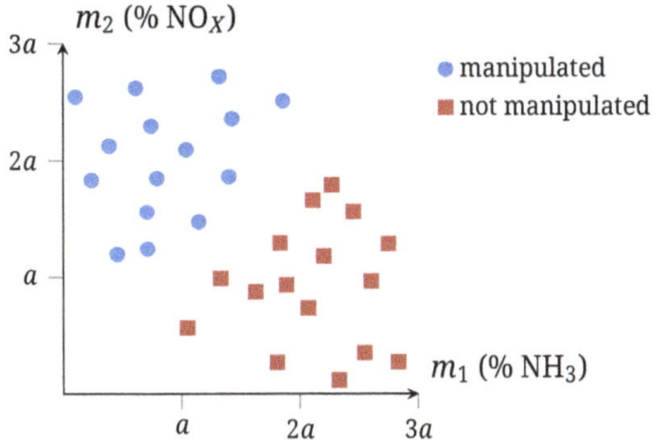

Construct a one-dimensional feature m' that can be used to classify cars into "manipulated" (ω_1) or "not manipulated" (ω_2). Give a formula to compute m'.

(2.7) The following feature is derived from a Fourier contour descriptor, i.e., the coefficients $Z_i \in \mathbb{C}, i \in \mathbb{Z}$ of the Fourier series expansion of the contour:

$$m := \frac{Z_8}{|Z_1|} - \frac{Z_6}{|Z_1|} + \frac{Z_4}{|Z_1|} - \frac{Z_2}{|Z_1|} + \frac{Z_0}{|Z_1|}.$$

Is m invariant under scaling, translation, and rotation? Why or why not?

(2.8) The following feature is derived from a Fourier contour descriptor, i.e., the coefficients $Z_i \in \mathbb{C}, i \in \mathbb{Z}$ of the Fourier series expansion of the contour:

$$m := \frac{\alpha|Z_0| + |Z_2|}{\sqrt{|Z_3|^2 + |Z_4|^\beta + \gamma}} \qquad \alpha,\beta,\gamma \in \mathbb{R}.$$

How must $\alpha, \beta, \gamma \in \mathbb{R}$ be chosen for m to be invariant under translation and scaling?

(2.9) Given are the following patterns:

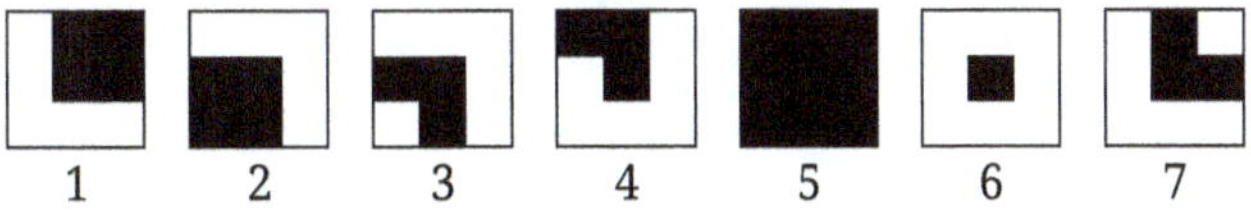

1. Which of the patterns are equivalent given a translation invariant feature?
2. Which of the patterns are equivalent given a translation and rotation invariant feature?

3. Which of the patterns are equivalent given a translation, rotation and scale invariant feature?

(2.10) In an optical inspection system the following variables are computed for each recorded object: center of mass $\mathbf{c}$, perimeter of the contour P, area of the object A, length of the main and secondary axes l_1 and l_2, and the rotation angle φ of the main axis w.r.t. the image coordinate system. The variables are shown in the following sketch:

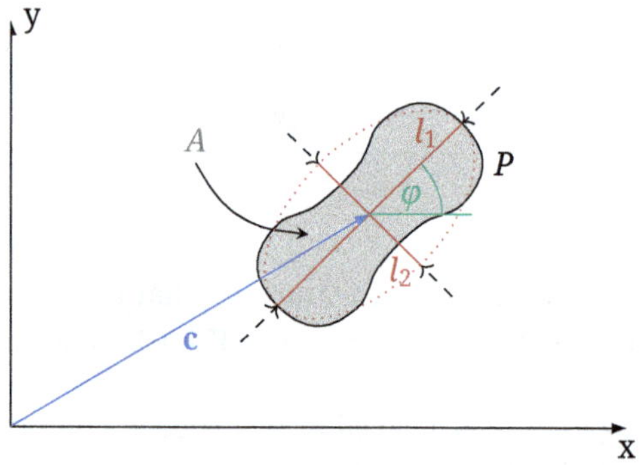

Five features, m_1 to m_5, are computed as defined below. Which of the features are invariant under translation, which are invariant under rotation, and which are invariant under scaling?

$$m_1 = \frac{\|\mathbf{c}\|^2}{A}$$

$$m_2 = \frac{l_1}{l_2} + \frac{P^2}{A}$$

$$m_3 = \varphi \frac{l_2}{P}$$

$$m_4 = \frac{A}{l_1} - \cos\varphi$$

$$m_5 = \frac{\varphi}{1\|\mathbf{c}\|^2} + \frac{l_1 - l_2}{A}$$

3 Bayesian decision theory

Up to now, this book has dealt with the question of how to select, define, and extract features from observed patterns of objects. In Figure 1.2, this is depicted as the first step of a pattern recognition system (first blue box) after the preparatory steps. From now on, our attention will be turned to the second step: how the features are used to assign objects to classes. Eventually, this task will be solved by classifiers.

Firstly, the entirety of classifiers can be divided into methods that use a probabilistic description of the problem and those which do not. The latter are considered in Chapter 7. The class of probabilistic methods is rather extensive, hence its discussion is divided into the Chapters 3 to 6. Under the assumption that the probabilistic description of the system is already given, the question of how the actual classification is to be performed needs to be answered. This is the subject of this chapter. The problem of how to obtain the probabilistic description temporarily remains open and will be answered later (Chapters 4 to 6).

In summary, the assumptions of this chapter are that the probabilistic description of the system is already given and that the features of the objects are already defined and extracted.

3.1 General considerations

The general idea is to think of the world as a random generator that outputs pairs of features with associated object classes $(\mathbf{m}, \omega)$. The collection of all pairs is described by a probability distribution. In addition, all elements of a sequence of pairs are pairwise independent, i.e., the joint distribution of N pairs $(\underline{\mathbf{m}}_1, \underline{\omega}_1), \ldots, (\underline{\mathbf{m}}_N, \underline{\omega}_N)$ equals the product of the individual distributions,

$$p((\mathbf{m}_1, \omega_1), \ldots, (\mathbf{m}_N, \omega_N)) = \prod_{j=1}^{N} p(\mathbf{m}_j, \omega_j). \tag{3.1}$$

The marginal distribution of the classes

$$P(\omega) = \int_{\mathcal{M}} P(\mathbf{m}, \omega)\, \mathrm{d}\mathbf{m} \tag{3.2}$$

is called an a priori distribution (of the classes) and the marginal distribution of the features

$$p(\mathbf{m}) = \sum_{\omega=\omega_1}^{\omega_c} P(\mathbf{m}, \omega) \tag{3.3}$$

is called the overall distribution of the features. The conditional distribution

$$p(\mathbf{m} \mid \omega) \tag{3.4}$$

https://doi.org/10.1515/9783111339207-003

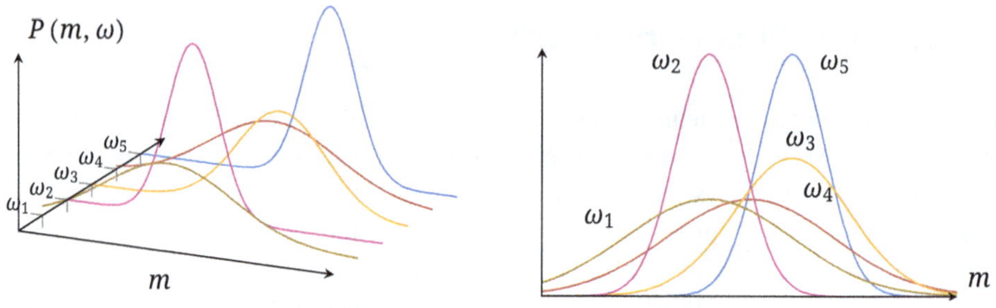

(a) Full joint distribution **(b)** Marginal distributions

Fig. 3.1: Example of a random distribution $P(\mathbf{m}, \omega)$ of mixed discrete and continuous quantities.

is called the class-specific feature distribution and the conditional distribution

$$P(\omega \mid \mathbf{m}) \tag{3.5}$$

is called the a posteriori distribution of the classes.

Throughout this chapter, we assume that the class-specific feature distribution and the a priori distribution are known. How to obtain those quantities from the data will be the subject of Chapters 4 and 5. After the observation of a specific feature $\mathbf{m}$, one is interested in the probability of the classes, that is, the a posteriori distribution. The joint distribution can be expressed as

$$p(\mathbf{m}, \omega) = p(\mathbf{m} \mid \omega)P(\omega) = P(\omega \mid \mathbf{m})p(\mathbf{m}) \tag{3.6}$$

which leads to Bayes' law

$$P(\omega \mid \mathbf{m}) = \frac{p(\mathbf{m} \mid \omega)P(\omega)}{p(\mathbf{m})}. \tag{3.7}$$

Note that $P(\mathbf{m}, \omega)$ is a mixed distribution of discrete and continuous random quantities. The set of classes $\{\omega_1, \ldots, \omega_c\}$ is finite but the range of the feature vector is usually a continuum. In terms of the continuous components of $\mathbf{m}$, $P(\mathbf{m}, \omega)$ is a probability density and in terms of ω and the discrete components of $\mathbf{m}$, $P(\mathbf{m},\omega)$ is a probability mass function. Figure 3.1 illustrates such a distribution.

As a preparatory step towards deriving an optimal classifier, we introduce the decision space $\mathcal{K}$ with $\dim(\mathcal{K}) = c$. This decision space is not strictly required, but it helps to unify the formal description of different classifiers and prevents a misinterpretation of the meaning of the classes. Originally, we had defined

$$\Omega/\!\sim\, = \{\omega_1, \ldots, \omega_c\} \tag{3.8}$$

where each $\omega_i \subseteq \Omega$ denoted one class and was formally defined as a subset of Ω. Now, we identify each of them with a unit vector in a c-dimensional space $\mathcal{K}$,

$$\boldsymbol{\omega}_i \in \mathcal{K} = \mathbb{R}^c \quad \text{with} \quad \omega_{ij} = \delta_i^j \tag{3.9}$$

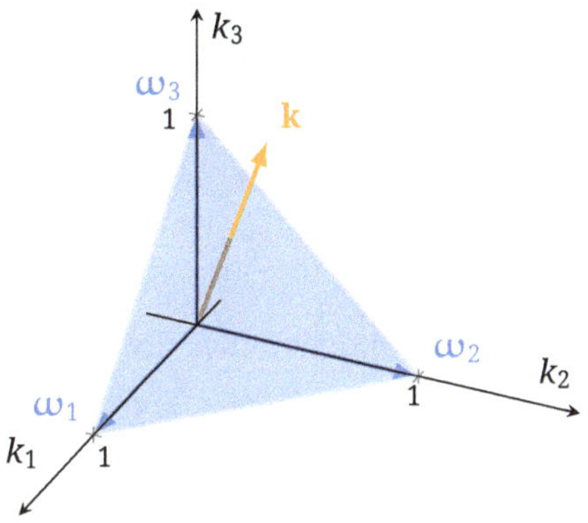

Fig. 3.2: The decision space $\mathcal{K}$ for $c = 3$ classes. The orange arrow shows a decision vector $\mathbf{k}$, the blue triangle shows the probability simplex, where $\sum_i k_i(\mathbf{m}) = 1$ and $k_i \geq 0$, $i = 1, \ldots, c$.

or, more elaborately (see Figure 3.2):

$$\omega_i = (0, \ldots, 0, \underbrace{1}_{i\text{th position}}, 0, \ldots, 0)^\top . \tag{3.10}$$

The unit vectors ω_i are also called target vectors, $\mathbf{k} \in \mathcal{K}$ decision vector, and its components k_i decision functions.

In light of this new perspective, a classifier follows a two-step construction scheme. In the first step, a feature vector is mapped to a value $\mathbf{k}$ in the decision space spanned by the ω_i (see Figure 3.2),

$$\mathbf{k} \in \operatorname{span}\{\omega_1, \ldots, \omega_c\} = \left\{ \sum_{i=1}^{c} \lambda_i \omega_i \,\middle|\, \lambda_i \in \mathbb{R} \right\} . \tag{3.11}$$

The second step is to take the target vector $\hat{\omega}$ as an estimation that has the shortest distance from $\mathbf{k}$, i.e.,

$$\hat{\omega} = \underset{\omega_i:\, i=1,\ldots,c}{\arg\min} \|\mathbf{k} - \omega_i\| . \tag{3.12}$$

While the second step is uniform for all classifiers, the actual logic of a classifier is merged into the mapping in the first step,

$$\mathbf{k} \colon \begin{cases} \mathcal{M} & \to \mathcal{K} \\ \mathbf{m} & \mapsto \mathbf{k}(\mathbf{m}) \end{cases} . \tag{3.13}$$

Beyond a uniform description of classifiers, the vectorized approach prevents a misconception of the classes' numbering. While the numbering misleadingly implies an ordering of the classes, the vectorized approach clearly shows that each pair of target vectors has the same distance. More concisely: the classes stem from a nominal, not an ordinal, scale and the vector description reflects this.

A parametric decision function is a mapping $\mathbf{k}(\mathbf{m}, \theta)$ that additionally depends on a parameter vector $\theta = (\theta_1, \ldots, \theta_k)^\top \in \Theta$ from a parameter space Θ. Learning by examples means to find the parameter vector $\hat{\theta}$ such that the mapping $\mathbf{k}(\mathbf{m}_i, \hat{\theta})$

approximates the true class $\omega(\mathbf{m}_i)$ for each of the training samples in some optimal way. Each decision region $\mathcal{R}_i \subseteq \mathcal{M}$ in the feature space corresponds to a subset $\{\mathbf{k} \mid \|\mathbf{k} - \omega_i\| < \|\mathbf{k} - \omega_j\| \; \forall j \neq i\} \in \mathcal{K}$ in the decision space. More precisely,

$$\mathcal{R}_i = \{\mathbf{m} \mid \|\mathbf{k}(\mathbf{m}, \theta) - \omega_i\| < \|\mathbf{k}(\mathbf{m}, \theta) - \omega_j\| \; \forall j \neq i\}, \tag{3.14}$$

$$\partial\mathcal{R}_i = \{\mathbf{m} \mid \|\mathbf{k}(\mathbf{m}, \theta) - \omega_i\| \leq \|\mathbf{k}(\mathbf{m}, \theta) - \omega_j\| \; \forall j \neq i\} \setminus \mathcal{R}_i. \tag{3.15}$$

This shows that the structure and parametrization of $\mathbf{k}(\cdot, \theta)$ determines the decision boundaries in $\mathcal{M}$.

3.2 The maximum a posteriori classifier

After these preliminary remarks, we now start with the first specific classifier. As the first objective, one can require that the expected squared Euclidean distance between the decision vector and the true target vector is minimal. Let $P(\mathbf{m}, \omega)$ denote the joint distribution of the feature vector and the target vectors. Then the objective function is

$$f(\mathbf{k}) = \mathrm{E}\{\|\mathbf{k}(\underline{\mathbf{m}}) - \underline{\omega}\|^2\} = \sum_{i=1}^{c} \int_{\mathcal{M}} \|\mathbf{k}(\mathbf{m}) - \omega_i\|^2 P(\mathbf{m}, \omega_i)\, \mathrm{d}\mathbf{m} \tag{3.16}$$

for all permissible functions $\mathbf{k}\colon \mathcal{M} \to \mathcal{K}$. In a strictly mathematical sense, one has to define the set of functions under consideration and ensure that the integrals are well defined. For the purposes of this book, we will ignore these technicalities. Let $\mathbf{k}$ denote an optimal solution and $\mathbf{k}_\Delta \neq \mathbf{0}$ a small deviation. It holds, that

$$f(\mathbf{k} + \mathbf{k}_\Delta) > f(\mathbf{k}). \tag{3.17}$$

In order to simplify notation in the following discussion, the argument to $\mathbf{k}$ is momentarily omitted. Keep in mind that the expectation is taken with respect to $\mathbf{m}$, which is an argument of $\mathbf{k}$. Consider the left side of Equation (3.17) more closely:

$$\begin{aligned} f(\mathbf{k} + \mathbf{k}_\Delta) &= \mathrm{E}\{\|\mathbf{k} + \mathbf{k}_\Delta - \underline{\omega}\|^2\} \\ &= \mathrm{E}\{\|\mathbf{k} - \underline{\omega}\|^2\} + 2\,\mathrm{E}\{\mathbf{k}_\Delta^\mathrm{T}(\mathbf{k} - \underline{\omega})\} + \mathrm{E}\{\|\mathbf{k}_\Delta\|^2\}. \end{aligned} \tag{3.18}$$

Hence, Equation (3.17) is equivalent to

$$\mathrm{E}\{\|\mathbf{k}_\Delta\|^2\} + 2\,\mathrm{E}\{\mathbf{k}_\Delta^\mathrm{T}(\mathbf{k} - \underline{\omega})\} > 0. \tag{3.19}$$

As $\mathbf{k}$ is an optimal solution, the inequality must hold for any $\mathbf{k}_\Delta$. This is surely true if there is a $\mathbf{k}$ such that the second term is identically zero. Note that it is not required that the second term be zero: this is a sufficient but not a necessary condition. Altogether, this assumption leads to

$$\begin{aligned} 0 = \mathrm{E}\{\mathbf{k}_\Delta^\mathsf{T}(\mathbf{k} - \underline{\omega})\} &= \int_{\mathcal{M}} \sum_{i=1}^{c} \mathbf{k}_\Delta^\mathsf{T}(\mathbf{k} - \omega_i) P(\mathbf{m}, \omega_i)\, \mathrm{d}\mathbf{m} \\ &= \int_{\mathcal{M}} \mathbf{k}_\Delta^\mathsf{T}(\mathbf{m}) \underbrace{\left(\sum_{i=1}^{c} (\mathbf{k}(\mathbf{m}) - \omega_i) P(\omega_i \mid \mathbf{m}) \right) p(\mathbf{m})\, \mathrm{d}\mathbf{m}}_{\stackrel{!}{=}\mathbf{0}}. \end{aligned} \tag{3.20}$$

In the last line, the dependence of $\mathbf{k}$ on $\mathbf{m}$ was made explicit again and the term was reordered so that $\mathbf{k}_\Delta^\mathsf{T}(\mathbf{m})$ was moved out of the brackets, because this term is outside of our control. But the overall formula becomes zero if the term in the brackets is zero and this term is only made up of known quantities and $\mathbf{k}$. One obtains

$$\begin{aligned} \mathbf{0} &= \sum_{i=1}^{c} (\mathbf{k}(\mathbf{m}) - \omega_i) P(\omega_i \mid \mathbf{m}) = \mathbf{k}(\mathbf{m}) \underbrace{\sum_{i=1}^{c} P(\omega_i \mid \mathbf{m})}_{=1} - \sum_{i=1}^{c} \omega_i P(\omega_i \mid \mathbf{m}) \\ &= \mathbf{k}(\mathbf{m}) - \mathrm{E}\{\underline{\omega} \mid \mathbf{m}\} \qquad \Rightarrow \qquad \mathbf{k}(\mathbf{m}) = \mathrm{E}\{\underline{\omega} \mid \mathbf{m}\}. \end{aligned} \tag{3.21}$$

The last line concludes the derivation. Going one step back and explicitly writing out the expectation gives a more illustrative representation of the optimal target vector:

$$\begin{aligned} \mathbf{k}(\mathbf{m}) &= \mathrm{E}\{\underline{\omega} \mid \mathbf{m}\} = \sum_{i=1}^{c} \omega_i P(\omega_i \mid \mathbf{m}) \\ &= \begin{pmatrix} 1 \\ 0 \\ \vdots \\ 0 \end{pmatrix} \cdot P(\omega_1 \mid \mathbf{m}) + \cdots + \begin{pmatrix} 0 \\ \vdots \\ 0 \\ 1 \end{pmatrix} \cdot P(\omega_c \mid \mathbf{m}) = \begin{pmatrix} P(\omega_1 \mid \mathbf{m}) \\ P(\omega_2 \mid \mathbf{m}) \\ \vdots \\ P(\omega_c \mid \mathbf{m}) \end{pmatrix}. \end{aligned} \tag{3.22}$$

The target vector is formed by the conditional distributions or a posteriori distributions of the classes, but this is only the first step of the classifier. Putting the last line into the second step (see Equation (3.12)) leads to

$$\hat{\omega} = \underset{\omega_i:\, i=1,\ldots,c}{\arg\min} \|\mathbf{k} - \omega_i\| = \underset{\omega_i:\, i=1,\ldots,c}{\arg\max} P(\omega_i \mid \mathbf{m}). \tag{3.23}$$

Intuitively, this result does not come as a surprise. The optimal classifier with respect to the least expected square error always takes the class with the highest a posteriori probability. Therefore, this classifier is called the maximum a posteriori (MAP) classifier. Every point that can be described by $\mathbf{k}$ shares the property that the sum of its components equals one. For this reason, the blue simplex in Figure 3.2 is also called the probability simplex. Figure 3.3 illustrates the workflow of the MAP classifier; Figure 3.4 depicts a 2-dimensional projection of the probability simplex from Figure 3.2.

Note that although this classifier is optimal in the sense of least square error in the decision space, this does not mean that it will classify without error. The error probability is given by

$$P_e := P(\hat{\omega} \neq \omega), \tag{3.24}$$

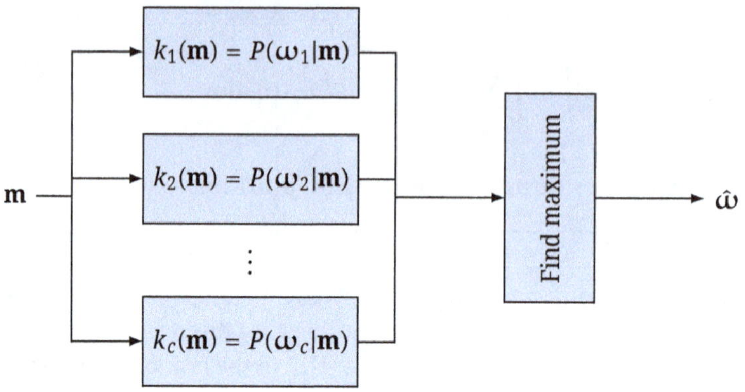

Fig. 3.3: Workflow of the MAP classifier.

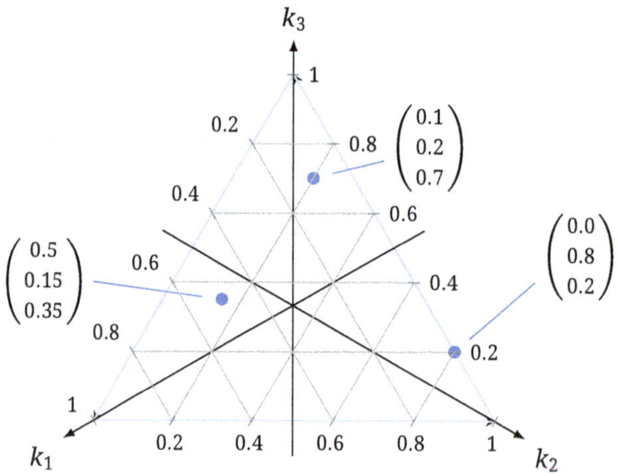

Fig. 3.4: 3-dimensional probability simplex in barycentric coordinates, i.e., projected onto the two-dimensional plane so that every point in the simplex is identified by three coordinates $k_1,k_2,k_3 \geq 0$ with $k_1 + k_2 + k_3 = 1$.

i.e., this is the probability that the estimated class ω does not match the true class ω. Since the a posteriori probability $P(\omega \mid \mathbf{m})$ depends on the class-specific feature distribution $p(\mathbf{m} \mid \omega)$, the asymptotic (i.e., the true theoretical) error probability can only be zero if the class-specific feature distributions do not overlap, or if the a priori probability of overlapping distributions is nonzero only for one of the classes. In practical applications, the former is almost never the case, while the latter is nonsensical as it prevents the classifier from deciding on one or more of the classes.

3.3 Bayesian classification

In the previous Section, the MAP classifier was derived as the optimal classifier with respect to the least expected square error. Of course, this criterion is only correct if each error is equally bad. Quite often one is faced with applications in which one kind of classification error is worse or more costly than an error of another kind. For example, an undetected cause for alarm might be worse than a false alarm.

Therefore, the goal of this chapter is to extend the Bayesian framework by a cost function:

$$l\colon \Omega^0/\!\sim \times\, \Omega/\!\sim\, \to \mathbb{R} \tag{3.25}$$

where $\Omega^0/\!\sim\, := \Omega/\!\sim \cup \{\omega_0\}$ and ω_0 denotes the rejection class, which will be discussed in more detail in Section 9.5. For now, it suffices to treat it as just another class without any deeper meaning. The cost function l expresses the costs of deciding on the class $\hat{\omega}$ if the true class is actually ω. In the case of a finite number of classes, the costs can also be expressed as the matrix

$$\mathbf{L} = \begin{pmatrix} l(\omega_0, \omega_1) & \dots & l(\omega_0, \omega_c) \\ l(\omega_1, \omega_1) & \dots & l(\omega_1, \omega_c) \\ \vdots & \ddots & \vdots \\ l(\omega_c, \omega_1) & \dots & l(\omega_c, \omega_c) \end{pmatrix} \in \mathbb{R}^{(c+1)\times c}. \tag{3.26}$$

Normally, one has $l(\omega_i,\omega_i) = 0$ and $l(\omega_i, \omega_j) \geq 0$ for $i \neq j$. Instead of reducing the average error the new objective is to reduce the expected cost

$$R = \mathrm{E}\{l(\hat{\omega}(\underline{\mathbf{m}}), \underline{\omega})\}\,. \tag{3.27}$$

This expected cost is also called the risk.

In the first subsection, we will develop the concept of the Bayesian classifier in a similar way as for the MAP classifier. The Bayesian classifier is the classifier with the minimal risk. In the light of this new concept, the MAP classifier will be subsumed as the special case when the cost function is the Kronecker delta, $l(\omega_i,\omega_j) = \delta_i^j$.

In the second subsection we will derive another special classifier, the so-called minimax classifier.

3.3.1 The Bayesian optimal classifier

The a posteriori risk or conditional risk is the risk of deciding on a class ω_i, $i = 1, \dots, c$ given a fixed feature vector $\mathbf{m}$,

$$r_i(\mathbf{m}) = R(\omega_i \mid \mathbf{m}) = \mathrm{E}\{l(\omega_i, \underline{\omega}) \mid \mathbf{m}\} = \sum_{j=1}^{c} l(\omega_i, \omega_j) P(\omega_j \mid \mathbf{m}), \tag{3.28}$$

which gives the a posteriori risk vector

$$\mathbf{r}(\mathbf{m}) = (r_1(\mathbf{m}), \dots, r_c(\mathbf{m}))^\mathsf{T}. \tag{3.29}$$

The objective is now to find a classifier $\hat{\omega} : \left\{ \begin{smallmatrix} \mathcal{M} \to \Omega/\sim \\ \mathbf{m} \mapsto \omega_i \end{smallmatrix} \right.$ that minimizes the total risk $R = \mathrm{E}\{l(\hat{\omega}(\underline{\mathbf{m}}), \underline{\omega})\}$. This leads to

$$\begin{aligned} R &= \mathrm{E}\{l(\hat{\omega}(\underline{\mathbf{m}}), \underline{\omega})\} \\ &= \int_{\mathcal{M}} \sum_{j=1}^{c} l(\hat{\omega}(\mathbf{m}), \omega_j) P(\mathbf{m}, \omega_j)\,\mathrm{d}\mathbf{m} \\ &= \int_{\mathcal{M}} \Big(\sum_{j=1}^{c} l(\hat{\omega}(\mathbf{m}), \omega_j) P(\omega_j \mid \mathbf{m}) \Big) p(\mathbf{m})\,\mathrm{d}\mathbf{m} \\ &= \int_{\mathcal{M}} R(\hat{\omega}(\mathbf{m}) \mid \mathbf{m}) p(\mathbf{m})\,\mathrm{d}\mathbf{m}. \end{aligned} \tag{3.30}$$

As $p(\mathbf{m})$ is a non-negative function, the risk R obviously becomes minimal if $\hat{\omega}$ is chosen such that $R(\hat{\omega}(\mathbf{m}) \mid \mathbf{m})$ is minimal point-wise for every $\mathbf{m}$. The estimator $\hat{\omega}$ can only have c distinct values and it is the same for $R(\hat{\omega}(\mathbf{m}) \mid \mathbf{m})$, hence the solution is to choose

$$\hat{\omega}(\mathbf{m}) = \omega_i \quad \text{with} \quad i = \underset{j=1,\dots,c}{\arg\min}\, r_j(\mathbf{m}). \tag{3.31}$$

In summary, the optimal decision is to choose the class with the smallest a posteriori risk.

Viewed from a distance, all this is not very surprising. Taking the less risky choice is probably what everyone would intuitively do, but now this is even mathematically proven to be the optimum. Hence, the optimal Bayesian decision is a benchmark for any other approach and defines an upper bound on the performance of any classifier. At this point, one could ask, why is there any reason to consider a different classifier if the optimum is already achieved? What is the rest of this book about? The answer is simple: the Bayesian classifier requires probability densities that are typically not (fully) known in real-world scenarios. The Bayesian classifier is optimal because it uses every piece of information that can eventually be known about the entirety of all the features and objects of the domain. Of course, an omniscient classifier has no difficulties in making the best decision.

Bayesian decision theory uses a cost function and the a posteriori probability of the classes. While the cost of a wrong decision can hopefully be determined with some degree of certainty, an accurate a posteriori probability or more precisely its determining pieces are hard to find. The class-specific distribution of the features and the a priori distribution of the classes are normally unknown. Note that the marginal distribution of the features $p(\mathbf{m})$ in the denominator of

$$P(\omega \mid \mathbf{m}) = \frac{p(\mathbf{m} \mid \omega) P(\omega)}{p(\mathbf{m})} \propto p(\mathbf{m} \mid \omega) P(\omega) \tag{3.32}$$

is not required to find the minimum or maximum with respect to ω for a fixed $\mathbf{m}$.

The quality of the Bayesian classifier stands or falls with the accuracy of these quantities. Therefore, the next two chapters will treat the question of how those quantities are estimated in practice. The next chapter will focus on the so-called parametrized methods and the following one will deal with parameter-free methods that work without a model assumption. Besides the technical challenge of how those distributions are mathematically obtained, another issue becomes evident, too. The question is what the concept “probability” actually means. This more philosophical matter will constitute the introductory part of the coming chapter.

But before that, the remainder of this chapter will discuss some simple examples and deduce another estimator that limits the maximal risk in the worst case.

Once again, consider a simple two-class scenario. Let $l_{ij} = l(\omega_i, \omega_j)$ be a shorter notation for the cost function. The a posteriori risk for a fixed feature $\mathbf{m}$ is

$$R(\omega_1 \mid \mathbf{m}) = l_{11}P(\omega_1 \mid \mathbf{m}) + l_{12}P(\omega_2 \mid \mathbf{m}), \tag{3.33}$$

$$R(\omega_2 \mid \mathbf{m}) = l_{21}P(\omega_1 \mid \mathbf{m}) + l_{22}P(\omega_2 \mid \mathbf{m}). \tag{3.34}$$

The estimator decides on the class ω_1 iff $R(\omega_1 \mid \mathbf{m}) < R(\omega_2 \mid \mathbf{m})$. This can be equivalently rewritten as

$$\begin{aligned} & R(\omega_1 \mid \mathbf{m}) < R(\omega_2 \mid \mathbf{m}) \\ \Leftrightarrow\ & (l_{21} - l_{11})P(\omega_1 \mid \mathbf{m}) > (l_{12} - l_{22})P(\omega_2 \mid \mathbf{m}) \\ \Leftrightarrow\ & (l_{21} - l_{11})p(\mathbf{m} \mid \omega_1)P(\omega_1) > (l_{12} - l_{22})p(\mathbf{m} \mid \omega_2)P(\omega_2) \\ \Leftrightarrow\ & \underbrace{\frac{p(\mathbf{m} \mid \omega_1)}{p(\mathbf{m} \mid \omega_2)}}_{\text{likelihood ratio}} > \frac{(l_{12} - l_{22})P(\omega_2)}{(l_{21} - l_{11})P(\omega_1)}. \end{aligned} \tag{3.35}$$

Hence, the Bayesian classifier decides according to

$$\hat{\omega}(\mathbf{m}) = \begin{cases} \omega_1 & \text{if } \frac{p(\mathbf{m} \mid \omega_1)}{p(\mathbf{m} \mid \omega_2)} > \frac{(l_{12}-l_{22})P(\omega_2)}{(l_{21}-l_{11})P(\omega_1)} \\ \omega_2 & \text{else} \end{cases}. \tag{3.36}$$

The last inequality is typical: the likelihood ratio is compared to a threshold. Figure 3.5 illustrates the relation between this ratio and the decision region in the feature space. Whenever the likelihood ratio is above the threshold (orange line) the corresponding subset of the feature space is assigned to the decision region $\mathcal{R}_1$ (violet intervals), otherwise to $\mathcal{R}_2$ (brown intervals).

Note that $p(\mathbf{m} \mid \omega)$ for fixed ω and variable $\mathbf{m}$ is the class-specific distribution of the feature vector $\mathbf{m}$. But as soon as a specific value of $\mathbf{m}$ is inserted into $p(\mathbf{m} \mid \omega)$ and $p(\mathbf{m} \mid \omega)$ is interpreted as a function of ω, the term $p(\mathbf{m} \mid \omega)$ is called the likelihood function.

Now consider the special cost function $l_{ij} = 1 - \delta_i^j$. This means that correct decisions are costless and every incorrect decision is penalized with a cost of 1 (this cost function

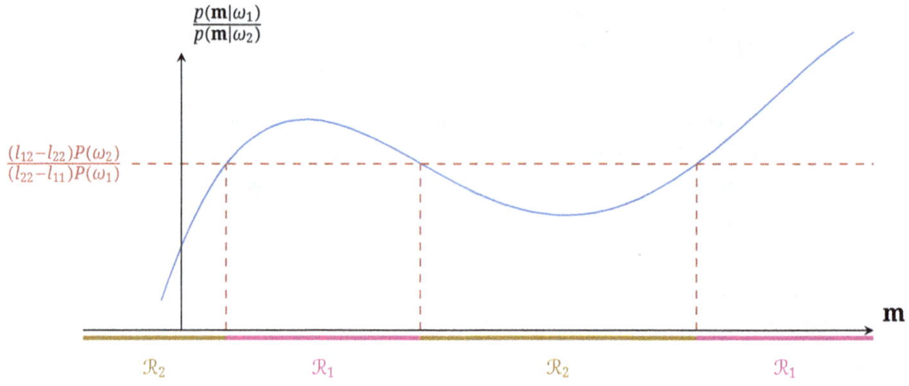

Fig. 3.5: Connection between the likelihood ratio and the optimal decision region.

is also known as *0–1 loss*). Then, the a posteriori risk

$$
\begin{aligned}
R(\omega_i \mid \mathbf{m}) &= \sum_{j=1}^{c} l_{ij} P(\omega_j \mid \mathbf{m}) = \sum_{j=1}^{c} (1 - \delta_i^j) P(\omega_j \mid \mathbf{m}) \\
&= 1 - \sum_{j=1}^{c} \delta_i^j P(\omega_j \mid \mathbf{m}) = 1 - P(\omega_i \mid \mathbf{m})
\end{aligned} \tag{3.37}
$$

equals the converse of the a posteriori probability. As the estimator chooses the least risk, this leads to the already known MAP classifier. Hence, the MAP classifier is a special case of the general Bayesian classifier with this particular cost function.

The overall risk is

$$
\begin{aligned}
R &= \int_{\mathcal{M}} R(\hat{\omega}(\mathbf{m}) \mid \mathbf{m}) p(\mathbf{m}) \, \mathrm{d}\mathbf{m} \\
&\overset{(3.37)}{=} \int_{\mathcal{M}} \big(1 - P(\hat{\omega}(\mathbf{m}) \mid \mathbf{m})\big) p(\mathbf{m}) \, \mathrm{d}\mathbf{m} \\
&= 1 - \int_{\mathcal{M}} P(\hat{\omega}(\mathbf{m}) \mid \mathbf{m}) p(\mathbf{m}) \, \mathrm{d}\mathbf{m} \\
&= 1 - \int_{\mathcal{M}} P(\hat{\omega}(\mathbf{m}), \mathbf{m}) \, \mathrm{d}\mathbf{m} .
\end{aligned} \tag{3.38}
$$

To interpret the last term, reconsider Figure 3.1 and compare it with Figure 3.6. Generally there will be tuples $(\omega(o_1),\mathbf{m}(o_1))$, $(\omega(o_2),\mathbf{m}(o_2))$ whose underlying objects belong to different classes, $\omega(o_1) \neq \omega(o_2)$, but show the same feature vector $\mathbf{m}(o_1) = \mathbf{m}(o_2)$. This always happens if the supports of the likelihoods overlap. The probability of the whole ensemble of all tuples $(\omega, \mathbf{m})$ is one by definition, i.e., $\sum_{j=1}^{c} \int_{\mathcal{M}} P(\omega_j, \mathbf{m}) \, \mathrm{d}\mathbf{m} = 1$, but the estimator $\hat{\omega}$ is a function of the feature vector and always chooses one (fixed) class, depending on the features. In the case of the MAP classifier, the estimator $\hat{\omega}$ always

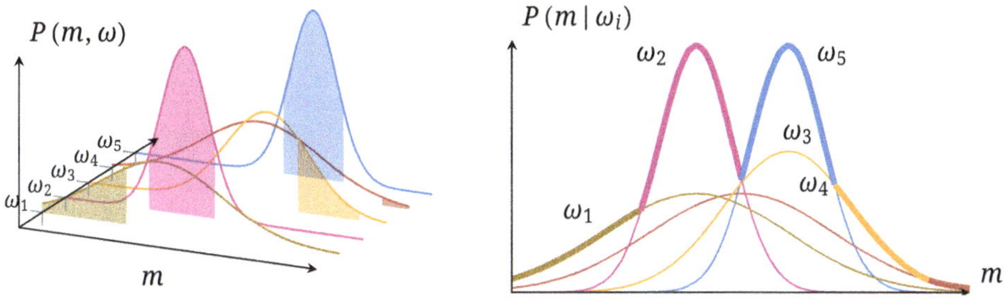

(a) Joint distribution **(b)** Marginal distributions

Fig. 3.6: Decision of a MAP classifier in relation to the a posteriori probabilities. The a priori probabilities are $P(\omega_i) = \frac{1}{5}$ for $i = 1, \ldots, 5$. The MAP classifier always chooses the class with the highest a posteriori probability (shaded areas/thick lines). The overall risk can be computed by summing the individual class-specific risks for each of these regions separately.

decides on the class with the highest a posteriori probability (thick lines in Figure 3.6b). Hence, the ensemble $(\hat{\omega}(\mathbf{m}), \mathbf{m})$ is only a subset of all the options. Rewriting the last term of Equation (3.38) in an artificially complicated manner leads to

$$\begin{aligned}\int_{\mathcal{M}} P(\hat{\omega}(\mathbf{m}), \mathbf{m})\,\mathrm{d}\mathbf{m} &= \sum_{j=1}^{c} \delta_{[\hat{\omega}(\mathbf{m})=\omega_j]} \int_{\mathcal{M}} P(\omega_j, \mathbf{m})\,\mathrm{d}\mathbf{m} \\ &= \sum_{j=1}^{c} \int_{\{\mathbf{m}\in\mathcal{M} \mid \hat{\omega}(\mathbf{m})=\omega_j\}} P(\omega_j, \mathbf{m})\,\mathrm{d}\mathbf{m} \\ &= \sum_{j=1}^{c} \int_{\mathcal{R}_j} P(\omega_j, \mathbf{m})\,\mathrm{d}\mathbf{m} \leq 1, \end{aligned} \quad (3.39)$$

where $\delta_{[\,\cdot\,]}$ denotes the generalized Kronecker symbol. This means that only those regions of the density are integrated that correspond to classes the estimator chooses (shaded areas in Figure 3.6a). In summary, the term in Equation (3.39) represents the probability that the classifier decides correctly. Conversely, Equation (3.38) is the probability of a wrong decision. This can also be rewritten as

$$R = 1 - \sum_{j=1}^{c} \int_{\mathcal{R}_j} P(\omega_j, \mathbf{m})\,\mathrm{d}\mathbf{m} = \sum_{j=1}^{c} \int_{\mathcal{M}\setminus\mathcal{R}_j} P(\omega_j, \mathbf{m})\,\mathrm{d}\mathbf{m}. \quad (3.40)$$

As a final result, one can state that the overall risk of the MAP classifier equals the probability of a wrong decision.

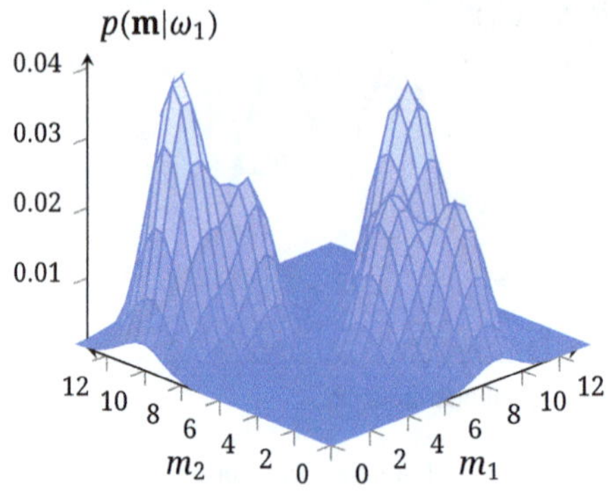

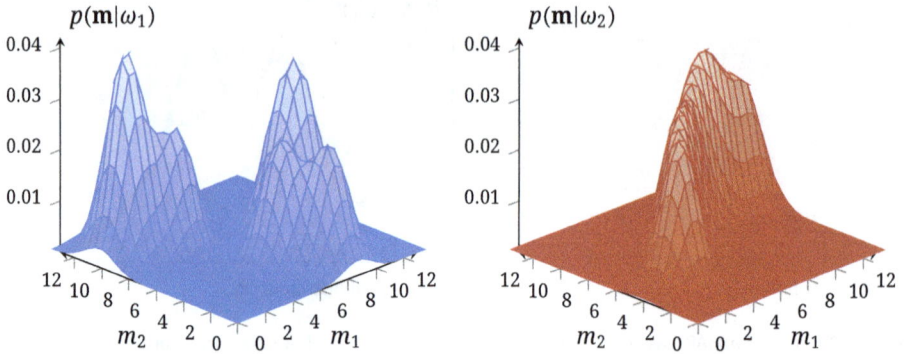

(a) Class-specific feature distributions

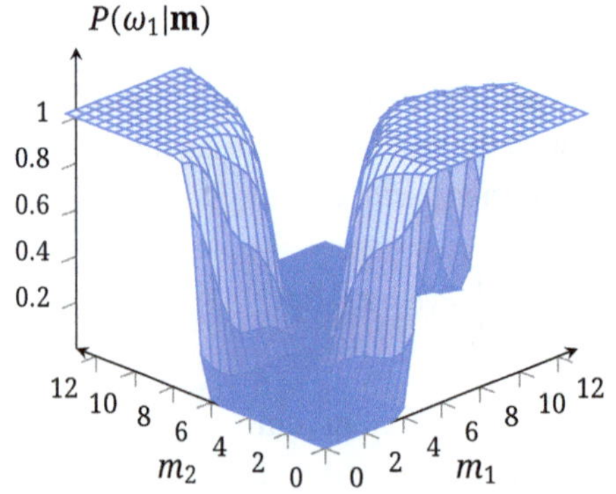

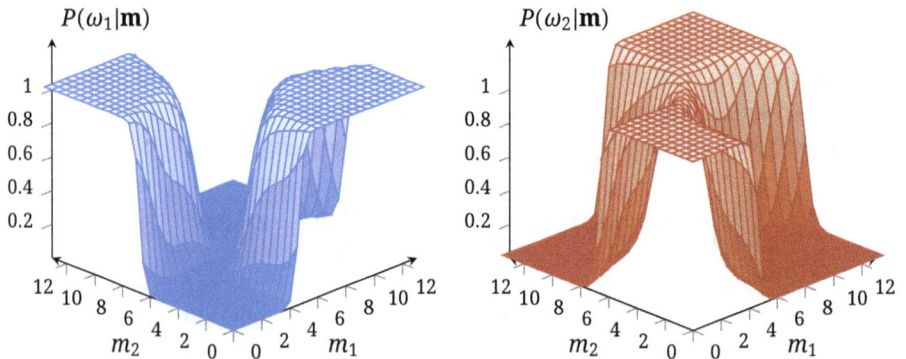

(b) A posteriori class probabilities

Fig. 3.7: Underlying densities in the reference example for classification.

3.3.2 Reference example: Optimal decision regions

We now turn our attention to a practical example of Bayesian classification. In particular, we consider a two-class problem ($c = 2$) with two-dimensional feature vectors $\mathbf{m} = (m_1,m_2)^\mathsf{T}$. Variations on this example will be found throughout this book.

The data is generated according to the joint probability density function $P(\mathbf{m},\omega) = p(\mathbf{m} \mid \omega)P(\omega)$. Both classes are equally likely, i.e., the prior class probabilities are $P(\omega_j) := \frac{1}{2}$ for $j = 1,2$. The likelihoods of the features given a class are modeled as mixtures of seven isotropic Gaussian densities:

$$p(\mathbf{m} \mid \omega_j) = \frac{1}{7}\sum_{i=1}^{7} \mathcal{N}\left(\mathbf{m}; \boldsymbol{\mu}_{ji}, \begin{pmatrix} \sigma_j^2 & 0 \\ 0 & \sigma_j^2 \end{pmatrix}\right) = \frac{1}{7}\sum_{i=1}^{7} \mathcal{N}\left(\mathbf{m}; \boldsymbol{\mu}_{ji}, \sigma_j^2\mathbf{I}\right), \tag{3.41}$$

where we further define $\sigma_1 = \sigma_2 := 1$. In other words, the components of both mixtures differ only in their mean $\boldsymbol{\mu}_{ji}$. Figure 3.7a shows the likelihoods $p(\mathbf{m} \mid \omega_j)$ of this example.

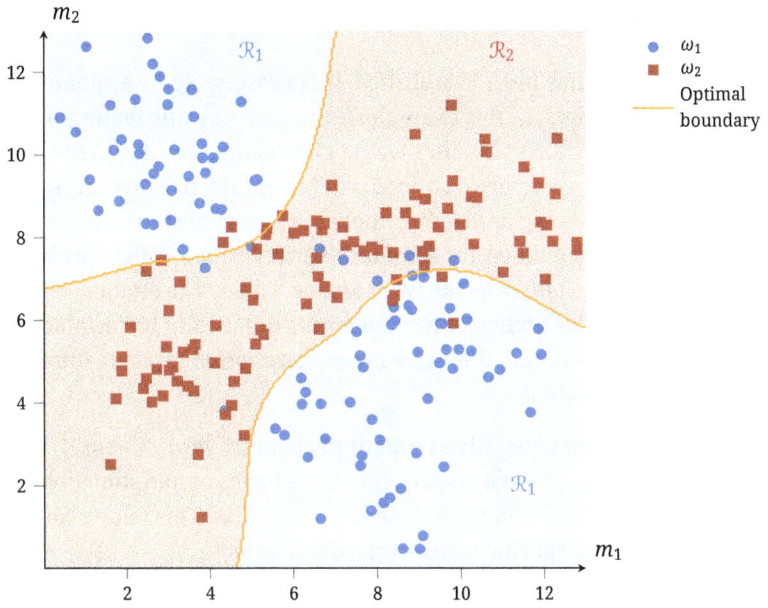

Fig. 3.8: Optimal decision regions derived from the true underlying distribution. In region $\mathcal{R}_1$, one has that $P(\omega_1 \mid \mathbf{m}) > P(\omega_2 \mid \mathbf{m})$. In region $\mathcal{R}_2$, the opposite is true. The dataset consists of 100 samples for each class (200 samples in all). The optimal classifier has an empirical classification error of $\frac{16}{200} = 8\,\%$ for the dataset shown.

The posterior class probabilities can be computed using the Bayes rule and are shown in Figure 3.7b.

The optimal decision boundaries correspond to the set of points where $P(\omega_1 \mid \mathbf{m}) = P(\omega_2 \mid \mathbf{m})$, as shown in Figure 3.8. Since the relevant distributions are known, it is possible to compute the theoretical classification error using Equation (3.40). The probability of error turns out to be approximately $P_e \approx 6.16\,\%$.

This model was also used to create a training sample $\mathcal{D}$ and a test sample $\mathcal{T}$ to use in the classifier examples throughout this book. Both $\mathcal{D}$ and $\mathcal{T}$ consist of 200 samples each, where 100 samples were drawn from $p(\mathbf{m} \mid \omega_1)$ and 100 samples were drawn from $p(\mathbf{m} \mid \omega_2)$. Figure 3.8 shows the training set $\mathcal{D}$ alongside the decision regions. It can be seen that some samples fall into the wrong decision regions. In fact, the empirical classification error is 8 %. Note that this error deviates from the theoretical classification error calculated above. The important lesson to take away from this is that the empirical classification error is subject to random perturbations, since it depends on the dataset. More reliable estimates of the true probability of error can be obtained by using a larger test sample, or by using the techniques described in Section 9.3.

3.3.3 The naive Bayes classifier

Now that a reference example has been established, it is reasonable to evaluate the performance of different classifiers on this example. If the class-specific feature distributions $p(\mathbf{m} \mid \omega)$ were known, the MAP classifier would give optimal results. In the next chapter, estimation methods for these distributions will be discussed. However, in the case of categorical classification, it is clear that the number of parameters to be estimated is quite large. If $\mathbf{m}$ is a d-dimensional vector of binary variables in a c-class problem, there are $2^d|c|$ parameters, as each m_i can take two possible values. The amount of data required to estimate such a high number of parameters is practically intractable. To minimize this computational overhead, the *naive Bayes assumption* treats features in $p(\mathbf{m} \mid \omega)$ as conditionally independent.

Definition 3.1 (Naive Bayes assumption). Given a joint probability $p(m_1, \ldots, m_d \mid \omega)$ of observing the features $m_1, \ldots m_d$ given a class ω, the naive Bayes assumption holds if the features are treated conditionally independent given the class. This allows for the simplified calculation of the class-specific feature distributions as:

$$p(\mathbf{m} \mid \omega) = \prod_{i=1}^{d} p(m_i \mid \omega). \tag{3.42}$$

Considering the same categorical classification task as above, only $2 \times d \times c$ parameters have to be estimated under this simplifying assumption. The MAP classifier from Equation (3.23) then resolves to

$$\begin{aligned}
\hat{\omega} &= \underset{\omega_j:\, j=1,\ldots,c}{\arg\max}\; P(\omega_j \mid \mathbf{m}) \\
&= \underset{\omega_j:\, j=1,\ldots,c}{\arg\max}\; \frac{p(\mathbf{m} \mid \omega_j)P(\omega_j)}{p(\mathbf{m})} \\
&= \underset{\omega_j:\, j=1,\ldots,c}{\arg\max}\; p(\mathbf{m} \mid \omega_j)P(\omega_j) \\
&= \underset{\omega_j:\, j=1,\ldots,c}{\arg\max}\; \prod_{i=1}^{d} p(m_i \mid \omega_j)P(\omega_j)
\end{aligned} \tag{3.43}$$

which considers each dimension of $\mathbf{m}$ seperately while it does not affect the estimation of $P(\omega)$. Based on the naming for its underlying assumption, this classifier is referred to as the *naive Bayes classifier*. Figure 3.9 illustrates the process of estimating the conditional probability $p(\mathbf{m} \mid \omega_j)$ by fitting a distribution for each dimension of $\mathbf{m}$ separately. Despite the fact that this approach often contradicts real-world scenarios where features are usually not independent, the classifier has demonstrated remarkable performance on many real-world datasets. Its primary advantage is the small amount of training data it requires.

When considering continuous features, the underlying class conditional distribution for each feature $p(m_i \mid \omega_j)$ often is assumed to be a Gaussian with a mean of μ_{ij}

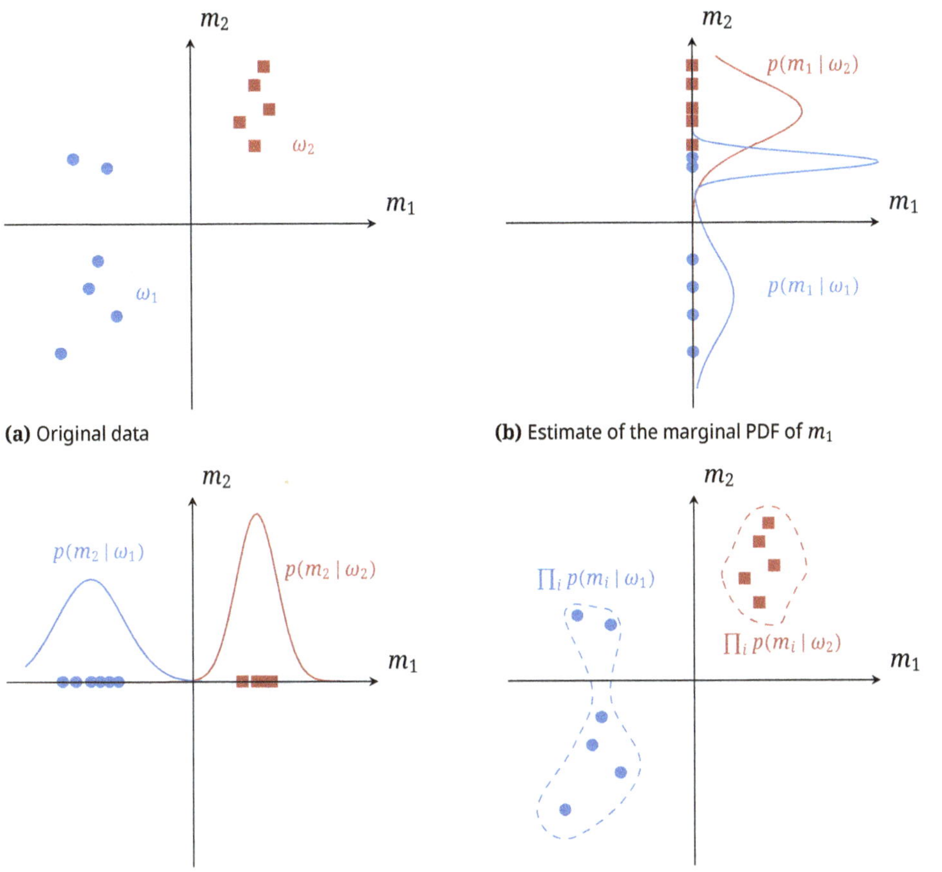

(a) Original data

(b) Estimate of the marginal PDF of m_1

(c) Estimate of the marginal PDF of m_2

(d) Contours of the resulting distribution

Fig. 3.9: Naive Bayes parameter estimation. The conditional $p(\mathbf{m} \mid \omega)$ is estimated separately for each dimension of $\mathbf{m}$. Note that the resulting distribution in (d) only indicates contours of the resulting probability density function (PDF), while the plots in (b) and (c) constitute the actual marginal PDFs.

and a variance of σ_{ij}^2 for each class ω_j. This additional assumption leads to the naming "Gaussian naive Bayes", which results in the following class-specific distributions:

$$p(\mathbf{m} \mid \omega_j) = \prod_{i=1}^{d} \frac{1}{\sqrt{2\pi\sigma_{ij}^2}} \mathrm{e}^{-\frac{(m_i-\mu_{ij})^2}{2\sigma_{ij}^2}} . \tag{3.44}$$

Since each class-specific distribution has its own parameters for mean μ and variance σ^2, the overall distribution $p(\mathbf{m} \mid \omega) = \mathcal{N}(\mathbf{m}; \boldsymbol{\mu}_\omega, \boldsymbol{\Sigma}_\omega)$ follows a Gaussian with a diagonal covariance matrix $\boldsymbol{\Sigma}_\omega$. However, selecting a Gaussian as underlying distribution is not the only option. Depending on the task at hand, one can use Bernoulli distributions or multinominal variants. In general, it is also possible to utilize various distributions within a single naive Bayes classifier. The naive Bayes classifier performs well in practice on datasets where the correlation between the features is not excessively high. On the real-world Iris Flower dataset introduced in Figure 2.1, the Naive Bayes classifier can bring down the classification error down to 0%, whereas it fails to perform well on the previously introduced reference example.

The primary cause for the poor performance on the reference example is the inflexibility of the decision boundaries. If the covariances vary across classes, the Gaussian naive Bayes will produce a conic decision boundary in the feature space, otherwise the decision boundary will be linear. The explanation for this will be presented in Section 3.3.5.

3.3.4 The minimax classifier

Now, another kind of classifier will be introduced, the minimax classifier. All of the concepts that have been discussed so far need either the a posteriori distribution $P(\omega \mid \mathbf{m})$ or must resort to the a priori distribution $P(\omega)$, thanks to Equation (3.32). Knowing the former is nearly a forlorn hope in real-world applications, but even having a feasible estimate of the latter can be a challenging task. Normally, one has to rely on expert knowledge of the application. Hence, a natural question to ask is what happens if one implements a pattern recognition system with a specific a priori distribution in mind that does not reflect the reality. To start, reconsider the risk

$$\begin{aligned} R &\overset{(3.30)}{=} \int_{\mathcal{M}} R(\hat{\omega}(\mathbf{m}) \mid \mathbf{m}) p(\mathbf{m})\,\mathrm{d}\mathbf{m} = \sum_{i=1}^{c} \int_{\mathcal{R}_i} R(\omega_i \mid \mathbf{m}) p(\mathbf{m})\,\mathrm{d}\mathbf{m} \\ &\overset{(3.28)}{=} \sum_{i=1}^{c} \int_{\mathcal{R}_i} \sum_{j=1}^{c} l_{ij} P(\omega_j \mid \mathbf{m}) p(\mathbf{m})\,\mathrm{d}\mathbf{m} \\ &\overset{(3.7)}{=} \sum_{i=1}^{c} \int_{\mathcal{R}_i} \sum_{j=1}^{c} l_{ij} p(\mathbf{m} \mid \omega_j) P(\omega_j)\,\mathrm{d}\mathbf{m}. \end{aligned} \tag{3.45}$$

The second line holds because the decision regions $\mathcal{R}_i$ are a partition of $\mathcal{M}$ and the classifier $\hat{\omega}$ is equivalent to those regions. Hence, $\hat{\omega}$ is constant on each of them and so $\hat{\omega}(\mathbf{m}) = \omega_i$ for $\mathbf{m} \in \mathcal{R}_i$.

Although the minimax classifier is not limited to the binary case, the following discussion will be restricted to this case in order to keep the formalism simple. With $c = 2$ and $P(\omega_2) = 1 - P(\omega_1)$, it follows that

$$\begin{aligned} R = {} & l_{11} P(\omega_1) \int_{\mathcal{R}_1} p(\mathbf{m} \mid \omega_1)\,\mathrm{d}\mathbf{m} + l_{12} P(\omega_2) \int_{\mathcal{R}_1} p(\mathbf{m} \mid \omega_2)\,\mathrm{d}\mathbf{m} \\ & + l_{21} P(\omega_1) \int_{\mathcal{R}_2} p(\mathbf{m} \mid \omega_1)\,\mathrm{d}\mathbf{m} + l_{22} P(\omega_2) \int_{\mathcal{R}_2} p(\mathbf{m} \mid \omega_2)\,\mathrm{d}\mathbf{m} \end{aligned}$$

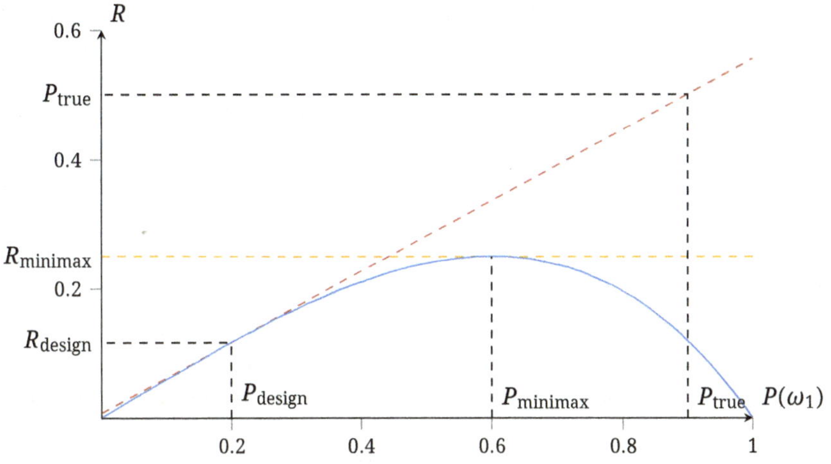

Fig. 3.10: Risk of the minimax classifier.

$$= \underbrace{l_{22} + (l_{12} - l_{22}) \int_{\mathcal{R}_1} p(\mathbf{m} \mid \omega_2)\,\mathrm{d}\mathbf{m}}_{R_{\text{minimax}}}$$
$$+ P(\omega_1) \underbrace{\left[(l_{11} - l_{22}) + (l_{21} - l_{11}) \int_{\mathcal{R}_2} p(\mathbf{m} \mid \omega_1)\,\mathrm{d}\mathbf{m} - (l_{12} - l_{22}) \int_{\mathcal{R}_1} p(\mathbf{m} \mid \omega_2)\,\mathrm{d}\mathbf{m}\right]}_{\overset{!}{=}0} . \tag{3.46}$$

One can see that the risk is a linear function of the a priori probability of ω_1. While the costs l_{ij} and the class-specific distributions of the feature vector are not adjustable, but rather determined by the environment, the decision regions $\mathcal{R}_i$ are the definable design parameters. They are made up of the ordinate and the slope. This can lead to a problem, as illustrated in Figure 3.10.

For every a priori probability, there exists a particular choice of the decision regions such that the risk is minimized. This relation is depicted by the blue curve in Figure 3.10, which represents the optimal Bayesian classifiers. This curve is always zero at the end points, because if any a priori probability equals 1, an error-free classification is possible. It attains its maximum somewhere in the middle, say at P_{minimax}, where the uncertainty is high. Of course, the uncertainty is maximized if all classes are equally distributed, but the maximal risk can lie somewhat off the mark if different costs are associated with each class.

Now, assume that the a priori probability was P_{design} and the classifier was originally constructed with this assumption in mind (hence, the classifier was thought to have a risk R_{design}). But if for some reason this assumption was wrong, and the classifier is deployed in a scenario in which the true a priori probability is $P_{\text{true}} \neq P_{\text{design}}$, then Equation (3.46) holds and the true risk R_{true} lies on the tangent (red line) that goes through the initially assumed point. This might even lead to a risk that is much higher than the minimal risk R_{minimax} in the worst case.

The idea is to construct a classifier such that Equation (3.46) becomes independent of the a priori distribution. This means the slope in Equation (3.46) must be set to zero by an appropriate choice of the decision regions $\mathcal{R}_i$. Implicitly, this choice belongs to the optimal Bayesian classifier for the worst-case a priori probability P_{minimax}. Then the classifier has the risk R_{minimax}, which remains constant even if the a priori probability diverges, because the tangent is constant (yellow line in Figure 3.10).

In summary, the objective is not to construct a classifier that has the minimal risk for a specific a priori probability, but a classifier that has the minimal risk in the worst case. The name "minimax" comes from the fact that one tries to minimize the maximal risk.

Anyway, in classical pattern recognition the optimal Bayesian classifier usually performs better than the minimax classifier. The situation in Figure 3.10 is somewhat far-fetched because P_{design} and P_{true} are extremely different. Though not optimal, a Bayesian classifier that is tuned for P_{design} will still be better for P_{true} than the minimax classifier if P_{true} is not too far away. (In Figure 3.10 these are all points where the red tangent is still below the yellow line.) The origin of the minimax approach is located in game theory. Here, two players have to consecutively perform moves and one has to decide what the best move is. The a priori probability describes what the adversary is likely to do for its next move, but of course the adversary can adapt its own strategy to the outcome of the decision one still has to make. Governed by the assumption that the adversary will always do what is worst for the player, the idea is to minimize the maximal risk. In classical pattern recognition, however, the adversary is the environment and is therefore passive. Under normal conditions, one should be able to make a reasonable assumption about the a priori distribution that is hopefully not too far off.

3.3.5 Normally distributed features

This Section repeats the previously introduced concepts in the case that the class-specific feature distributions are Gaussian. Besides the goal of presenting some concrete examples, this section also serves to reinforce the understanding of the previous information.

Definition 3.2 (Univariate normal distribution). A random variable $\underline{m}$ is said to be *normally (or Gaussian) distributed* with expectation μ and variance σ^2 if its distribution can be described by the density

$$p(m) = \frac{1}{\sqrt{2\pi}\sigma} \exp\left(-\frac{1}{2}\left(\frac{m-\mu}{\sigma}\right)^2\right). \tag{3.47}$$

The importance of the normal distribution is explained by the central limit theorem. In its most basic and classical form, it states that the normalized sum of a sequence of independent and identically distributed random variables with existing expectation and variance converges almost surely to a normally distributed random variable. Stated more precisely:

Theorem 3.3 ((Classical) central limit theorem (Billingsley [1995])). Let $\underline{m}_1, \underline{m}_2, \ldots$ be a sequence of independent and identically distributed random variables with $\mathrm{E}\{\underline{m}_j\} = \mu$ and $\mathrm{Var}\{\underline{m}_j\} = \sigma^2$. Let

$$\underline{s}_n = \frac{1}{n} \sum_{i=1}^{n} \underline{m}_i \tag{3.48}$$

denote the average partial sum. Then

$$\sqrt{n}\,\frac{\underline{s}_n - \mu}{\sigma} \xrightarrow{\mathrm{P}} \underline{z} \sim \mathcal{N}(0, 1). \tag{3.49}$$

Note that this theorem does not explain the ubiquity of the normal distribution. Indeed, there are more sophisticated and generalized variants of the central limit theorem, but convergence (in distribution) can still be guaranteed under very mild assumptions. Without going into too much detail, the individual random variables do not even need to be identically distributed, and the pairwise independence can be replaced by some limit value condition that ensures that no subsequence has too much influence. Hence, the normal distribution gained extreme popularity in modeling naturally occurring phenomena.

Loosely formulated, one could say that if a feature is generated by summing many independent contributions, it is reasonable to approximate the distribution of the feature by a Gaussian distribution. This holds in one dimension, but also in many dimensions. The multi-dimensional normal distribution is given by

Definition 3.4 (Multivariate normal distribution). A random vector $\underline{\mathbf{m}} \in \mathbb{R}^d$ is called *normally (or Gaussian) distributed* with expectation vector $\boldsymbol{\mu} \in \mathbb{R}^d$ and covariance matrix $\boldsymbol{\Sigma} \in \mathbb{R}^{d \times d}$ if its distribution can be described by the density

$$p(\mathbf{m}) = \frac{1}{(2\pi)^{\frac{d}{2}} |\boldsymbol{\Sigma}|^{\frac{1}{2}}} \exp\left(-\frac{1}{2} (\mathbf{m} - \boldsymbol{\mu})^\mathsf{T} \boldsymbol{\Sigma}^{-1} (\mathbf{m} - \boldsymbol{\mu}) \right). \tag{3.50}$$

Note that $\boldsymbol{\Sigma}$ is a covariance matrix and therefore symmetric and positive definite. In the 2-dimensional case, $\boldsymbol{\Sigma}$ can be decomposed as

$$\boldsymbol{\Sigma} = \begin{pmatrix} \sigma_1^2 & \rho\sigma_1^2\sigma_2^2 \\ \rho\sigma_1^2\sigma_2^2 & \sigma_2^2 \end{pmatrix} \tag{3.51}$$

for $\sigma_1, \sigma_2 > 0$ and $\rho \in [-1, 1]$. Here, ρ is called the correlation coefficient.

For the rest of this chapter, we will assume that the class-specific feature distributions of the features are normal. This means for each $j = 1, \ldots, c$,

$$p(\mathbf{m} \mid \omega_j) = \mathcal{N}(\mathbf{m}; \boldsymbol{\mu}_j, \boldsymbol{\Sigma}_j). \tag{3.52}$$

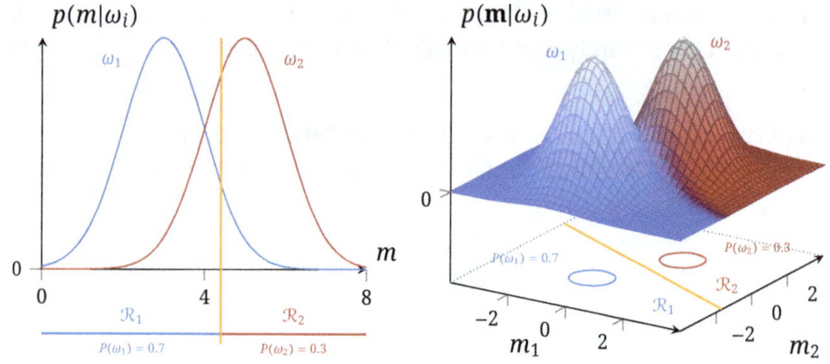

(a) Decision boundaries for $P(\omega_1) = 0.7$ and $P(\omega_2) = 0.3$

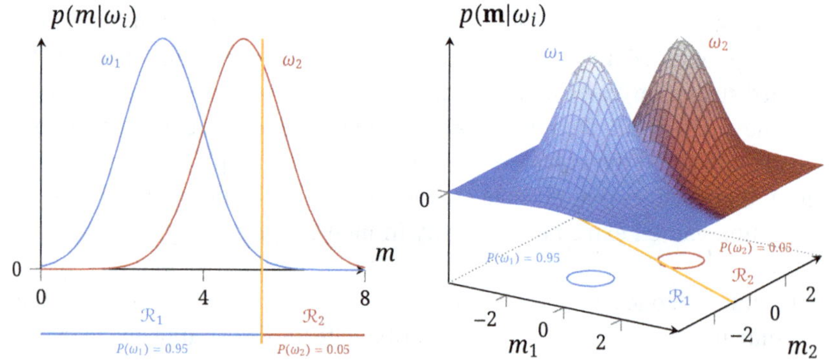

(b) Decision boundaries for $P(\omega_1) = 0.95$ and $P(\omega_2) = 0.05$

Fig. 3.11: Decision boundary of a two-class Gaussian classifier with unequal a priori probabilities. Left: one-dimensional feature space; right: two-dimensional feature space.

Note that such an assumption is only reasonable if the features are at least on an interval scale.

Because the MAP classifier chooses the class with the highest a posteriori probability and due to the strict monotonicity of the logarithm and Equation (3.32), the following holds for every fixed feature vector $\mathbf{m}$:

$$\begin{aligned} & P(\omega_i \mid \mathbf{m}) > P(\omega_j \mid \mathbf{m}) \\ \Leftrightarrow\quad & \ln P(\omega_i \mid \mathbf{m}) > \ln P(\omega_j \mid \mathbf{m}) \\ \Leftrightarrow\quad & \ln p(\mathbf{m} \mid \omega_i) + \ln P(\omega_i) > \ln p(\mathbf{m} \mid \omega_j) + \ln P(\omega_j). \end{aligned} \tag{3.53}$$

As the logarithmic representation is easier and numerically more stable with respect to normally distributed features, the decision function is reformulated as

$$k_i(\mathbf{m}) = \ln p(\mathbf{m} \mid \omega_i) + \ln P(\omega_i)$$

$$= -\frac{1}{2}(\mathbf{m}-\boldsymbol{\mu}_i)^{\mathsf{T}}\boldsymbol{\Sigma}_i^{-1}(\mathbf{m}-\boldsymbol{\mu}_i) - \frac{d}{2}\ln 2\pi - \frac{1}{2}\ln|\boldsymbol{\Sigma}_i| + \ln P(\omega_i). \quad (3.54)$$

The goal is to discuss the shape of the decision boundaries. For this purpose, the scenario will now be specialized even more, to the easiest case, and then generalized again step by step. If all likelihoods share the same covariance matrix $\boldsymbol{\Sigma}_i = \sigma^2\mathbf{I}$, the decision functions become

$$k_i(\mathbf{m}) = -\frac{1}{2\sigma^2}(\mathbf{m}-\boldsymbol{\mu}_i)^{\mathsf{T}}(\mathbf{m}-\boldsymbol{\mu}_i) - \underbrace{\frac{d}{2}\ln 2\pi - d\ln\sigma}_{\text{independent of } i} + \ln P(\omega_i). \quad (3.55)$$

After reduction to the crucial components, $k_i(\mathbf{m})$ can be redefined as

$$k_i(\mathbf{m}) = -\frac{1}{2\sigma^2}\|\mathbf{m}-\boldsymbol{\mu}_i\|^2 + \ln P(\omega_i) \quad (3.56)$$

and the decision boundaries are given by

$$\begin{aligned} 0 &= k_i(\mathbf{m}) - k_j(\mathbf{m}) \\ &= \frac{1}{\sigma^2}(\boldsymbol{\mu}_i-\boldsymbol{\mu}_j)^{\mathsf{T}}\mathbf{m} - \frac{1}{2\sigma^2}(\|\boldsymbol{\mu}_i\|^2 - \|\boldsymbol{\mu}_j\|^2) + \ln\frac{P(\omega_i)}{P(\omega_j)}. \end{aligned} \quad (3.57)$$

The decision boundaries are hyperplanes that are perpendicular to the connection lines between the expectation. If the a priori probabilities of the classes are equal, i.e., $\ln\frac{P(\omega_i)}{P(\omega_j)} = 0$, then the hyperplanes lie at the center points between the expectation vectors. If the a priori probabilities are not equal, the hyperplanes move toward the component with lower a priori probability. Examples with one and two features are shown in Figure 3.11.

For the second step, continue assuming an equal covariance matrix $\boldsymbol{\Sigma}_i = \boldsymbol{\Sigma}$ for all classes, but no longer are they necessarily diagonal. Again, the crucial part of the decision function is

$$k_i(\mathbf{m}) = -\frac{1}{2}(\mathbf{m}-\boldsymbol{\mu}_i)^{\mathsf{T}}\boldsymbol{\Sigma}^{-1}(\mathbf{m}-\boldsymbol{\mu}_i) + \ln P(\omega_i) \quad (3.58)$$

and the decision boundary is given by

$$0 = (\boldsymbol{\mu}_i-\boldsymbol{\mu}_j)^{\mathsf{T}}\boldsymbol{\Sigma}^{-1}\mathbf{m} - \frac{1}{2}(\boldsymbol{\mu}_i^{\mathsf{T}}\boldsymbol{\Sigma}^{-1}\boldsymbol{\mu}_i - \boldsymbol{\mu}_j^{\mathsf{T}}\boldsymbol{\Sigma}^{-1}\boldsymbol{\mu}_j) + \ln\frac{P(\omega_i)}{P(\omega_j)}. \quad (3.59)$$

Again, the decision boundaries are hyperplanes. But unlike before (see Equation (3.57)), the hyperplanes are rotated by $\boldsymbol{\Sigma}^{-1}$ and thus are not perpendicular to the connection lines between the centers of the Gaussians $\boldsymbol{\mu}_i$. In the most general case k_i can be redefined as

$$\begin{aligned} k_i(\mathbf{m}) &= \mathbf{m}^{\mathsf{T}}\mathbf{W}_i\mathbf{m} + \mathbf{w}_i^{\mathsf{T}}\mathbf{m} + w_{i0} \quad \text{with} && (3.60) \\ \mathbf{W}_i &= -\frac{1}{2}\boldsymbol{\Sigma}_i^{-1} && (3.61) \\ \mathbf{w}_i &= \boldsymbol{\Sigma}_i^{-1}\boldsymbol{\mu}_i && (3.62) \\ w_{i0} &= -\frac{1}{2}\boldsymbol{\mu}_i^{\mathsf{T}}\boldsymbol{\Sigma}_i^{-1}\boldsymbol{\mu}_i - \frac{1}{2}\ln|\boldsymbol{\Sigma}_i| + \ln P(\omega_i), && (3.63) \end{aligned}$$

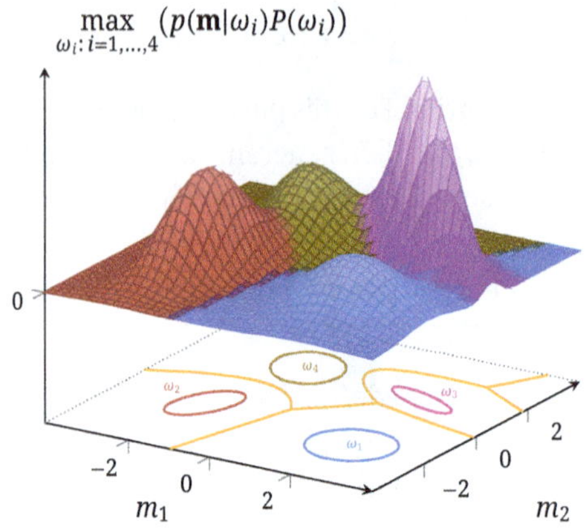

Fig. 3.12: Decision regions of a generic Gaussian classifier (i.e., full covariances) with $c = 4$ classes and two features ($d = 2$). The diagram shows $\max_{\omega_i:\, i=1,\dots,4} (p(\mathbf{m}|\omega_i)P(\omega_i))$, where the regions are colored according to the decision made in the region.

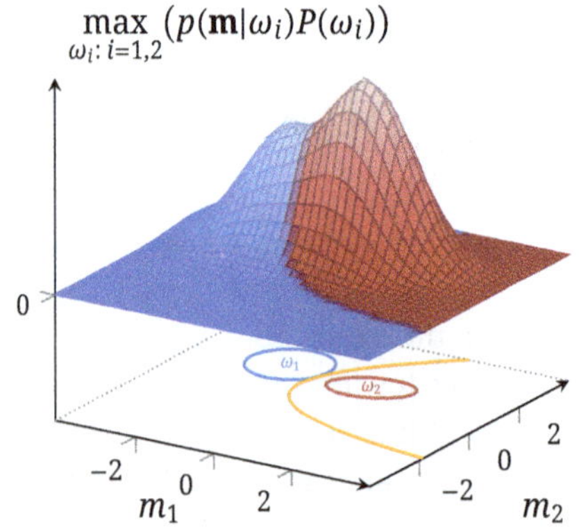

(a) Parabolic decision boundary

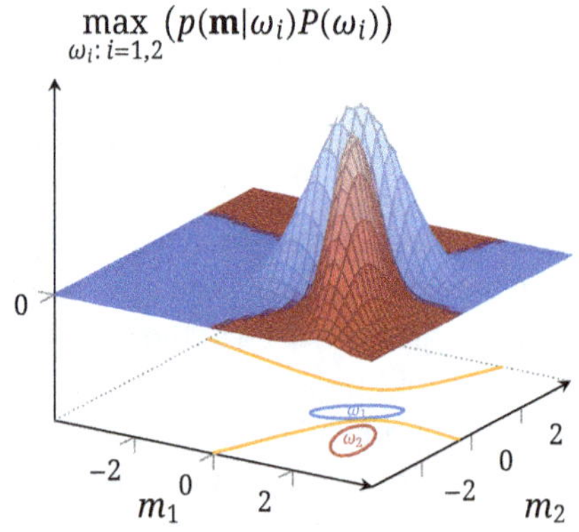

(b) Hyperbolic decision boundaries

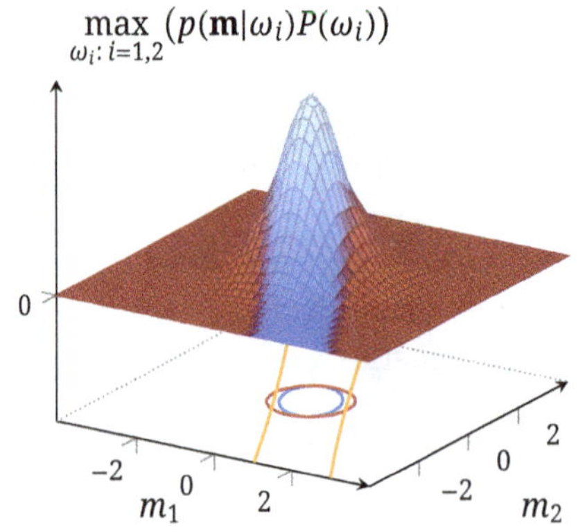

(c) Linear decision boundaries

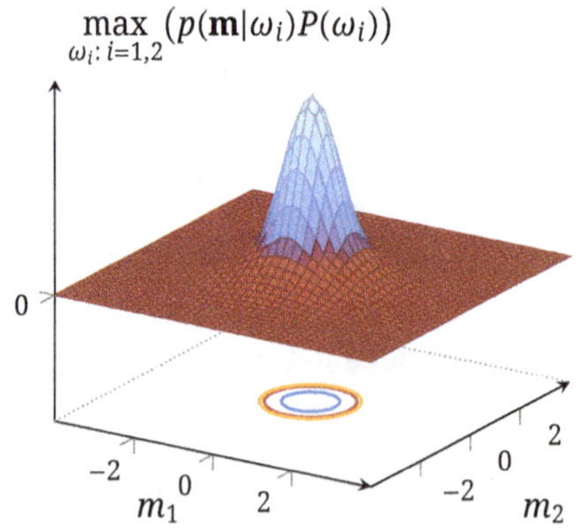

(d) Elliptic decision boundary

Fig. 3.13: Decision regions of a generic Gaussian classifier (i.e., full covariances) with $c = 2$ classes and two features ($d = 2$) are conic sections. The diagram shows $\max_{\omega_i:\, i=1,2} (p(\mathbf{m}|\omega_i)P(\omega_i))$, where the regions are colored according to the decision made in the region.

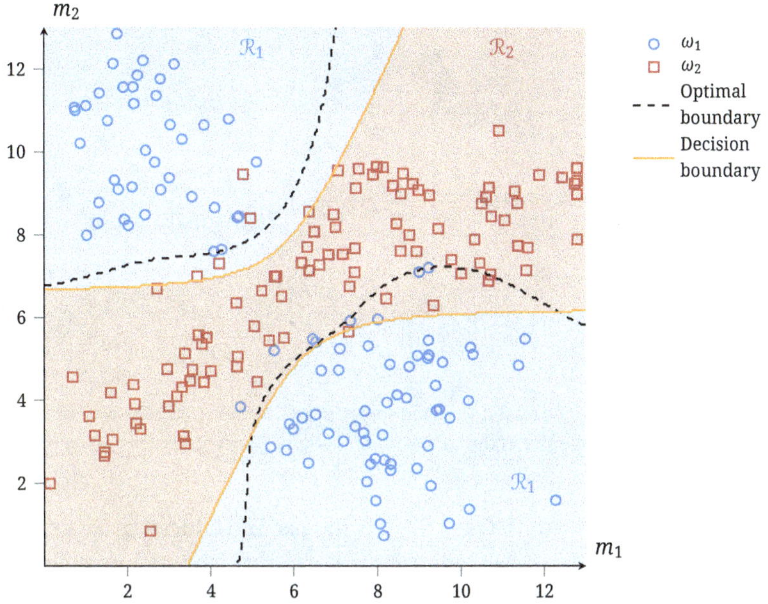

Fig. 3.14: Application to the reference example of Section 3.3.2. Decision regions of a classifier with Gaussian densities $p(\mathbf{m} \mid \omega_j) = \mathcal{N}(\mathbf{m}; \boldsymbol{\mu}_j, \boldsymbol{\Sigma}_j)$ ($j = 1,2$). The parameters $\boldsymbol{\mu}_j$ and $\boldsymbol{\Sigma}_j$ are estimated from a training sample. The training error is $e_{\text{train}} = 12.5\,\%$; the testing error is $e_{\text{test}} = 7\,\%$, but asymptotically approaches $e_{\text{test}} \approx 9.6\,\%$. The training set is the same as in Figure 3.8. Test samples are shown with hollow marks.

if only the crucial parts remain. Then the decision boundary equals

$$0 = \mathbf{m}^{\mathsf{T}}(\mathbf{W}_i - \mathbf{W}_j)\mathbf{m} + (\mathbf{w}_i - \mathbf{w}_j)^{\mathsf{T}}\mathbf{m} + w_{i0} - w_{j0}. \tag{3.64}$$

In summary, for normally distributed class-specific densities, the decision boundaries are hyperquadrics in the feature space. An example of a four-class ($c = 4$) Gaussian classifier in a two-dimensional feature space is shown in Figure 3.12. Note that the decision boundaries between the classes are piecewise conic sections. Further examples are shown in Figure 3.13.

Next, Figure 3.14 shows the decision regions of a Gaussian classifier with the reference example dataset from Section 3.3.2. Again, the decision boundary is a conic section, which approximates the optimal decision boundary. However, conic sections are not flexible enough to exactly match the optimal decision boundary, which results in a 3.5 percentage point increase in the error probability.

3.3.6 Arbitrarily distributed features

When modeling arbitrarily distributed features, using a single Gaussian might not be sufficient to fit the data. This could occur if the feature distribution is skewed or has a

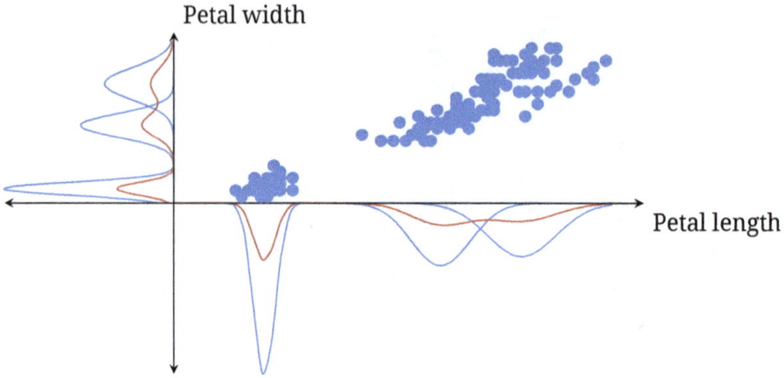

Fig. 3.15: GMM modeling of the Iris flower dataset. The blue curves depict a single Gaussian distribution for each class in the dataset. The red curve shows the convex combination of those Gaussians as a mixture.

multimodal shape. Examining the data from the Iris dataset introduced in Figure 2.1, deriving parameters for each class-specific density is necessary from the perspective of a classification task. To accomplish this, we often assume that the underlying feature distribution can be modeled using a Gaussian. However, if another classification task was to distinguish between classes of *Iridaceae*, such as *Iris* and *Gladiolus*, then a single distribution must accommodate all the data. Without specific knowledge about the Iris class, it might not be possible to identify the latent factors that induce different sub-species within the distribution. In order to still fit the data, a single Gaussian is insufficient. Instead, one can use a linear combination of multiple Gaussians known as a Gaussian mixture model (GMM). GMMs allow to model arbitrarily complex data distributions, which often occur in real-world datasets. In Figure 3.15, the Iris dataset is modeled utilizing a GMM that comprises three components.

In the following Section 3.4, the construction, properties and applications of GMMs will be derived in greater detail.

3.4 Gaussian mixtures

A Gaussian mixture is a random variable whose density equals the convex combination of Gaussian densities. More formally:

Definition 3.5 (Gaussian mixture). A random variable $\underline{\mathbf{m}} \in \mathbb{R}^d$ is a Gaussian mixture with K components if its density is of the form

$$p(\mathbf{m}) = \sum_{k=1}^{K} \pi_k \mathcal{N}(\mathbf{m}; \boldsymbol{\mu}_k, \boldsymbol{\Sigma}_k) \tag{3.65}$$

with $0 \leq \pi_k \leq 1$ and $\sum_{k=1}^{K} \pi_k = 1$.

Note that the term "Gaussian mixture" is misleading because $\underline{\mathbf{m}}$ is not a mixture of Gaussian random variables (which would itself be a Gaussian random variable, see Theorem 3.3). Instead, its probability density function is a mixture of Gaussian probability density functions. Each of the Gaussian probability density functions appears as a so-called *component* k of the mixture and is associated with its respective mean $\boldsymbol{\mu}_k$ and covariance $\boldsymbol{\Sigma}_k$. Each component is weighted by a factor π_k which determines its influence on the resulting density. The properties $0 \leq \pi_k \leq 1$ and $\sum_{k=1}^{K} \pi_k = 1$ stated in Definition 3.5 hold, as each component satisfies the formal requirements of a density function. This interpretation of π_k as probability will be used to establish a Bayesian perspective on Gaussian mixtures.

The reference example for this book, introduced in Section 3.3.2, originated from a Gaussian mixture with seven components, each of which had equal weight.

In general, Gaussian mixtures are widely used as they are easy to handle and enjoy a powerful approximation property: any density (within reason) can be approximated by Gaussian mixtures with arbitrary precision. More precisely, let f be a density with a finite number of discontinuities on each compact subset of its support. Let

$$f_n = \sum_{k=1}^{K_n} \pi_{nk} \mathcal{N}(\mathbf{m}; \boldsymbol{\mu}_{nk}, \boldsymbol{\Sigma}_{nk}) \tag{3.66}$$

be a sequence of Gaussian mixtures. Then there are K_n, π_{nk}, $\boldsymbol{\mu}_{nk}$, $\boldsymbol{\Sigma}_{nk}$ such that f_n converges uniformly to f except at the points of discontinuity (see Maz'ya and Schmidt [1996]). Note that K_n is the number of components of the nth member f_n and that, in general, each f_n has different components. Furthermore, the $\boldsymbol{\mu}_{nk}$ and $\boldsymbol{\Sigma}_{nk}$ are not necessarily a superset of the components of the previous member.

Despite the fact that GMMs can fit arbitrarily complex data distributions, estimating good parameter sets is still a difficult task. It is not possible to establish a general rule for the number of components K required to achieve a specific approximation error. Moreover, there is no closed-form solution to obtain optimal parameter values. Assume that the task is to approximate each d-dimensional class-specific feature distribution $p(m_j), j = 1, \ldots, c$ with K components. Then, for each class, $\frac{1}{2}K(d^2+3d+2)-1$ parameters must be estimated:

$$\underbrace{\pi_{j1}, \ldots, \pi_{jK}}_{K-1} \quad \underbrace{\boldsymbol{\mu}_{j1}, \ldots, \boldsymbol{\mu}_{jK}}_{Kd} \quad \underbrace{\boldsymbol{\Sigma}_{j1}, \ldots, \boldsymbol{\Sigma}_{jK}}_{\frac{1}{2}Kd(d+1)} . \tag{3.67}$$

Luckily, various approaches and frameworks exist to determine suitable parameters, one of which is the *expectation maximization* (EM) algorithm, which will be discussed in detail in Section 4.5.

As an intuition in the probabilistic Bayesian framework, the mixing-coefficients π_k can be interpreted as an a priori distribution $\pi_k = P(k)$ for the different mixture components k, whereas the component-specific feature distribution $p(\mathbf{m} \mid k) = \mathcal{N}(\mathbf{m}; \boldsymbol{\mu}_k, \boldsymbol{\Sigma}_k)$ is given by the Gaussian for each component. The question arises how to determine the posterior probabilities $p(k \mid \mathbf{m})$, which represent the likelihood of a datapoint emerging from a specific component. This posterior can be derived using Bayes' law (see Equation (3.7)):

$$\begin{aligned} p(k \mid \mathbf{m}) &= \frac{P(k)p(\mathbf{m} \mid k)}{\sum_l P(l)p(\mathbf{m} \mid l)} \\ &= \frac{\pi_k \mathcal{N}(\mathbf{m}; \boldsymbol{\mu}_k, \boldsymbol{\Sigma}_k)}{\sum_l \pi_l \mathcal{N}(\mathbf{m}; \boldsymbol{\mu}_l, \boldsymbol{\Sigma}_l)} \end{aligned} \tag{3.68}$$

An alternative representation of this posterior uses a latent variable interpretation, which later allows to apply standard approaches for the recovery of latent variables and simplifies the derivation of the EM algorithm in Section 4.5. Let $\mathbf{z}$ denote a K-dimensional binary random variable with $z_k \in \{0,1\}$ and $\sum_k z_k = 1$. In other words, $\mathbf{z}$ serves as a one-hot encoded vector that assigns a data point to a particular (unknown) distribution, thereby imposing a causal relationship of $\mathbf{z} \to \mathbf{m}$. In practice, this assignment of a data point to a particular component is not given. Therefore, instead of having a one-hot encoded vector $\mathbf{z}$, a distribution $P(z_k = 1) = \pi_k$ is specified using the mixing-coefficients π_k. Again, the properties $0 \leq \pi_k \leq 1$ and $\sum_{k=1}^{K} \pi_k = 1$ hold so that the π_k constitute a valid probability mass function and the prior extends to

$$P(\mathbf{z}) = \prod_{k=1}^{K} \pi_k^{z_k}. \tag{3.69}$$

Given a specific component z_k, the conditional distribution is simply the Gaussian

$$p(\mathbf{m} \mid z_k = 1) = \mathcal{N}(\mathbf{m}; \boldsymbol{\mu}_k, \boldsymbol{\Sigma}_k) \tag{3.70}$$

with the parametrization for the component k. With the same reasoning as above, the conditional on $\mathbf{z}$ expands to

$$p(\mathbf{m} \mid \mathbf{z}) = \prod_{k=1}^{K} \mathcal{N}(\mathbf{m}; \boldsymbol{\mu}_k, \boldsymbol{\Sigma}_k)^{z_k}. \tag{3.71}$$

These two distributions now form a joint distribution

$$p(\mathbf{m}, \mathbf{z}) = P(\mathbf{z})p(\mathbf{m} \mid \mathbf{z}) \tag{3.72}$$

which explicitly uses the latent variable representation $\mathbf{z}$. By marginalizing over all components $\mathbf{z}$, the original form of the Gaussian mixture evolves:

$$\begin{aligned} p(\mathbf{m}) &= \sum_{\mathbf{z}} p(\mathbf{m},\mathbf{z}) \\ &= \sum_{\mathbf{z}} P(\mathbf{z})p(\mathbf{m} \mid \mathbf{z}) \\ &= \sum_{k=1}^{K} \pi_k \mathcal{N}(\mathbf{m}; \boldsymbol{\mu}_k, \boldsymbol{\Sigma}_k). \end{aligned} \tag{3.73}$$

This formulation allows to get some insight into the contribution of each mixture component to a specific observation. The contribution $p(z_k = 1 \mid \mathbf{m})$ is simply the posterior within the latent variable representation. It can be evaluated using Bayes' law

$$p(z_k = 1 \mid \mathbf{m}) = \frac{P(z_k = 1)p(\mathbf{m} \mid z_k = 1)}{p(\mathbf{m})} \tag{3.74}$$

and playes an important role in the derivation of the parameter estimation for Gaussian mixtures, which we further explore in Section 4.5.

3.5 Exercises

(3.1) Show: If two random variables $\underline{a}$,$\underline{b}$ are stochastically independent under the event a, $P(a \mid b) = P(a)$, then they are also independent under the opposite event $\overline{a}$, i.e., $P(\overline{a} \mid b) = P(\overline{a})$.

(3.2) Let $\mathbf{m}_1$ and $\mathbf{m}_2$ be two feature vectors that are to be classified using a maximum posteriori (MAP) classifier.
1. When is the result of classification according to maximum a posteriori probability, $\hat{\omega} = \arg\max_\omega P(\omega \mid \mathbf{m})$, the same as that of classification according to maximum likelihood, $\hat{\omega} = \arg\max_\omega p(\mathbf{m} \mid \omega)$?
2. Under what conditions will the result of the MAP classifier only depend on the a priori probabilities $P(\omega)$?

(3.3) In a classification problem with three classes $\omega_1, \omega_2, \omega_3$, with $P(\omega_1) = 0.1$ and $P(\omega_2) = 0.6$, the following is known about the feature vectors $\mathbf{m} \in \mathbb{R}^d$:

$$p(\mathbf{m} \mid \omega_1) \equiv p(\mathbf{m} \mid \omega_2) \qquad \text{and}$$
$$\mathrm{supp}_{\mathbf{m}}\{p(\mathbf{m} \mid \omega_3)\} \cap \mathrm{supp}_{\mathbf{m}}\{p(\mathbf{m} \mid \omega_2)\} = \emptyset.$$

Maximum a posteriori classification is used for classification.
1. How well can ω_1 be separated from ω_2 using $\mathbf{m}$?
2. How well can ω_3 be separated from ω_1 and ω_2 using $\mathbf{m}$?
3. How large is the error probability P_e?

(3.4) Suppose given two classes ω_1 and ω_2 and a feature $m \in \mathbb{R}$ with the following class-dependent feature distributions:

$$p(m \mid \omega_1) = \begin{cases} m & \text{if } 0 < m \leq 1 \\ 2 - m & \text{if } 1 < m \leq 2 \\ 0 & \text{else} \end{cases}$$

and

$$p(m \mid \omega_2) = \begin{cases} m-1 & \text{if } 1 < m \le 2 \\ 3-m & \text{if } 2 < m \le 3\,. \\ 0 & \text{else} \end{cases}$$

1. Sketch the class-dependent feature distributions $p(m \mid \omega_1)$ and $p(m \mid \omega_2)$ in a single diagram.
2. Calculate the decision boundary of the Bayesian optimal classifier under the assumption that $P(\omega_1) = P(\omega_2) = 0.5$. Mark the boundary in your diagram.
3. Calculate the decision boundary for $P(\omega_1) = 0.25$ and mark it in your diagram.
4. Calculate the error probabilities in both cases.

(3.5) Let ω_1 to ω_4 be four classes with $P(\omega_1) = P(\omega_2)$, $P(\omega_3) = 0.3$ and $P(\omega_4) = 0.5$. For a feature m_1, the following are to hold:

$$p(m_1 \mid \omega_1) = p(m_1 \mid \omega_4) > p(m_1 \mid \omega_3) \quad \text{and} \quad p(m_1 \mid \omega_4) > p(m_1 \mid \omega_2).$$

1. Compute the probabilities $P(\omega_1)$ and $P(\omega_2)$.
2. What class will be chosen if one classifies according to the maximum a priori probability?
3. What class will be chosen if one classifies according to the maximum a posteriori probability?
4. Let m_2 be a second feature with $p(m_2 \mid \omega_1) = p(m_2 \mid \omega_2) > p(m_2 \mid \omega_3) > p(m_2 \mid \omega_4)$. Which classes are separated the best using (only) m_2? Which classes cannot be separated?
5. Which of the features m_1 and m_2 is better suited for maximum a posteriori classification on its own?

(3.6) Let ω be a class and let $\mathbf{m} = (m_1, m_2)^\mathsf{T}$ be a feature vector of stochastically independent features m_1 and m_2. Give the a posteriori class probability $P(\omega \mid \mathbf{m})$ using Bayes' law. Simplify as much as possible.

(3.7) A bulk material sorter is used to separate healthy wheat grains (ω_1) from grains infected with ergot (a fungus that produces a very potent toxin, ω_2) and assorted foreign bodies like dirt and the grains of other plants (ω_3). If an infected grain remains undetected, the infection will spread and 100,000 grains with a value of 1 EUR will have to be discarded, on average. If a foreign body remains undetected, the damage will only be to the brand image, which is calculated at 0.01 EUR.
The sorting system uses a Bayesian classifier to classify each individual grain, where only the length of the object is used as a feature. The sensor used can only detect whether a grain is longer or shorter than 7 mm.
It is known that the material stream consists of 97 % healthy grains, 2 % infected grains, and 1 % foreign materials. The manufacturer of the length sensor gives

the following performance characteristics:

$$\Pr(\text{length} < 7\,\text{mm} \mid \omega_1) = \frac{90}{100},$$
$$\Pr(\text{length} < 7\,\text{mm} \mid \omega_2) = \frac{3}{100},$$
$$\Pr(\text{length} < 7\,\text{mm} \mid \omega_3) = \frac{5}{100}.$$

1. Construct the cost matrix **L** for classification according to minimal a posteriori risk.
2. Which class will be chosen by a maximum a posteriori classifier if the sensor signals length $< 7\,\text{mm}$?
3. Which class will be chosen by a risk minimizing classifier if the sensor signals length $< 7\,\text{mm}$?

4 Parameter estimation

The previous Chapter assumed that the quantities

$$P(\omega|\mathbf{m}) = \frac{p(\mathbf{m}|\omega)P(\omega)}{p(\mathbf{m})} \propto p(\mathbf{m}|\omega)P(\omega) \tag{4.1}$$

(see Equation (3.32)) are known or at least the very right side (without $p(\mathbf{m})$). As already stated, the methods for determining these quantities can basically be divided into two groups, parametric and non-parametric methods. This chapter deals with the parametric approach.

In principle, one already has a mathematical model $p(\mathbf{m}|\omega_i, \theta)$ of the distribution that bounds the major traits and restricts the degrees of freedom to a finite-dimensional parameter vector $\theta \in \mathbb{R}^q$. For example, $p(\mathbf{m}|\omega_i, \theta)$ might be the family of normal densities with unknown expectation and variance, i.e., $\theta = \left(\mu, \sigma^2\right)^\mathsf{T}$. Furthermore, a dataset $\mathcal{D} = \{\mathbf{m}_1, \ldots, \mathbf{m}_N\}$ is given which is assumed to have been generated by $p(\mathbf{m}|\omega_i, \theta)$ for a fixed, but unknown θ. The goal is to find the "true" value of the parameter vector θ given the samples $\mathcal{D}$.

Definition 4.1 (Statistic, estimator). Let $\mathcal{S} = \mathcal{M} \cup \mathcal{M}^2 \cup \mathcal{M}^3 \cup \cdots$ denote the space of all finite samples and $\Theta \ni \theta$ the set of parameters.

1. A measurable function $s : \mathcal{S} \to \mathbb{R}$ is called a (real) statistic.
2. A statistic $s : \mathcal{S} \to \Theta$ that maps into the parameter space is called an estimator.

This definition of an estimator is rather broad and especially does not make any statement about quality. A constant function is an estimator, too, but intuition tells us that this should not be a good one. Performance indicators, which help to decide if an estimator is reasonable with respect to the application, are a subsequent topic of this chapter.

First, however, there will be a short excursus on two doctrines about the meaning (semantics) of probability. Both philosophies share the same syntactical foundation: the axiom system of Kolmogorov governs how to calculate with probabilities.

Definition 4.2 (Axiom system of Kolmogorov). Let $(\mathcal{A}, \mathfrak{A})$ be a measurable space ($\mathcal{A}$ can be thought of as the set of all possible elementary events). The triple $(\mathcal{A}, \mathfrak{A}, P)$ is called a probability space if

1. $\Pr(\mathcal{B}) \geq 0$ for all $\mathcal{B} \in \mathfrak{A}$ (non-negativity),
2. $\Pr(\mathcal{A}) = 1$ (the certain event has probability 1), and
3. for all $\mathcal{B}_1, \mathcal{B}_2, \ldots$ with $\mathcal{B}_j \cap \mathcal{B}_i = \emptyset$, we have

$$\Pr(\mathcal{B}_1 \cup \mathcal{B}_2 \cup \cdots) = \sum_{j=1}^{\infty} \Pr(\mathcal{B}_i). \tag{4.2}$$

Moreover, the probability $\Pr(\mathcal{B}|\mathcal{C}) = \frac{\Pr(\mathcal{B} \cap \mathcal{C})}{\Pr(\mathcal{C})}$ is called the conditional probability of $\mathcal{B}$ given $\mathcal{C}$.

https://doi.org/10.1515/9783111339207-004

Although these definitions explain the syntactical handling of probability, they do not explain what is actually meant by an expression like "The probability of tossing a head is $\frac{1}{2}$" or "Tomorrow it is going to rain with a chance of 1 out of 3". There are two doctrines. On the one hand, there is frequentism, which is also called the objective concept of probability. On the other hand, there is the Bayesian interpretation, which is also called the subjective concept of probability. Note that the Bayesian interpretation is not directly linked to the Bayes law except that both concepts are named in honor of the same person. The Bayes law also holds in frequentism and is not less correct under this doctrine.

The frequentist interpretation tries to interpret the probability as some kind of physical property like weight, size, etc. Hence, a probability can only be assigned to events for which a well-defined (theoretical) experiment can be designed. Then the probability of the event is the limit of infinitely many repetitions of the experiment. Coin tossing is the classical example of such an experiment. As the existence of a well-defined experiment is a fundamental requirement, the frequentist view is also called objective.

Although there are good reasons for frequentism, reasons that are not going to be elaborated further here, this concept quickly meets its limits. No probability can be assigned to infrequent events, such as volcanic eruptions. Even an statement about tomorrow's weather is difficult. Nonetheless, in many cases, one can imagine some kind of (theoretical) experiment in an ideal environment.

The Bayesian approach avoids those difficulties by interpreting probability as a *degree of belief*. As belief is very subjective, there is no need for an experiment. The principal idea is to trace back probability to a fair bet. For example, the probability of rain equals $\frac{1}{3}$ if one believes it is a fair bet (for both sides) to put 1 money unit on rain if the prize is 2 in case of rain and a total loss in case of sunshine. Though this concept opens the world of probability to a much wider field of applications, this approach has the drawback that probability becomes somewhat intangible. The Bayesian interpretation has a branch that is called "objective Bayesianism" which perhaps could be explained by "degree of belief after good reasoning".

This textbook is not going to resolve the conflict between frequentism and Bayesianism, but this topic is mentioned to draw attention to some important differences with respect to parameter estimation. For a deeper look into the different interpretations of probability, see, for example, the work of Robert [2011] or Efron and Hastie [2016].

In frequentism (classical statistics), the parameter vector θ is assumed to be a fixed though unknown quantity. The parameter vector is not a random variable. On the notational level, this leads to $p(\mathbf{m} \mid \theta)$ for the density, i.e., the density of a random quantity $\underline{\mathbf{m}}$ evaluated at $\mathbf{m}$ that is additionally controlled by a (non-random) parameter θ. Note that although from the frequentist point of view, θ is *not* a random variable, the density $p(\mathbf{m} \mid \theta)$ is written like a conditional density. This is usual within the engineering literature and should not cause any confusion within this book. Given a dataset $\mathcal{D} = \{\mathbf{m}_1, \ldots, \mathbf{m}_N\}$, a reasonable approach is to choose $\hat{\theta}(\mathcal{D})$ such that the observation

$\mathcal{D}$ becomes most likely,

$$\hat{\theta}(\mathcal{D}) = \arg\max_{\theta \in \Theta} \prod_{\mathbf{m} \in \mathcal{D}} p(\mathbf{m} \mid \theta). \tag{4.3}$$

This approach is called the likelihood method and this estimator is called the maximum likelihood estimator (see Definition 4.7).

In the Bayesian framework, the parameter vector is assumed to be a random quantity $\underline{\theta}$, too, so that there is a joint distribution of $(\underline{\mathbf{m}}, \underline{\theta})$. For practical purposes the joint distribution is rather uninteresting, but the distribution assumption is expressed as a conditional distribution $p(\mathbf{m}|\theta)$. Applying Bayes' law, this can be rewritten to a conditional distribution of $\underline{\theta}$ given an observation $\mathcal{D}$. Then $\hat{\theta}(\mathcal{D})$ is assigned the value for which the a posteriori probability attains a maximum. Unsurprisingly, this is called the maximum a posteriori estimator.

Let us repeat this so that the difference between both philosophies becomes abundantly clear. In classical statistics, the distribution of the features is given by $p(\mathbf{m} \mid \theta)$. The parameter vector θ is an unknown, but constant quantity. On the Bayesian view, the distribution of the features is a conditional distribution given by $p(\mathbf{m}|\theta)$. The parameter vector θ is a random variable itself. Moreover, in classical statistics the parameter is chosen such that the observation becomes most likely. There is no point in speaking about something like the probability of the parameter. In the Bayesian world, on the contrary, the parameter vector θ is chosen to have the maximum probability conditioned by the observation.

In maximum a posteriori pattern classification, there is one feature distribution per class ω_i. Consequently, one needs to estimate one parameter vector θ_i per class. More precisely, one seeks the distributions $p(\mathbf{m}|\omega_i, \theta_i)$ with $i = 1, \ldots, c$. The general assumption is that one uses supervised learning. This means that the number of classes and the class assignment of samples is given beforehand. The whole dataset can be decomposed into a partition $\mathcal{D} = \mathcal{D}_1 \uplus \cdots \uplus \mathcal{D}_c$ and furthermore samples from $\mathcal{D}_i$ bear information about the unknown parameter vector θ_i, but do not have any influence on the parameter vectors θ_j $(j \neq i)$ of the other classes. In conclusion, the task of parameter estimation can be independently repeated for each class and one can assume without loss of generality that only one class exists. Here and below, the explicit notation of a class will be suppressed.

Definition 4.1 stated that any statistic that maps into the parameter space is an estimator. One of the minimal requirements for a reasonable estimator is its unbiasedness. The principal idea is to put random variables (instead of observations) into the estimator, so that the estimator becomes a random quantity on its own. The estimator is called unbiased if its expectation equals the parameter being estimated.

Definition 4.3 (Unbiased estimator). Depending on the interpretation of statistics:

In classical statistics Let $\underline{\mathbf{m}}_1, \ldots, \underline{\mathbf{m}}_N$ denote independently and identically distributed random variables with distribution $p(\mathbf{m} \mid \theta)$ and let $\hat{\theta} : \mathcal{M}^N \to \Theta$ be an estimator of θ. $\underline{\hat{\theta}} = \hat{\theta}(\underline{\mathbf{m}}_1, \ldots, \underline{\mathbf{m}}_N)$ will denote the estimator considered as a random variable.

Then $\hat{\theta}$ is an unbiased estimator iff the following holds:

$$\begin{aligned}\mathrm{E}\{\underline{\hat{\theta}}\} &= \mathrm{E}\{\hat{\theta}(\underline{\mathbf{m}}_1, \ldots, \underline{\mathbf{m}}_N)\} \\ &= \int\limits_{\mathcal{M}^N} \cdots \int \hat{\theta}(\mathbf{m}_1, \ldots, \mathbf{m}_N)\, p(\mathbf{m}_1 \mid \theta) \cdots p(\mathbf{m}_N \mid \theta)\, \mathrm{d}\mathbf{m}_1\, \mathrm{d}\mathbf{m}_2 \ldots \mathrm{d}\mathbf{m}_N \\ &\stackrel{!}{=} \theta. \end{aligned} \tag{4.4}$$

In Bayesian statistics Let $\underline{\mathbf{m}}_1, \ldots, \underline{\mathbf{m}}_N$ denote independently and identically distributed random variables with conditional distribution $p(\mathbf{m}|\theta)$ and let $\hat{\theta} : \mathcal{M}^N \to \Theta$ be an estimator of $\underline{\theta}$. $\underline{\hat{\theta}} = \hat{\theta}(\underline{\mathbf{m}}_1, \ldots, \underline{\mathbf{m}}_N)$ will denote the estimator considered as a random variable. Then $\hat{\theta}$ is an unbiased estimator iff the following holds:

$$\begin{aligned}\mathrm{E}\{\underline{\hat{\theta}}\} &= \mathrm{E}\{\hat{\theta}(\underline{\mathbf{m}}_1, \ldots, \underline{\mathbf{m}}_N)\} \\ &= \int\limits_{\mathcal{M}^N} \cdots \int \hat{\theta}(\mathbf{m}_1, \ldots, \mathbf{m}_N)\, p(\mathbf{m}_1 \mid \underline{\theta}) \cdots p(\mathbf{m}_N \mid \underline{\theta})\, \mathrm{d}\mathbf{m}_1\, \mathrm{d}\mathbf{m}_2 \ldots \mathrm{d}\mathbf{m}_N \\ &\stackrel{!}{=} \mathrm{E}\{\underline{\theta}\}. \end{aligned} \tag{4.5}$$

As illustrated by the definition above, it is necessary to distinguish between classical and Bayesian statistics in order to get the details right, although the principal ideas (like unbiasedness) exist in both worlds.

We will discuss the viewpoint of classical statistics first. If one already considers the expectation of an estimator, it seems reasonable to discuss the variance $\mathrm{Var}\{\underline{\hat{\theta}}\} = \mathrm{E}\left\{\left(\underline{\hat{\theta}} - \mathrm{E}\{\underline{\hat{\theta}}\}\right)^2\right\} = \mathrm{E}\{\underline{\hat{\theta}}^2\} - \mathrm{E}\{\underline{\hat{\theta}}\}^2$ as well. Here, the variance of a random vector distributes over the elements of the vector, i.e., $\mathrm{Var}\{\underline{\alpha}\} := (\mathrm{Var}\{\underline{\alpha}_1\}, \mathrm{Var}\{\underline{\alpha}_2\}, \ldots)^\mathsf{T}$; similarly, the square distributes over the elements, too. As a minimal requirement, the discussion must be restricted to unbiased estimators with $\mathrm{Var}\{\underline{\hat{\theta}}\} = \mathrm{E}\{\underline{\hat{\theta}}^2\} - \mathrm{E}\{\underline{\hat{\theta}}\}^2 = \mathrm{E}\{\underline{\hat{\theta}}^2\} - \theta^2$, because one is interested in how much the estimator fluctuates away from the true value. Without this requirement, it would be easy to construct estimators with zero variance by taking a constant estimator. Nonetheless, a small variance is desirable. If one cannot expect to obtain zero variance as long as the estimator is based on the observations, one can still ask, if there is some lower bound among all reasonable estimators. This lower bound is given by the Cramér–Rao bound (CRB). For the sake of simplicity, the theorem is only given for a real, scalar parameter $\theta \in \Theta = \mathbb{R}$.

Definition 4.4 (Cramér–Rao bound). Let the hypotheses be the same as in Definition 4.3, first item, with the simplification $\Theta = \mathbb{R}$ and the following additions:

1. $\hat{\theta}$ is unbiased,
2. $\hat{\theta} : \mathcal{M}^N \to \Theta$ does not depend on the unknown value θ,
3. $\mathrm{E}\{\underline{\hat{\theta}}^2\} = \mathrm{E}\{\hat{\theta}(\underline{\mathbf{m}}_1, \ldots, \underline{\mathbf{m}}_N)^2\} < \infty$,
4. the density $p(\mathbf{m} \mid \theta)$ is differentiable with respect to θ, and
5. $\frac{\partial}{\partial\theta} \int_{\mathcal{M}} p(\mathbf{m} \mid \theta)\, \mathrm{d}\mathbf{m} = \int_{\mathcal{M}} \frac{\partial}{\partial\theta} p(\mathbf{m} \mid \theta)\, \mathrm{d}\mathbf{m}$.

Then the following holds:

$$\mathrm{Var}\{\hat{\underline{\theta}}\} \geq \frac{1}{N\,\mathrm{E}\Big\{\big(\frac{\partial}{\partial\theta}\ln p(\underline{\mathbf{m}}\,|\,\Theta)\big)^2\Big\}} = \frac{1}{N\int_{\mathcal{M}}\big(\frac{\partial}{\partial\theta}\ln p(\mathbf{m}\,|\,\Theta)\big)^2 p(\mathbf{m}\,|\,\Theta)\,\mathrm{d}\mathbf{m}}. \tag{4.6}$$

An estimator whose variance equals this bound for all θ is called a CR-efficient estimator. Before a sketch of the proof is given, let us discuss the prerequisites. The first two are rather natural. The need for unbiasedness was already explained. An estimator that would depend on the value being estimated is not forbidden by definition, but rather pointless for practical purposes. Hence, this is no real limitation. The last three requirements are rather technical and normally summarized by the term *regularity conditions*. Distributions that comply with these are called *regular distributions*. For all practical (engineering) purposes, one can take them as granted.

The proof is mainly an application of the Cauchy–Schwarz inequality

$$\int x(\mathbf{m})^2\,\mathrm{d}\mathbf{m}\int y(\mathbf{m})^2\,\mathrm{d}\mathbf{m} \geq \Big(\int x(\mathbf{m})y(\mathbf{m})\,\mathrm{d}\mathbf{m}\Big)^2 \tag{4.7}$$

for square-integrable functions. The sketch of the proof is only presented for $N = 1$ to avoid cumbersome integrals. Because $\hat{\theta}$ is unbiased, $\mathrm{E}\{\hat{\underline{\theta}} - \theta\} = 0$ for all θ, and likewise for the partial derivative $\frac{\partial}{\partial\theta}\mathrm{E}\{\hat{\underline{\theta}} - \theta\} = 0$. Altogether, this leads to

$$\begin{aligned}
0 &= \frac{\partial}{\partial\theta}\mathrm{E}\{\hat{\underline{\theta}} - \theta\} = \frac{\partial}{\partial\theta}\int(\hat{\theta}(\underline{\mathbf{m}}) - \theta)p(\mathbf{m}\,|\,\theta)\,\mathrm{d}\mathbf{m} \\
&= \int\frac{\partial}{\partial\theta}(\hat{\theta}(\underline{\mathbf{m}}) - \theta)\cdot p(\mathbf{m}\,|\,\theta)\,\mathrm{d}\mathbf{m} + \int(\hat{\theta}(\underline{\mathbf{m}}) - \theta)\frac{\partial}{\partial\theta}p(\mathbf{m}\,|\,\theta)\,\mathrm{d}\mathbf{m} \\
&\overset{(4.9)}{=} -\underbrace{\int p(\mathbf{m}\,|\,\theta)\,\mathrm{d}\mathbf{m}}_{=1} + \int(\hat{\theta}(\underline{\mathbf{m}}) - \theta)p(\mathbf{m}\,|\,\theta)\frac{\partial}{\partial\theta}\ln p(\mathbf{m}\,|\,\theta)\,\mathrm{d}\mathbf{m}.
\end{aligned} \tag{4.8}$$

The last line holds because

$$p(\mathbf{m}\,|\,\theta)\frac{\partial}{\partial\theta}\ln p(\mathbf{m}\,|\,\theta) = p(\mathbf{m}\,|\,\theta)\frac{1}{p(\mathbf{m}\,|\,\theta)}\frac{\partial}{\partial\theta}p(\mathbf{m}\,|\,\theta) = \frac{\partial}{\partial\theta}p(\mathbf{m}\,|\,\theta), \tag{4.9}$$

which follows from the chain rule of differentiation, where $p(\mathbf{m}\,|\,\theta)$ is viewed as a function of θ. Reorganizing and simplifying Equation (4.8) yields

$$\begin{aligned}
1 &= \Big(\int(\hat{\theta}(\underline{\mathbf{m}}) - \theta)p(\mathbf{m}\,|\,\theta)\frac{\partial}{\partial\theta}\ln p(\mathbf{m}\,|\,\theta)\,\mathrm{d}\mathbf{m}\Big)^2 \\
&= \Big(\int(\hat{\theta}(\underline{\mathbf{m}}) - \theta)\sqrt{p(\mathbf{m}\,|\,\theta)}\sqrt{p(\mathbf{m}\,|\,\theta)}\frac{\partial}{\partial\theta}\ln p(\mathbf{m}\,|\,\theta)\,\mathrm{d}\mathbf{m}\Big)^2 \\
&\overset{(4.7)}{\leq} \int(\hat{\theta}(\underline{\mathbf{m}}) - \theta)^2 p(\mathbf{m}\,|\,\theta)\,\mathrm{d}\mathbf{m}\cdot\int\Big(\frac{\partial}{\partial\theta}\ln p(\mathbf{m}\,|\,\theta)\Big)^2 p(\mathbf{m}\,|\,\theta)\,\mathrm{d}\mathbf{m} \\
&= \mathrm{Var}\{\hat{\underline{\theta}}\}\cdot\underbrace{\mathrm{E}\Big\{\big(\tfrac{\partial}{\partial\theta}\ln p(\underline{\mathbf{m}}\,|\,\theta)\big)^2\Big\}}_{=J(\theta)\text{ (Fisher information)}},
\end{aligned} \tag{4.10}$$

which concludes the sketch of the proof.

The expectation in Equations (4.6) and (4.10) is called the Fisher information $J(\theta)$: it is the variance of the *score*. The score estimates how much the parameter θ influences the density of the random variable $\underline{\mathbf{m}}$ at every point $\mathbf{m}$, normalized to the absolute value of the density at the point $\mathbf{m}$. So, the score is

$$\frac{1}{p(\mathbf{m}\,|\,\theta)}\frac{\partial}{\partial\theta}p(\mathbf{m}\,|\,\theta) \overset{(4.9)}{=} \frac{\partial}{\partial\theta}\ln p(\mathbf{m}\,|\,\theta), \tag{4.11}$$

which, again, follows from the chain rule of differentiation.

This information can be further concentrated by not looking at the change for every point $\mathbf{m}$, but by looking at the *average* change. If one puts the random variable $\underline{\mathbf{m}}$ into its own density, one can regard the score as a random quantity and calculate its expectation. Under the regulatory conditions of Definition 4.4

$$\mathrm{E}\left\{\frac{\partial}{\partial\theta}\ln p(\underline{\mathbf{m}}\,|\,\theta)\right\} = 0. \tag{4.12}$$

This makes intuitive sense, because the integral of every member of the family $p(\mathbf{m}\,|\,\theta)$ equals 1. So, if a change of θ increases the density $p(\mathbf{m}\,|\,\theta)$ at some point $\mathbf{m}$, the density must decrease at some other point $\mathbf{m}'$. With respect to the weighted average over all $\mathbf{m}$, the changes being induced by θ must be balanced. But the squared average does not disappear, because both a decrease and an increase count as positive. Due to the fact that the mean is zero, this equals the variance,

$$\mathrm{E}\left\{\left(\frac{\partial}{\partial\theta}\ln p(\underline{\mathbf{m}}\,|\,\theta)\right)^2\right\} = \mathrm{Var}\left\{\frac{\partial}{\partial\theta}\ln p(\underline{\mathbf{m}}\,|\,\theta)\right\} = J(\theta). \tag{4.13}$$

Without any proof, we just state the multi-dimensional generalization of Equations (4.6) and (4.10):

$$\mathrm{Cov}\{\underline{\hat{\boldsymbol{\theta}}}\} \geq \frac{1}{N}J^{-1}(\boldsymbol{\theta}) \tag{4.14}$$

where the scalar Fisher information is replaced by the Fisher information matrix:

$$J^{-1}(\theta) = \mathrm{E}\left\{(\nabla_{\boldsymbol{\theta}}\ln p(\underline{\mathbf{m}}\,|\,\boldsymbol{\theta}))(\nabla_{\boldsymbol{\theta}}\ln p(\underline{\mathbf{m}}\,|\,\boldsymbol{\theta}))^{\mathsf{T}}\right\}. \tag{4.15}$$

Of course, Equation (4.14) requires the explanation of what "$\succeq$" means in the context of matrices. We say $\mathrm{Cov}\{\underline{\hat{\boldsymbol{\theta}}}\} \succeq \frac{1}{N}J^{-1}(\boldsymbol{\theta})$ iff $\left(\mathrm{Cov}\{\underline{\hat{\boldsymbol{\theta}}}\} - \frac{1}{N}J^{-1}(\boldsymbol{\theta})\right)$ is a positive semi-definite matrix. This is equivalent to

$$\boldsymbol{\alpha}^{\mathsf{T}}\,\mathrm{Cov}\{\underline{\hat{\boldsymbol{\theta}}}\}\,\boldsymbol{\alpha} \geq \frac{1}{N}\boldsymbol{\alpha}^{\mathsf{T}}J^{-1}(\boldsymbol{\theta})\boldsymbol{\alpha} \qquad \text{for all } \boldsymbol{\alpha} \in \mathbb{R}^k \tag{4.16}$$

and

$$\mathrm{tr}\,\mathrm{Cov}\{\underline{\hat{\boldsymbol{\theta}}}\} \geq \frac{1}{N}\,\mathrm{tr}\,J^{-1}(\boldsymbol{\theta}). \tag{4.17}$$

After unbiased estimators and CR-efficient estimators, the third type of estimator we consider is the consistent estimator.

Definition 4.5 (Consistent estimator). Again, the setting is the same as in Definition 4.3, first item. An estimator is called a consistent estimator iff

$$\lim_{N\to\infty} \Pr\left(\|\hat{\underline{\theta}} - \theta\| \geq \varepsilon\right) = 0 \qquad \text{for all } \varepsilon > 0 \tag{4.18}$$

with $\hat{\underline{\theta}} = \hat{\theta}(\underline{\mathbf{m}}_1, \ldots, \underline{\mathbf{m}}_N)$.

This means an estimator is consistent if its value converges almost surely to the true value. Actually, this should be a minimal requirement of a reasonable estimator. One should realize that neither is an unbiased estimator necessarily consistent, nor is a consistent estimator necessarily unbiased. These properties are independent of each other.

Before looking at some examples of estimators that illustrate the above concepts, the terms will be discussed and weighed against each other. From a purely theoretical perspective, the unbiasedness of an estimator is a minimal requirement. The gap between the expectation of the estimator and the true value

$$b(\hat{\theta}) = \mathrm{E}\{\hat{\underline{\theta}}\} - \theta \tag{4.19}$$

is called the bias of the estimator. If an estimator is truly biased (and not only biased, but even asymptotically biased), then the bias will remain no matter how many samples are used. For this reason a bias is also called a systematic error. In contrast, the variance $\mathrm{Var}\{\hat{\underline{\theta}}\}$ of an estimator typically diminishes when more samples are considered. Hence, the variance is called the stochastic error of the estimator. As the theory can use as many samples as needed, the stochastic error is not crucial. From a practical perspective this is not entirely true, because the number of samples is limited and cannot be arbitrarily increased. The mean squared error (MSE) of an estimator

$$\mathrm{E}\{(\hat{\underline{\theta}} - \theta)^2\} = b(\hat{\theta})^2 + \mathrm{Var}\{\hat{\underline{\theta}}\} \tag{4.20}$$

equals the sum of the squared bias and the variance. Given two estimators $\hat{\theta}_1$ and $\hat{\theta}_2$ with $\mathrm{Var}\{\hat{\underline{\theta}}_2\} \ll \mathrm{Var}\{\hat{\underline{\theta}}_1\}$, a biased estimator $\hat{\theta}_2$ can have a much smaller mean squared error than an unbiased estimator $\hat{\theta}_1$. This is depicted in Figure 4.1.

As a starting example, consider an estimator of the expectation. For this purpose, $\underline{m}_i$ $(i = 1, \ldots, N)$ will be independently and identically distributed with $p(m \mid \mu)$ with a parameter $\theta = \mu$. Moreover, $\mathrm{E}\{\underline{m}_i\} = \mu$. Then $\sigma^2 = \mathrm{Var}\{\underline{m}_i\} = \mathrm{E}\{(\underline{m} - \mu)^2\}$ follows. Note that although the distribution is assumed to be parametrized by its expectation μ, the distribution is *not* assumed to be Gaussian. The expectation and variance are called μ and σ^2 for convenience only.

The empirical mean suggests itself as an estimator for the expectation:

$$\hat{\mu} = \frac{1}{N} \sum_{i=1}^{N} m_i. \tag{4.21}$$

The expectation of this estimator is

$$\mathrm{E}\{\hat{\underline{\mu}}\} = \mathrm{E}\left\{\frac{1}{N} \sum_{i=1}^{N} \underline{m}_i\right\} = \frac{1}{N} \sum_{i=1}^{N} \mathrm{E}\{\underline{m}_i\} = \frac{1}{N} \sum_{i=1}^{N} \mu = \mu. \tag{4.22}$$

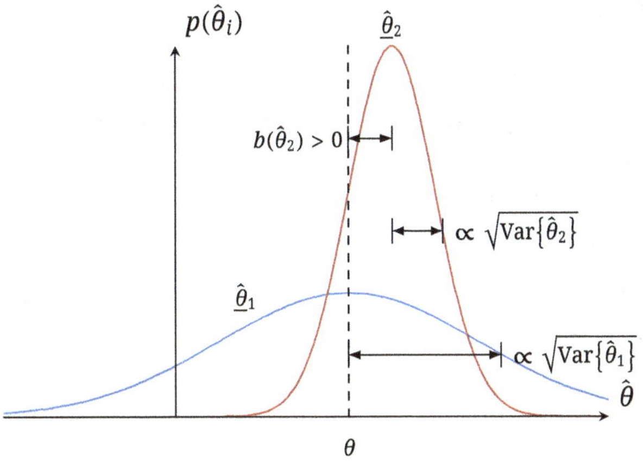

Fig. 4.1: Comparison of an unbiased estimator with large variance ($\hat{\theta}_1$, blue) with a biased estimator with small variance ($\hat{\theta}_2$, red).

This proves, that for any distribution, the empirical average is an unbiased estimator for the expectation. The variance of this estimator equals

$$
\begin{aligned}
\mathrm{Var}\{\underline{\hat{\mu}}\} &= \mathrm{E}\{(\underline{\hat{\mu}} - \mu)^2\} = \mathrm{E}\left\{\left(\frac{1}{N}\sum_{i=1}^{N} \underline{m}_i - \mu\right)^2\right\} \\
&= \frac{1}{N^2}\sum_{k=1}^{N}\sum_{i=1}^{N} \mathrm{E}\{(\underline{m}_k - \mu)(\underline{m}_i - \mu)\} \\
&= \frac{1}{N^2}\sum_{k=1}^{N} \mathrm{E}\{(\underline{m}_k - \mu)^2\} + \frac{1}{N^2}\sum_{k=1}^{N}\sum_{\substack{i=1\\ i\neq k}}^{N} \mathrm{E}\{(\underline{m}_k - \mu)(\underline{m}_i - \mu)\} \\
&= \frac{1}{N^2}\sum_{k=1}^{N} \sigma^2 + \frac{1}{N^2}\sum_{k=1}^{N}\sum_{\substack{i=1\\ i\neq k}}^{N} \mathrm{E}\{\underline{m}_k - \mu\}\,\mathrm{E}\{\underline{m}_i - \mu\} \\
&\overset{\text{i.i.d.}}{=} \frac{\sigma^2}{N}\,.
\end{aligned}
\tag{4.23}
$$

The variance of the estimator linearly vanishes with respect to the sample size. In other words, the standard deviation $\sqrt{\mathrm{Var}\{\underline{\hat{\mu}}\}}$ decreases with $\frac{1}{\sqrt{N}}$. This asymptotic behavior is usual for most applications. Applying Chebyshev's law

$$
\Pr\left(\left|\underline{\hat{\mu}} - \mathrm{E}\{\underline{\hat{\mu}}\}\right| \geq \varepsilon\right) \leq \frac{\mathrm{Var}\{\underline{\hat{\mu}}\}}{\varepsilon^2} \qquad \forall \varepsilon > 0 \tag{4.24}
$$

yields

$$
\Pr\left(\left|\underline{\hat{\mu}} - \mu\right| \geq \varepsilon\right) \leq \frac{\sigma^2}{N\varepsilon^2}, \tag{4.25}
$$

which shows that the estimator is also consistent.

4.1 Maximum likelihood estimation

The empirical mean as an estimator for the expectation can more or less be found through an educated guess. We now turn to a more systematic approach to finding estimators. The maximum likelihood estimator has already been mentioned in the introduction of this chapter (see Equation (4.3)). For many distribution assumptions, the maximum likelihood estimator of the expectation value equals the average mean.

Definition 4.6 (Likelihood function, log-likelihood function). Let $\underline{\mathbf{m}}_i$ $(i = 1, \dots, N)$ be independently and identically distributed with $p(\mathbf{m} \mid \theta)$ and $\mathcal{D} = \{\mathbf{m}_1, \dots, \mathbf{m}_N\}$ a set of realized samples.

1. The function
$$L(\theta) = \prod_{\mathbf{m} \in \mathcal{D}} p(\mathbf{m} \mid \theta) \tag{4.26}$$
is called the *likelihood* function.
2. The function
$$l(\theta) = \ln \prod_{\mathbf{m} \in \mathcal{D}} p(\mathbf{m} \mid \theta) = \sum_{\mathbf{m} \in \mathcal{D}} \ln p(\mathbf{m} \mid \theta) \tag{4.27}$$
is called the *log-likelihood* function.

Due to the strict monotonicity of the logarithm, the likelihood function and the log-likelihood function share the same extremal points. In practice, the log-likelihood function is often easier to use, since it involves sums instead of products. The maximum likelihood estimator determines the parameter that maximizes the likelihood given the observation. In other words, maximum likelihood estimation chooses the value $\theta = \hat{\theta}$ which makes the given observation $\mathcal{D}$ maximally probable under the model.

Definition 4.7 (Maximum likelihood estimator). The hypothesis will be the same as in Definition 4.6. Then

$$\hat{\theta}_{\mathrm{ML}}(\mathcal{D}) = \operatorname*{arg\,max}_{\theta \in \Theta} \prod_{\mathbf{m} \in \mathcal{D}} p(\mathbf{m} \mid \theta) = \operatorname*{arg\,max}_{\theta \in \Theta} \sum_{\mathbf{m} \in \mathcal{D}} \ln p(\mathbf{m} \mid \theta) \tag{4.28}$$

is called the *maximum likelihood estimator* (ML estimator).

Under the usual implicit assumption that all functions are sufficiently smooth,

$$\mathbf{0} \stackrel{!}{=} \nabla_\theta l(\theta) = \sum_{\mathbf{m} \in \mathcal{D}} \nabla_\theta \ln(p(\mathbf{m} \mid \theta)) \quad \text{with } \nabla_\theta = \left(\tfrac{\partial}{\partial \theta_1} \quad \cdots \quad \tfrac{\partial}{\partial \theta_q}\right)^{\mathsf{T}} \tag{4.29}$$

is a necessary condition.

The first example will be to find the ML estimator for the expectation value of a d-dimensional normal distribution. Let $\underline{\mathbf{m}}_k \sim \mathcal{N}(\boldsymbol{\mu}, \boldsymbol{\Sigma})$ with $\boldsymbol{\mu}$ unknown but known $\boldsymbol{\Sigma}$. It follows that

$$\ln p(\mathbf{m}_k) = -\frac{1}{2}(\mathbf{m}_k - \boldsymbol{\mu})^{\mathsf{T}} \boldsymbol{\Sigma}^{-1} (\mathbf{m}_k - \boldsymbol{\mu}) - \frac{d}{2} \ln 2\pi - \frac{1}{2} \ln \det \boldsymbol{\Sigma}$$
$$\Rightarrow \nabla_{\boldsymbol{\mu}} \ln p(\mathbf{m}_k) = \boldsymbol{\Sigma}^{-1}(\mathbf{m}_k - \boldsymbol{\mu}). \tag{4.30}$$

Applying the necessary condition, Equation (4.29) leads to

$$\begin{aligned}
\sum_{k=1}^{N} \nabla_{\hat{\boldsymbol{\mu}}} \ln p(\mathbf{m}_k) &= \sum_{k=1}^{N} \boldsymbol{\Sigma}^{-1}(\mathbf{m}_k - \hat{\boldsymbol{\mu}}) \stackrel{!}{=} \mathbf{0} \\
\Leftrightarrow \sum_{k=1}^{N} (\mathbf{m}_k - \hat{\boldsymbol{\mu}}) &= \mathbf{0} \\
\Leftrightarrow \hat{\boldsymbol{\mu}}_{\mathrm{ML}} &= \frac{1}{N}\sum_{k=1}^{N} \mathbf{m}_k. \qquad (4.31)
\end{aligned}$$

In summary, the ML estimator of the expectation is the empirical average, as already has been noted earlier. This result holds for many distributions.

The next example extends the estimator to the case that the variance is unknown, too. To avoid some tedious matrix calculus, the scenario is restricted to the scalar case, but the result is equally true in a multi-dimensional setting. Hence, let $\underline{m}_k \sim \mathcal{N}(\mu, \sigma^2)$ with $\boldsymbol{\theta} = (\theta_1, \theta_2)^\mathsf{T} := (\mu, \sigma^2)^\mathsf{T}$ as the parameter vector. It follows that

$$\begin{aligned}
\ln p(m_k) &= -\frac{1}{2\theta_2}(m_k - \theta_1)^2 - \frac{1}{2}\ln\theta_2 - \frac{1}{2}\ln 2\pi \\
\Rightarrow \nabla_{\boldsymbol{\theta}} \ln p(m_k) &= \begin{pmatrix} \frac{1}{\theta_2}(m_k - \theta_1) \\ \frac{1}{2\theta_2^2}(m_k - \theta_1)^2 - \frac{1}{2\theta_2} \end{pmatrix} \stackrel{!}{=} \mathbf{0}. \qquad (4.32)
\end{aligned}$$

The solution of the system of equations is

$$\theta_1 = \hat{\mu}_{\mathrm{ML}} = \frac{1}{N}\sum_{k=1}^{N} m_k \qquad (4.33)$$

$$\theta_2 = \hat{\sigma}^2_{\mathrm{ML}} = \frac{1}{N}\sum_{k=1}^{N} (m_k - \hat{\mu}_{\mathrm{ML}})^2. \qquad (4.34)$$

Accordingly, in the multi-dimensional case, the ML estimator is

$$\hat{\boldsymbol{\mu}}_{\mathrm{ML}} = \frac{1}{N}\sum_{k=1}^{N} \mathbf{m}_k \qquad (4.35)$$

$$\hat{\boldsymbol{\Sigma}}_{\mathrm{ML}} = \frac{1}{N}\sum_{k=1}^{N} (\mathbf{m}_k - \hat{\boldsymbol{\mu}}_{\mathrm{ML}})(\mathbf{m}_k - \hat{\boldsymbol{\mu}}_{\mathrm{ML}})^\mathsf{T}. \qquad (4.36)$$

Note that the ML estimator for the variance is biased. It would be unbiased if the true expectation μ was known. But as the ML estimator $\hat{\mu}$ is put into the estimator for the variance, this estimator underestimates the variance systematically due to an additional uncertainty coming from $\hat{\mu}$. It can be shown that the unbiased estimator is $\frac{N}{N-1}\hat{\sigma}^2_{\mathrm{ML}}$. In any case, both estimators are consistent.

4.2 Bayesian estimation of the class-specific distributions

In this section, the estimation problem is reconsidered under the Bayesian framework. Unlike the former approach, the parameter vector $\underline{\theta}$ is also regarded as a random quantity. Moreover, the classical approach introduced the parameter right from the start and aimed to estimate the parameter directly from the given dataset. In the Bayesian concept, the parameter vector fades a little bit into the background, because here the starting point is the original aim of estimating the class of an unknown object given the training samples. The parameter is introduced as an intermediate link between the training samples and the unknown object.

The fundamental quantity of the Bayesian classification is the a posteriori distribution of the classes

$$P(\omega_i|\mathbf{m}) = \frac{p(\mathbf{m}|\omega_i)P(\omega_i)}{p(\mathbf{m})} \qquad i = 1, \dots, c. \tag{4.37}$$

As usual, the dataset is $\mathcal{D} = \mathcal{D}_1 \uplus \cdots \uplus \mathcal{D}_c$ with $\mathbf{m} \in \mathcal{D}_i \Leftrightarrow \omega(\mathbf{m}) = \omega_i$. Taking into account that all quantities in Equation (4.37) are based on the data $\mathcal{D}$, the formula can be extended to

$$P(\omega_i|\mathbf{m}, \mathcal{D}) = \frac{p(\mathbf{m}|\omega_i, \mathcal{D})P(\omega_i|\mathcal{D})}{p(\mathbf{m}|\mathcal{D})}. \tag{4.38}$$

The conceptual difference of the Bayesian view is to actually regard every probability as a conditional probability. Any unconditional distribution is just a convenient utility, if the condition is negligible. This means one actually wants to know the probability that a realized feature $\mathbf{m}$ of an random feature $\underline{\mathbf{m}}$ belongs to class ω_i, given that the concrete dataset $\mathcal{D}$ out of the entirety of $\underline{\mathcal{D}}$ has been observed before. In this sense, $P(\omega|\mathbf{m})$ is only an abbreviation for $P(\omega|\mathbf{m}, \mathcal{D})$ given that $P(\cdot, \cdot, \mathcal{D}') \approx P(\cdot, \cdot, \mathcal{D}'')$, if the datasets $\mathcal{D}'$ and $\mathcal{D}''$ are large enough.

Equation (4.38) can immediately be simplified again, because supervised sampling is assumed. This means that the membership of a sample $\mathbf{m}$ in one of the partitions $\mathcal{D}_i$ is controlled, because its class is known. This has two consequences:

First, though the a priori distribution of the classes $P(\omega|\mathcal{D})$ depends on $\mathcal{D}$, one must not use a realization of the random variable $\underline{\mathcal{D}}$, because the realization is generally not truly sampled but artificially composed. This means that the proportions of the partition $\mathcal{D}_1 \uplus \cdots \uplus \mathcal{D}_c$ do not reflect the distribution of the classes. Hence, the assumption is that an a priori distribution $P(\omega)$ is known.

Second, one assumes that the class-specific feature distribution does not depend on samples of a different class. This means that

$$P(\mathbf{m}|\omega_i, \mathcal{D}) = P(\mathbf{m}|\omega_i, \mathcal{D}_i). \tag{4.39}$$

Applying these considerations to Equation (4.38) and replacing the denominator by a summation over all classes yields

$$P(\omega_i|\mathbf{m}, \mathcal{D}) = \frac{p(\mathbf{m}|\omega_i, \mathcal{D}_i)P(\omega_i)}{\sum_{j=1}^{c} p(\mathbf{m}|\omega_j, \mathcal{D}_j)P(\omega_j)}. \tag{4.40}$$

Note the additional indices in the right side of the above equation. Hence, the only quantity to be determined is the class-specific feature distribution $p(\mathbf{m}|\omega_i, \mathcal{D}_i)$ given the matching partition of a specific dataset. This quantity can be calculated independently for each of the c classes and it is only required for matching indices of ω_i and $\mathcal{D}_i$. For this reason, the explicit notation of the class is omitted,

$$p(\mathbf{m}|\omega_i, \mathcal{D}_i) = p(\mathbf{m}|\mathcal{D}), \tag{4.41}$$

but it is implicitly stipulated that $\mathbf{m}$ is conditioned on the same class as the samples in $\mathcal{D}_i$.

Until now, no parameter vector θ has been introduced so far, but everything was based on the data $\mathcal{D}$. We now assume that the feature distribution has a known parametric form with an unknown parameter θ that is a random quantity. Then one can write

$$\begin{aligned} p(\mathbf{m}|\mathcal{D}) &= \int_{\Theta} p(\mathbf{m}, \theta|\mathcal{D})\, \mathrm{d}\theta \\ &= \int_{\Theta} p(\mathbf{m}|\theta, \mathcal{D}) p(\theta|\mathcal{D})\, \mathrm{d}\theta \\ &= \int_{\Theta} p(\mathbf{m}|\theta) p(\theta|\mathcal{D})\, \mathrm{d}\theta. \end{aligned} \tag{4.42}$$

The latter equation assumes that the distribution of the feature is conditionally independent of the data $\mathcal{D}$ given the parameter vector θ.

The open question is whether the last line of Equation (4.42) must be calculated every time and for each class when a new feature $\mathbf{m}$ is to be classified. (Recall that the indices were suppressed and that Equation (4.42) is only a sub-term in Equation (4.40).) Under certain conditions the answer is that this is not ultimately necessary and the calculation can be decoupled into two steps.

Assume the data $\mathcal{D}$ imply strong evidence for one singular parameter, i.e., the density $p(\theta|\mathcal{D})$ has a sharp and singular maximum at

$$\hat{\theta}(\mathcal{D}) = \underset{\theta \in \Theta}{\arg\max}\, p(\theta|\mathcal{D}). \tag{4.43}$$

Then the density can be approximated by $p(\theta|\mathcal{D}) \approx \delta(\theta - \hat{\theta})$, where δ denotes the Dirac distribution. It follows that

$$\begin{aligned} p(\mathbf{m}|\mathcal{D}) &= \int_{\Theta} p(\mathbf{m}|\theta) p(\theta|\mathcal{D})\, \mathrm{d}\theta \\ &\approx \int_{\Theta} p(\mathbf{m}|\theta) \delta(\theta - \hat{\theta})\, \mathrm{d}\theta = p(\mathbf{m}|\hat{\theta}) \end{aligned} \tag{4.44}$$

and the integral calculation can be avoided. In summary, the conditional feature distribution with respect to the dataset can be approximately replaced by a conditional feature distribution with respect to the parameter vector with the highest a posteriori distribution given the data.

The first example considers a univariate normal distribution with random expectation $\underline{\mu}$ but known variance σ^2, i.e., $\underline{m}_k \sim \mathcal{N}(\underline{\mu}, \sigma^2)$. The expectation is also normally distributed with $\underline{\mu} \sim \mathcal{N}(\mu_0, \sigma_0^2)$. We start with the calculation of the a posteriori distribution of $\underline{\mu}$ given the data,

$$\begin{aligned}
p(\mu|\mathcal{D}) &= \frac{p(\mathcal{D}|\mu)p(\mu)}{\int p(\mathcal{D}|\mu)p(\mu)\,\mathrm{d}\mu} \\
&\propto p(\mu)\prod_{k=1}^{N} p(m_k|\mu) \\
&\propto \exp\left\{-\frac{1}{2}\left(\frac{\mu-\mu_0}{\sigma_0}\right)^2\right\}\cdot\prod_{k=1}^{N}\exp\left\{-\frac{1}{2}\left(\frac{m_k-\mu}{\sigma}\right)^2\right\} \\
&\propto \exp\left\{-\frac{1}{2}\left[\left(\frac{\mu-\mu_0}{\sigma_0}\right)^2+\sum_{k=1}^{N}\left(\frac{m_k-\mu}{\sigma}\right)^2\right]\right\} \\
&\propto \exp\left\{-\frac{1}{2}\left[\left(\frac{N}{\sigma^2}+\frac{1}{\sigma_0^2}\right)\mu^2-2\left(\frac{1}{\sigma^2}\sum_{k=1}^{N}m_k+\frac{\mu_0}{\sigma_0^2}\right)\mu\right]\right\} \\
\Rightarrow p(\mu|\mathcal{D}) &= \alpha\exp\left\{-\frac{1}{2}\left(\frac{\mu-\mu_N}{\sigma_N}\right)^2\right\},
\end{aligned} \tag{4.45}$$

where the quantities in the last line are

$$\mu_N = \left(\frac{N\sigma_0^2}{N\sigma_0^2+\sigma^2}\right)\hat{\mu}_N + \frac{\sigma^2}{N\sigma_0^2+\sigma^2}\mu_0, \quad \text{with} \quad \hat{\mu}_N = \frac{1}{N}\sum_{k=1}^{N} m_k, \tag{4.46}$$

$$\sigma_N^2 = \frac{\sigma_0^2\sigma^2}{N\sigma_0^2+\sigma^2}, \quad \text{and} \tag{4.47}$$

$$\alpha = \frac{1}{\sqrt{2\pi}\sigma_N}. \tag{4.48}$$

The quantities μ_N, σ_N^2, and $\hat{\mu}_N$ can be found by comparing the coefficients of the last and the second last line. The factor α can be easily determined, because the last line shows that $p(\mu \mid \mathcal{D})$ is a Gaussian density.

Before we go on to finally calculate the feature distribution $p(m|\mathcal{D})$, let us discuss this intermediate result. The estimate of $\underline{\mu}$ given the data $\mathcal{D}$ is a Gaussian density on its own. We consider the two extreme cases with respect to the sample number N. If there is no sample, $N = 0$, then

$$\mu_N = \underbrace{\left(\frac{N\sigma_0^2}{N\sigma_0^2+\sigma^2}\right)}_{=0}\hat{\mu}_N + \underbrace{\frac{\sigma^2}{N\sigma_0^2+\sigma^2}}_{=1}\mu_0 = \mu_0 \quad \text{and} \tag{4.49}$$

$$\sigma_N^2 = \frac{\sigma_0^2\sigma^2}{N\sigma_0^2+\sigma^2} = \sigma_0^2. \tag{4.50}$$

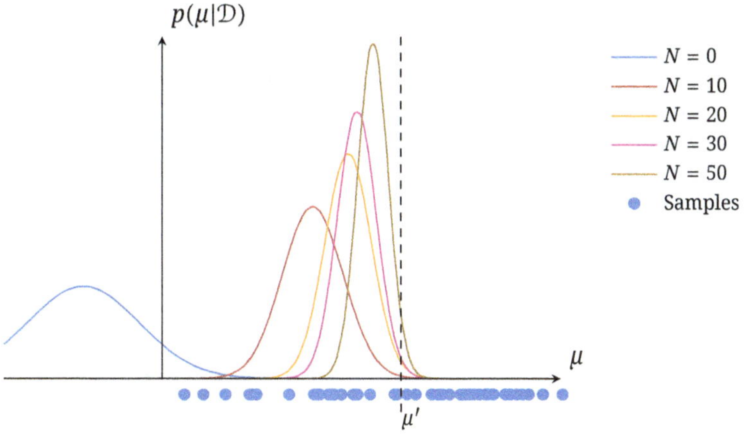

Fig. 4.2: Sequence of Bayesian a posteriori densities estimating the mean μ of a Gaussian distribution; the true Gaussian has $\mu' = 3$, $\sigma^2 = 2$, the prior distribution of $\underline{\mu}$ was assumed to be distributed with $\mu_0 = -1$ and $\sigma_0^2 = 0.5$.

This is a reasonable result, as the distribution of the best estimate equals the prior if no data is given. In contrast, as $N \to \infty$,

$$\lim_{N\to\infty} \mu_N = \lim_{N\to\infty} \underbrace{\left(\frac{N\sigma_0^2}{N\sigma_0^2 + \sigma^2}\right)}_{\to 1} \hat{\mu}_N + \lim_{N\to\infty} \underbrace{\frac{\sigma^2}{N\sigma_0^2 + \sigma^2}}_{\to 0} \mu_0 = \lim_{N\to\infty} \hat{\mu}_N = \lim_{N\to\infty} \frac{1}{N} \sum_{k=1}^{N} m_k, \tag{4.51}$$

$$\lim_{N\to\infty} \sigma_N^2 = \lim_{N\to\infty} \frac{\sigma_0^2 \sigma^2}{N\sigma_0^2 + \sigma^2} = 0. \tag{4.52}$$

In conclusion, for infinitely many samples, the uncertainty of the estimation vanishes and the a posteriori distribution converges to a Dirac distribution at the empirical mean of the samples. This means that any resemblance to the a priori assumption vanishes and the result depends solely on the data and actually equals the ML estimator. An example of such a sequence of a posteriori distributions is depicted in Figure 4.2.

To conclude the example, we must still calculate the conditional feature distribution given the dataset $p(m|\mathcal{D})$. As all densities are Gaussian, the calculation of Equation (4.42) needs little effort. Again, α denotes a universal normalizing constant in

$$\begin{aligned} p(m|\mathcal{D}) &= \int p(m|\mu)p(\mu|\mathcal{D})\,\mathrm{d}\mu \\ &= \alpha \int \exp\left\{-\frac{1}{2}\left(\frac{m-\mu}{\sigma}\right)^2\right\} \exp\left\{-\frac{1}{2}\left(\frac{\mu-\mu_N}{\sigma_N}\right)^2\right\} \mathrm{d}\mu \\ &= \alpha \exp\left\{-\frac{1}{2}\frac{(m-\mu_N)^2}{\sigma^2+\sigma_N^2}\right\}. \end{aligned} \tag{4.53}$$

In summary, $\underline{m} \sim \mathcal{N}(\mu_N, \sigma^2 + \sigma_N^2)$.

The next example is the multivariate case. It is a straight-forward generalization of the previous example. Let the samples be $\underline{\mathbf{m}}_k \sim \mathcal{N}(\boldsymbol{\mu}, \boldsymbol{\Sigma})$ and the mean be $\underline{\boldsymbol{\mu}} \sim \mathcal{N}(\boldsymbol{\mu}_0, \boldsymbol{\Sigma}_0)$. Similar to Equation (4.45), one obtains

$$
\begin{aligned}
p(\boldsymbol{\mu}|\mathcal{D}) &\propto p(\boldsymbol{\mu}) \prod_{k=1}^{N} p(\mathbf{m}_k|\boldsymbol{\mu}) \\
&\propto \exp\Big\{-\frac{1}{2}(\boldsymbol{\mu}-\boldsymbol{\mu}_0)^{\mathsf{T}}\boldsymbol{\Sigma}_0^{-1}(\boldsymbol{\mu}-\boldsymbol{\mu}_0)\Big\} \cdot \prod_{k=1}^{N} \exp\Big\{-\frac{1}{2}(\mathbf{m}_k-\boldsymbol{\mu})^{\mathsf{T}}\boldsymbol{\Sigma}^{-1}(\mathbf{m}_k-\boldsymbol{\mu})\Big\} \\
&\propto \exp\left\{-\frac{1}{2}\left[(\boldsymbol{\mu}-\boldsymbol{\mu}_0)^{\mathsf{T}}\boldsymbol{\Sigma}^{-1}(\boldsymbol{\mu}-\boldsymbol{\mu}_0) + \sum_{k=1}^{N}(\mathbf{m}_k-\boldsymbol{\mu})^{\mathsf{T}}\boldsymbol{\Sigma}^{-1}(\mathbf{m}_k-\boldsymbol{\mu})\right]\right\} \\
&\propto \exp\left\{-\frac{1}{2}\left[\boldsymbol{\mu}^{\mathsf{T}}(N\boldsymbol{\Sigma}^{-1}+\boldsymbol{\Sigma}_0^{-1})\boldsymbol{\mu} - 2\boldsymbol{\mu}^{\mathsf{T}}\Big(\boldsymbol{\Sigma}^{-1}\sum_{k=1}^{N}\mathbf{m}_k + \boldsymbol{\Sigma}_0^{-1}\boldsymbol{\mu}_0\Big)\right]\right\} \\
p(\boldsymbol{\mu}|\mathcal{D}) &= \alpha \exp\Big\{-\frac{1}{2}(\boldsymbol{\mu}-\boldsymbol{\mu}_N)^{\mathsf{T}}\boldsymbol{\Sigma}_N^{-1}(\boldsymbol{\mu}-\boldsymbol{\mu}_N)\Big\},
\end{aligned} \tag{4.54}
$$

where the quantities in the last line are

$$
\boldsymbol{\mu}_N = N\boldsymbol{\Sigma}_0(N\boldsymbol{\Sigma}_0+\boldsymbol{\Sigma})^{-1}\hat{\boldsymbol{\mu}}_N + \boldsymbol{\Sigma}(N\boldsymbol{\Sigma}_0+\boldsymbol{\Sigma})\boldsymbol{\mu}_0, \quad \text{with} \quad \hat{\boldsymbol{\mu}}_N = \frac{1}{N}\sum_{k=1}^{N}\mathbf{m}_k, \tag{4.55}
$$

$$
\boldsymbol{\Sigma}_N = \boldsymbol{\Sigma}_0\boldsymbol{\Sigma}(N\boldsymbol{\Sigma}_0+\boldsymbol{\Sigma})^{-1}, \quad \text{and} \tag{4.56}
$$

$$
\alpha = \frac{1}{\sqrt{2\pi \det \boldsymbol{\Sigma}_N}}. \tag{4.57}
$$

Analogously to Equation (4.53), the feature distribution equals

$$
\begin{aligned}
p(\mathbf{m}|\mathcal{D}) &= \int p(\mathbf{m}|\boldsymbol{\mu})p(\boldsymbol{\mu}|\mathcal{D})\,\mathrm{d}\boldsymbol{\mu} \\
&\propto \int \exp\Big\{-\frac{1}{2}(\mathbf{m}-\boldsymbol{\mu})^{\mathsf{T}}\boldsymbol{\Sigma}^{-1}(\mathbf{m}-\boldsymbol{\mu})\Big\} \cdot \exp\Big\{-\frac{1}{2}(\mu-\boldsymbol{\mu}_N)^{\mathsf{T}}\boldsymbol{\Sigma}_N^{-1}(\mathbf{m}-\boldsymbol{\mu})\Big\}\,\mathrm{d}\boldsymbol{\mu} \\
&\propto \exp\Big\{-\frac{1}{2}(\mathbf{m}-\boldsymbol{\mu}_N)^{\mathsf{T}}(\boldsymbol{\Sigma}+\boldsymbol{\Sigma}_N)^{-1}(\mathbf{m}-\boldsymbol{\mu}_N)\Big\}.
\end{aligned} \tag{4.58}
$$

In summary, the multivariate result is $\underline{\mathbf{m}} \sim \mathcal{N}(\boldsymbol{\mu}_N, \boldsymbol{\Sigma}+\boldsymbol{\Sigma}_N)$.

At the end of this section, the following list recapitulates the principal steps of the Bayesian approach to estimating the feature distribution. The implicit suppressed notation of the classes is re-introduced again. These steps must be performed for each class $\omega_i, i = 1, \ldots, c$:

1. $p(\mathbf{m}|\boldsymbol{\theta}_i, \omega_i)$ is assumed to be structurally known; the parameter vector $\underline{\boldsymbol{\theta}}_i$ is a random quantity, too.
2. $p(\boldsymbol{\theta}_i \mid \omega_i)$ includes the a priori knowledge about the $\underline{\boldsymbol{\theta}}_i$.

3. The dataset $\mathcal{D}_i$ bears the additional knowledge about $\underline{\theta}_i$; $\mathcal{D}_i$ is assumed to be a set of i.i.d. feature vectors $\underline{\mathbf{m}}_1, \ldots, \underline{\mathbf{m}}_{N_i} \sim p(\mathbf{m} \mid \theta_i, \omega_i)$,

$$p(\mathcal{D}_i|\theta_i, \omega_i) = \prod_{k=1}^{N_i} p(\mathbf{m}_k|\theta_i, \omega_i). \tag{4.59}$$

4. Then the class-specific feature distribution is calculated by

$$p(\theta_i|\mathcal{D}_i, \omega_i) = \frac{p(\mathcal{D}_i|\theta_i, \omega_i)p(\theta_i|\omega_i)}{\int p(\mathcal{D}_i|\theta_i, \omega_i)p(\theta_i|\omega_i)\,\mathrm{d}\theta_i} \tag{4.60}$$

   followed by

$$p(\mathbf{m}|\mathcal{D}, \omega_i) = \int p(\mathbf{m}|\theta_i, \omega_i)p(\theta_i|\mathcal{D}_i, \omega_i)\,\mathrm{d}\theta_i. \tag{4.61}$$

In comparison to maximum likelihood estimation, two essential differences can be captured as such:

- The Bayesian technique allows bringing in prior knowledge about the parameter vector; the maximum likelihood technique does not.
- An ML estimator is usually simpler to compute, because one needs to find an extremum only; the Bayesian technique requires the numerical calculation of multidimensional integrals.

4.3 Bayesian parameter estimation

The goal of the ML estimator is to find the best value of $\hat{\theta}$ that can be plugged into the parametric density $p(\mathbf{m} \mid \theta)$. In contrast, the Bayesian technique does not yield a single value of $\hat{\theta}$, but a whole a posteriori distribution $p(\theta|\mathcal{D})$. Hence, the classical approach and the Bayesian approach are not directly comparable. The class-specific feature distribution is calculated by

$$p(\mathbf{m}|\mathcal{D}) = \int p(\mathbf{m}|\theta)p(\theta|\mathcal{D})\,\mathrm{d}\theta \tag{4.62}$$

(see Equation (4.42)). As already stated in Section 4.2, this computational effort can be avoided if the a posteriori distribution can be approximated by a Dirac distribution. In this case, the additional integration degenerates into a simple replacement of θ by the value of $\hat{\theta}$ with the highest a posteriori probability. But setting $\hat{\theta} = \arg\max_{\theta \in \Theta} p(\theta|\mathcal{D})$ (see Equation (4.43)) is only one option for condensing a full density into a single value of the parameter. This section will present the two most important ways of Bayesian parameter estimation.

The basic approach is to find the estimate of $\hat{\theta}(\mathcal{D})$ such that the expectation

$$\mathrm{E}\left\{l(\hat{\theta}(\mathcal{D}), \underline{\theta})\right\} \tag{4.63}$$

is minimized, where $\underline{\theta}$ is the a posteriori parameter and l is a cost function.

4.3.1 Least squared estimation error

The cost function will be

$$l(\hat{\theta}, \underline{\theta}) = (\hat{\theta} - \underline{\theta})^{\mathsf{T}}(\hat{\theta} - \underline{\theta}). \tag{4.64}$$

Here and below, $\mathrm{d}\mathcal{D}$ abbreviates $\mathrm{d}\mathbf{m}_1 \dots \mathrm{d}\mathbf{m}_N$. The expected square error

$$\mathrm{E}\left\{\|\hat{\theta}(\mathcal{D}) - \underline{\theta}\|^2\right\} = \int\limits_{\mathcal{M}^N} \int\limits_{\Omega} (\hat{\theta}(\mathcal{D}) - \theta)^{\mathsf{T}}(\hat{\theta}(\mathcal{D}) - \theta) p(\theta|\mathcal{D}) \,\mathrm{d}\theta \,\mathrm{d}\mathcal{D} \tag{4.65}$$

is minimized if the integrand

$$I(\mathcal{D}) = \int\limits_{\Omega} (\hat{\theta}(\mathcal{D}) - \theta)^{\mathsf{T}}(\hat{\theta}(\mathcal{D}) - \theta) p(\theta|\mathcal{D}) \,\mathrm{d}\theta \tag{4.66}$$

is minimized point-wise for every $\mathcal{D} = \{\mathbf{m}_1, \dots, \mathbf{m}_N\}$. Under the assumption that integration and differentiation can be interchanged, the necessary condition for the minimum becomes

$$\mathbf{0} \stackrel{!}{=} \frac{\partial}{\partial \hat{\theta}(\mathcal{D})} I(\mathcal{D}) = \int\limits_{\Omega} 2(\hat{\theta}(\mathcal{D}) - \theta) p(\theta|\mathcal{D}) \,\mathrm{d}\theta \tag{4.67}$$

$$\Leftrightarrow \hat{\theta}(\mathcal{D}) = \mathrm{E}\{\underline{\theta} \,|\, \mathcal{D}\}\,. \tag{4.68}$$

This is a minimizing point if the second derivative is a positive definite matrix,

$$\frac{\partial}{\partial^2 \hat{\theta}(\mathcal{D})} I(\mathcal{D}) = \int\limits_{\Omega} 2\mathbf{U} p(\theta|\mathcal{D}) \,\mathrm{d}\theta = 2\mathbf{U}, \tag{4.69}$$

where $\mathbf{U}$ denotes the unit matrix, i.e., the matrix all of whose entries are unity. In summary, the estimator with the least quadratic error is

$$\hat{\theta}(\mathcal{D}) = \mathrm{E}\{\underline{\theta} \,|\, \mathcal{D}\}\,, \tag{4.70}$$

the a posteriori expectation of $\underline{\theta}$.

4.3.2 Constant penalty for failures

Now, the cost function will be

$$l(\hat{\theta}, \underline{\theta}) = \begin{cases} 0 & \text{if } \|\hat{\theta} - \underline{\theta}\| < \Delta \\ 1 & \text{else} \end{cases} \tag{4.71}$$

for an arbitrary but fixed $\Delta > 0$. An interesting special case is surely $\Delta = 0$. But as $\{\hat{\theta} = \underline{\theta}\}$ is a null set, the direct approach does not lead to any result.

Again, the expectation $\mathrm{E}\{l(\hat{\theta}, \underline{\theta})\}$ is minimized if the integrand

$$I(\mathcal{D}) = \int_{\Omega} l(\hat{\theta}, \theta) p(\theta|\mathcal{D}) \, \mathrm{d}\theta = \int_{\{\theta \mid \|\hat{\theta}(\mathcal{D}) - \theta\| > \Delta\}} p(\theta|\mathcal{D}) \, \mathrm{d}\theta = 1 - \int_{\{\theta \mid \|\hat{\theta}(\mathcal{D}) - \theta\| \leq \Delta\}} p(\theta|\mathcal{D}) \, \mathrm{d}\theta \tag{4.72}$$

is minimized point-wise. The last line is minimal if the integration is over a region where $p(\theta|\mathcal{D})$ is large. If Δ becomes small enough and if the density is sufficiently smooth, this is achieved for $\arg\max_{\theta} p(\theta \mid \mathcal{D})$. Hence, it follows that

$$\hat{\theta}(\mathcal{D}) = \arg\max_{\theta} p(\theta \mid \mathcal{D}) \tag{4.73}$$

for $\Delta \to 0$. This is called the maximum a posteriori estimator.

4.4 Additional remarks on Bayesian classification

Now, we briefly turn our attention back to Bayesian classification. With the results of Chapter 4, it is possible to discuss in greater depth the errors that arise in Bayesian classification.

Although Bayesian classification is the optimal classification, it is not free of errors. Basically, three different sources of errors can be distinguished:

Bayesian or indistinguishability error This error arises because of overlapping class-specific feature distributions. This means there are at least two classes whose objects might have the same feature values. Then the features are not perfectly discriminative. The solution to this problem is to find better features that are more powerful.

Model error In this case, an unsuitable parametric model was chosen to describe the class-specific feature distributions. For example, the features were assumed to be Gaussian distributed, but actually they are uniformly distributed. Of course, the parameter estimation techniques of this chapter generate a result, but for a non-fitting model. Finding a suitable model requires expert knowledge. There are options to test if a model seems to fit the data, but these tests cannot find a better model. Anyway, these tests are outside the scope of this textbook.

Estimation error This error occurs because there are too few samples compared to the number of parameters to give a reliable estimation. This kind of error can be diminished by either gathering more samples or by reducing the number of parameters. Normally, the former is naturally limited by costs, effort, or time. The latter can be limited by the quality of the new features.

4.5 Parameter estimation for Gaussian mixtures

For more complex distributions, estimating the parameters can be challenging. In Section 3.4, Gaussian mixtures have been presented as universal approximators. The density of a Gaussian mixture

$$p(\mathbf{m}) = \sum_{k=1}^{K} \pi_k \mathcal{N}(\mathbf{m}; \mu_k, \Sigma_k) \tag{4.74}$$

involves the weighted summation of individual Gaussians over all components of the mixture. As we will see in this section, this introduces a difficulty, when applied to the standard maximum likelihood approach as presented earlier in the chapter. Given a dataset $\mathcal{D} = \{\mathbf{m}_1, \ldots, \mathbf{m}_N\}$, the likelihood function conditioned on the mixing coefficients π, the component means μ, and the component variances Σ is given by

$$p(\mathcal{D} \mid \pi, \mu, \Sigma) = \prod_{n=1}^{N} p(\mathbf{m}_n \mid \pi, \mu, \Sigma), \tag{4.75}$$

leading to

$$l(\theta) = l(\mathcal{D} \mid \pi, \mu, \Sigma) = \sum_{n=1}^{N} \ln \sum_{k=1}^{K} \pi_k \mathcal{N}(\mathbf{m}_n \mid \mu_k, \Sigma_k) \tag{4.76}$$

as log-likelihood function. Directly minimizing this function poses several challenges. Firstly, the ML term from Equation (4.76) is analytically intractable due to the logarithm of a sum. Additionally, singularities arise when one of the means μ_k exactly falls on one of the datapoints $\mathbf{m}_i$, while the variance collapses for the respective component. As a consequence, the component's likelihood

$$\lim_{\sigma_k \to 0} \mathcal{N}\left(\mathbf{m}_i; \mathbf{m}_i, \sigma_k^2 \mathbf{I}\right) = \lim_{\sigma_k \to 0} \frac{1}{(2\pi)^{\frac{1}{2}}} \frac{1}{\sigma_k} = \infty \tag{4.77}$$

blows up the overall likelihood, as a single datapoint "explains" the whole mixture component. Note that this cannot happen in the case of a single Gaussian, as the variance won't collapse unless the entire dataset falls on the exact same point.

In general, the Bayesian approach discussed in Section 4.2 is also challenging to implement, since the number of terms in the likelihood undergo a combinatorial explosion. One possible solution to these difficulties is the expectation maximization (EM) algorithm, which optimizes the likelihood iteratively in a probabilistically sound manner.

The EM algorithm was originally introduced by Dempster et al. [1977]. It serves as a generic framework for iteratively fitting latent variables to a complex distribution. To do so, it alternates between estimating the current *expected* likelihood (E step) and adapting the distribution parameters θ to maximize the corresponding likelihood (M step). This process is iteratively repeated until no improvement on the parameters θ is observed anymore. The EM framework is widely used as a tool for estimating mixture parameters, as it provides numerical stability, as well as certain convergence and efficiency guarantees.

Given a fixed number of components, the main objective of the EM algorithm is to determine the parameter set $\theta = (\pi, \mu, \Sigma)$ that yields a maximum for the log-likelihood function $l(\theta)$ from Equation (4.76). This requires the calculation of the gradient

$$\nabla_\theta \equiv \left[\frac{\partial}{\partial \pi_k}, \frac{\partial}{\partial \mu_k}, \frac{\partial}{\partial \Sigma_k}\right] \tag{4.78}$$

and solving for

$$\nabla_\theta \ln p(\mathcal{D} \mid \theta) \stackrel{!}{=} 0. \tag{4.79}$$

As discussed above, the direct optimization of that term is not straightforward.

In abstract terms, when the EM algorithm is applied to mixture models, each observed datapoint is given an unobserved, latent counterpart representing its respective mixture component. This latent variable view on mixture models has already been established in Section 3.4, where a binary vector $\mathbf{z}$ of size K represents the assignment of the corresponding datapoint to the k'th mixture component. Within the latent variable view, the density of a mixture evaluates to

$$p(\mathbf{m}) = \sum_{\mathbf{z}} P(\mathbf{z}) p(\mathbf{m} \mid \mathbf{z}) = \sum_{k=1}^{K} \pi_k \mathcal{N}(\mathbf{m} \mid \mu_k, \Sigma_k). \tag{4.80}$$

Using this formulation, the a priori probability of choosing a particular component $P(\mathbf{z}_k = 1)$ corresponds to the mixing coefficient π_k of that component. The posterior of the k'th component being responsible for the datapoint $\mathbf{m}_n$ is also called the *responsibility* and is denoted by $\gamma(z_{nk}) = P(z_k = 1 \mid \mathbf{m}_n)$. These responsibilities are now used to derive an *expected likelihood* function $Q(\theta)$ as a lower bound for the log-likelihood function from Equation (4.76). Instead of maximizing $l(\theta)$ directly, the maximization is performed on

$$\begin{aligned} l(\theta) &= \sum_{n=1}^{N} \ln p(\mathbf{m}_n \mid \theta) \\ &= \sum_{n=1}^{N} \ln \sum_{\mathbf{z}_n} p(\mathbf{m}_n, \mathbf{z}_n \mid \theta) \\ &= \sum_{n=1}^{N} \ln \sum_{k=1}^{K} \gamma(z_{nk}) \frac{p(\mathbf{m}_n, \mathbf{z}_n \mid \theta)}{\gamma(z_{nk})} \\ &\geq \sum_{n=1}^{N} \sum_{k=1}^{K} \gamma(z_{nk}) \ln \frac{p(\mathbf{m}_n, \mathbf{z}_n \mid \theta)}{\gamma(z_{nk})} \\ &=: Q(\theta) \end{aligned} \tag{4.81}$$

as lower bound. The above inequality holds as consequence of *Jensen's inequality* for the logarithm as concave function.

Theorem 4.8 (Jensen's inequality (Jensen [1906])). Let $\underline{x}$ be an integrable random variable and $f : \mathbb{R} \to \mathbb{R}$ a concave function. Then, the inequality

$$f(\mathrm{E}\{\underline{x}\}) \geq \mathrm{E}\{f(\underline{x})\} \tag{4.82}$$

holds, if both the expectations are finite.

Note that Jensen's inequality is applicable since its argument

$$\sum_{k=1}^{K} \gamma(z_{nk}) \frac{p(\mathbf{m}_n, \mathbf{z}_n \mid \theta)}{\gamma(z_{nk})} = \sum_{\mathbf{z}_n} \gamma(\mathbf{z}_n) \frac{p(\mathbf{m}_n, \mathbf{z}_n \mid \theta)}{\gamma(\mathbf{z}_n)} = \mathrm{E}_{\mathbf{z}_n \sim \gamma(\mathbf{z}_n)} \left\{ \frac{p(\mathbf{m}_n, \mathbf{z}_n \mid \theta)}{\gamma(\mathbf{z}_n)} \right\} \tag{4.83}$$

is simply an expectation of $\frac{p(\mathbf{m}_n, \mathbf{z}_n \mid \theta)}{\gamma(z_{nk})}$ with respect to $\mathbf{z}_n$ drawn from the responsibilities $\gamma(\mathbf{z}_n)$.

Since $Q(\theta)$ acts as a lower bound on the likelihood function $l(\theta)$ for different values of $\gamma(\mathbf{z}_n)$, the question remains how to choose these. The main idea of the E step is to use the currently best parameter estimate $\hat{\theta}$ in order to find an estimate for $\gamma(\mathbf{z}_n)$ that makes the above inequality as tight as possible. It turns out that this is exactly the case if the responsibilities are set to be the posterior distribution of $\mathbf{z}_n$ given the samples $\mathbf{m}_n$ and the current parameter estimate $\hat{\theta}$, which appears

$$\gamma(\mathbf{z}_n) = p(\mathbf{z}_n \mid \mathbf{m}_n, \hat{\theta}) \tag{4.84}$$

as main result for the E step in the EM algorithm. With that choice for $\gamma(\mathbf{z}_n)$, the expectation in Equation (4.83) is taken over a constant quantity

$$c = \frac{p(\mathbf{m}_n, \mathbf{z}_n \mid \theta)}{\gamma(\mathbf{z}_n)} = \frac{p(\mathbf{m}_n, \mathbf{z}_n \mid \theta)}{p(\mathbf{z}_n \mid \mathbf{m}_n, \theta)} = p(\mathbf{m}_n \mid \theta) \tag{4.85}$$

which does not depend on $\mathbf{z}$. As a result, Jensen's inequality holds with equality, which is one of the key ideas behind the convergence properties of the EM algorithm. This is illustrated in Figure 4.3, where the likelihood function's lower bound equals the actual likelihood $l(\theta)$ for a specific choice of θ^t.

During the M step, the actual maximization on the lower bound $Q(\theta)$ is performed, resulting in a new set of parameters

$$\begin{aligned} \hat{\theta}^{t+1} &= \arg\max_{\hat{\theta}^t} Q(\hat{\theta}^t) \\ &= \arg\max_{\hat{\theta}^t} \sum_{n=1}^{N} \sum_{k=1}^{K} \gamma(z_{nk}) \ln \frac{p\left(\mathbf{m}_n, \mathbf{z}_n \mid \hat{\theta}^t\right)}{\gamma(z_{nk})}, \end{aligned}$$

where the superscript t denotes the iteration. In Figure 4.3, this corresponds to the gradient ascent on the lower bound, until a maximum is reached. The parameters $\hat{\theta}^{t+1}$ obtained at this particular maximum are used to construct an improved lower bound in the E step of the next iteration.

Maximizing the lower bound to obtain updated parameters $\hat{\theta}^{t+1}$ can be challenging as well. However, in the case of GMMs, it is possible to find closed-form solutions for the means μ, covariances $\boldsymbol{\Sigma}$ and mixture coefficients π. To find the updated means, the partial derivative

$$\frac{\partial}{\partial \mu_c} Q(\pi, \mu, \boldsymbol{\Sigma}) \stackrel{!}{=} \mathbf{0} \tag{4.86}$$

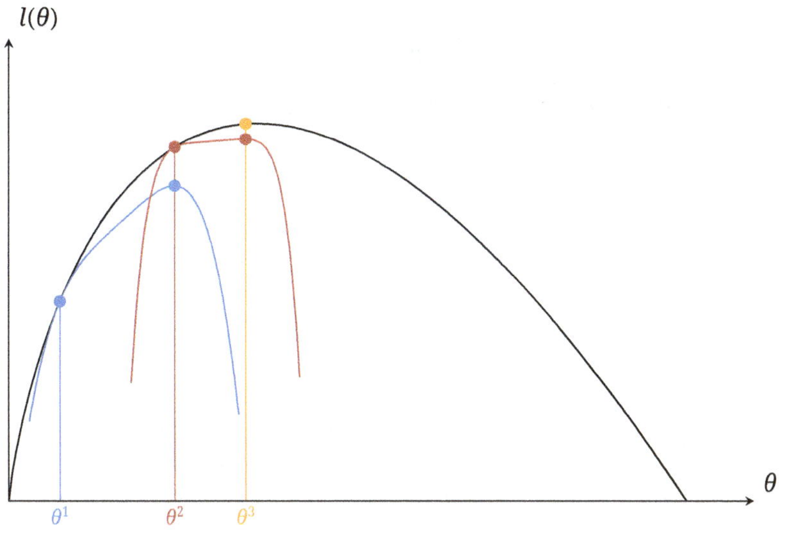

Fig. 4.3: Illustration of the EM algorithm. Each colored curve is a lower bound on $l(\theta^t)$ for different iterations. The choice of the responsibilities in the E step makes sure that the lower bounds align with $l(\theta^t)$ for the current choice of θ^t. The M step then finds the maximum of the lower bound in order to get a new set of parameters θ^{t+1}.

with respect to μ_c for some component c must equal zero.

Explicitly writing out the Gaussians yields

$$
\begin{aligned}
\frac{\partial}{\partial \mu_c} Q(\pi,\mu,\Sigma) &= \frac{\partial}{\partial \mu_c} \sum_{n=1}^{N} \sum_{\mathbf{z}_n} \gamma(\mathbf{z}_n) \ln \frac{p(\mathbf{m}_n,\mathbf{z}_n \mid \pi, \mu, \Sigma)}{\gamma(\mathbf{z}_n)} \\
&= \frac{\partial}{\partial \mu_c} \sum_{n=1}^{N} \sum_{k=1}^{K} \gamma(z_{nk}) \ln \frac{\pi_k \mathcal{N}(\mathbf{m}_n; \mu_k, \Sigma_k)}{\gamma(z_{nk})} \\
&= \frac{\partial}{\partial \mu_c} \sum_{n=1}^{N} \sum_{k=1}^{K} \gamma(z_{nk}) \\
&\qquad \ln \frac{\pi_k \left((2\pi)^{\frac{d}{2}} |\Sigma_k|^{\frac{1}{2}} \right)^{-1} \exp\left(-\frac{1}{2}(\mathbf{m}_n - \mu_k)^{\mathsf{T}} \Sigma_k^{-1} (\mathbf{m}_n - \mu_k)\right)}{\gamma(z_{nk})} \\
&= \frac{\partial}{\partial \mu_c} \sum_{n=1}^{N} \sum_{k=1}^{K} \gamma(z_{nk}) \left(-\frac{1}{2} (\mathbf{m}_n - \mu_k)^{\mathsf{T}} \Sigma_k^{-1} (\mathbf{m}_n - \mu_k) \right) \\
&\overset{(4.88)}{=} \sum_{n=1}^{N} \gamma(z_{nc}) \Sigma_c^{-1} (\mathbf{m}_n - \mu_c) \overset{!}{=} \mathbf{0},
\end{aligned}
\tag{4.87}
$$

which can now be solved for μ_c. Note that these calculations basically follow the standard maximum likelihood approach outlined in Equation (4.31). The last step results from the fact that

$$\frac{\partial}{\partial \mathbf{s}}(\mathbf{m}-\mathbf{s})^{\mathsf{T}}\mathbf{W}(\mathbf{m}-\mathbf{s}) = -2\mathbf{W}(\mathbf{m}-\mathbf{s}) \tag{4.88}$$

holds for symmetric matrices **W**. Similar to Equation (4.31), one arrives at

$$\begin{aligned}
\mathbf{0} &= \sum_{n=1}^{N} \gamma(z_{nc})\boldsymbol{\Sigma}_c^{-1}(\mathbf{m}_n - \boldsymbol{\mu}_c)\\
\mathbf{0} &= \sum_{n=1}^{N} \gamma(z_{nc})(\mathbf{m}_n - \boldsymbol{\mu}_c)\\
\Rightarrow \hat{\boldsymbol{\mu}}_c &= \frac{1}{\sum_{n=1}^{N}\gamma(z_{nc})}\sum_{n=1}^{N}\gamma(z_{nc})\mathbf{m}_n
\end{aligned} \tag{4.89}$$

as ML estimation for $\boldsymbol{\mu}_c$. This outcome closely resembles the standard maximum likelihood estimator for the mean of a single Gaussian. The key difference is that the averaging occurs over the effective number of datapoints

$$N_c := \sum_{n=1}^{N} \gamma(z_{nc}) \tag{4.90}$$

which are "involved" in explaining the c-th component. Subsequently, the estimation

$$\hat{\boldsymbol{\mu}}_c = \frac{1}{N_c}\sum_{n=1}^{N}\gamma(z_{nc})\mathbf{m}_n \tag{4.91}$$

basically becomes a weighted mean over all involved datapoints, where the weight is given by the respective responsibilities $\gamma(z_{nc})$.

Solving for the covariances $\hat{\boldsymbol{\Sigma}}_c$ that maximize the lower bound follows a similar line. Again, the partial derivative

$$\frac{\partial}{\partial \boldsymbol{\Sigma}_c} Q(\boldsymbol{\pi},\boldsymbol{\mu},\boldsymbol{\Sigma}) \stackrel{!}{=} \mathbf{0} \tag{4.92}$$

with respect to $\boldsymbol{\Sigma}_c$ for some component c must equal zero:

$$\begin{aligned}
\frac{\partial}{\partial \boldsymbol{\Sigma}_c} Q(\boldsymbol{\pi},\boldsymbol{\mu},\boldsymbol{\Sigma}) &= \frac{\partial}{\partial \boldsymbol{\Sigma}_c}\sum_{n=1}^{N}\sum_{\mathbf{z}_n}\gamma(\mathbf{z}_n)\ln\frac{p(\mathbf{m}_n,\mathbf{z}_n \mid \boldsymbol{\pi},\boldsymbol{\mu},\boldsymbol{\Sigma})}{\gamma(\mathbf{z}_n)}\\
&= \sum_{n=1}^{N}\gamma(z_{nc})\frac{\partial}{\partial \boldsymbol{\Sigma}_c}\ln \pi_c \mathcal{N}(\mathbf{m}_n;\boldsymbol{\mu}_c,\boldsymbol{\Sigma}_c)\\
&= \sum_{n=1}^{N}\gamma(z_{nc})\left[-\frac{1}{2}\frac{\partial}{\partial \boldsymbol{\Sigma}_c}\ln|\boldsymbol{\Sigma}_c| - \frac{1}{2}\frac{\partial}{\partial \boldsymbol{\Sigma}_c}(\mathbf{m}_n-\boldsymbol{\mu}_c)^{\mathsf{T}}\boldsymbol{\Sigma}_c^{-1}(\mathbf{m}_n-\boldsymbol{\mu}_c)\right]\\
&\stackrel{(4.96)}{=} -\frac{1}{2}\left(\sum_{n=1}^{N}\gamma(z_{nc})\right)\boldsymbol{\Sigma}_c^{-1} + \frac{1}{2}\boldsymbol{\Sigma}_c^{-1}\left(\sum_{n=1}^{N}\gamma(z_{nc})\,(\mathbf{m}_n-\boldsymbol{\mu}_c)\,(\mathbf{m}_n-\boldsymbol{\mu}_c)^{\mathsf{T}}\right)\boldsymbol{\Sigma}_c^{-1}\\
&\stackrel{!}{=} \mathbf{0}.
\end{aligned} \tag{4.93}$$

This is essentially a more general matrix form of Equation (4.32). The last step makes use of the fact that

$$\frac{\partial}{\partial \mathbf{W}} \ln |\mathbf{W}| = \left(\mathbf{W}^{\mathsf{T}}\right)^{-1} \tag{4.94}$$

applies for the derivative of the covariance matrix and that $\boldsymbol{\Sigma}^{\mathsf{T}} = \boldsymbol{\Sigma}$ holds because of its symmetry. In addition,

$$\frac{\partial}{\partial \mathbf{W}} \mathbf{a}^{\mathsf{T}} \mathbf{W}^{-1} \mathbf{b} = -(\mathbf{W}^{-1})^{\mathsf{T}} \mathbf{a} \mathbf{b}^{\mathsf{T}} (\mathbf{W}^{-1})^{\mathsf{T}} \tag{4.95}$$

applies for the derivative for an inverse and hence

$$\frac{\partial}{\partial \boldsymbol{\Sigma}_c} (\mathbf{m}_n - \boldsymbol{\mu}_c)^{\mathsf{T}} \boldsymbol{\Sigma}_c^{-1} (\mathbf{m}_n - \boldsymbol{\mu}_c) = -\boldsymbol{\Sigma}_c^{-1} (\mathbf{m}_n - \boldsymbol{\mu}_c)(\mathbf{m}_n - \boldsymbol{\mu}_c)^{\mathsf{T}} \boldsymbol{\Sigma}_c^{-1} \tag{4.96}$$

follows for the symmetric covariance $\boldsymbol{\Sigma}_c$. Solving Equation (4.93) for the covariance yields

$$\begin{aligned}
\mathbf{0} &\overset{!}{=} -\frac{1}{2} \left(\sum_{n=1}^{N} \gamma(z_{nc}) \right) \boldsymbol{\Sigma}_c^{-1} + \frac{1}{2} \boldsymbol{\Sigma}_c^{-1} \left(\sum_{n=1}^{N} \gamma(z_{nc}) (\mathbf{m}_n - \boldsymbol{\mu}_c)(\mathbf{m}_n - \boldsymbol{\mu}_c)^{\mathsf{T}} \right) \boldsymbol{\Sigma}_c^{-1} \\
\mathbf{0} &= \left(\sum_{n=1}^{N} \gamma(z_{nc}) \right) \mathbf{I} - \left(\sum_{n=1}^{N} \gamma(z_{nc}) (\mathbf{m}_n - \boldsymbol{\mu}_c)(\mathbf{m}_n - \boldsymbol{\mu}_c)^{\mathsf{T}} \right) \boldsymbol{\Sigma}_c^{-1} \\
\Rightarrow \hat{\boldsymbol{\Sigma}}_c &= \frac{1}{N_c} \sum_{n=1}^{N} \gamma(z_{nc}) (\mathbf{m}_n - \boldsymbol{\mu}_c)(\mathbf{m}_n - \boldsymbol{\mu}_c)^{\mathsf{T}},
\end{aligned} \tag{4.97}$$

which once more resembles the outcome obtained in Equation (4.36), but again weighted with the responsibilities.

The maximization of the mixing coefficients $\boldsymbol{\pi}$ is slightly more difficult as they undergo the additional constraint $\sum_{k=1}^{K} \pi_k = 1$. Therefore, the Lagrangian form is utilized, resulting in the optimization objective

$$L(\boldsymbol{\pi}) = \sum_{n=1}^{N} \sum_{k=1}^{K} \gamma(z_{nk}) \ln \frac{\pi_k \mathcal{N}(\mathbf{m}_n; \boldsymbol{\mu}_k, \boldsymbol{\Sigma}_k)}{\gamma(z_{nk})} + \lambda \left(\sum_{k=1}^{K} \pi_k - 1 \right) \tag{4.98}$$

with λ as Lagrange multiplier. Setting the derivative of the Lagrangian with respect to some π_c equal to zero

$$\begin{aligned}
\frac{\partial}{\partial \pi_c} L(\boldsymbol{\pi}) &= \frac{\partial}{\partial \pi_c} \sum_{n=1}^{N} \sum_{k=1}^{K} \gamma(z_{nk}) \ln \pi_k + \lambda \left(\sum_{k=1}^{K} \pi_k - 1 \right) \\
&= \sum_{n=1}^{N} \frac{\gamma(z_{nc})}{\pi_c} + \lambda \\
&\overset{!}{=} 0
\end{aligned} \tag{4.99}$$

gives

$$\hat{\pi}_c = -\frac{1}{\lambda}\sum_{n=1}^{N}\gamma(z_{nc}) \tag{4.100}$$

as solution to the Lagrangian. Plugging this result back into the constraint yields

$$\begin{aligned}\sum_{c=1}^{K} -\frac{1}{\lambda}\sum_{n=1}^{N}\gamma(z_{nc}) &\stackrel{!}{=} 1\\ -\frac{1}{\lambda}\sum_{n=1}^{N}\sum_{c=1}^{K}\gamma(z_{nc}) &= 1\\ \Rightarrow -N &= \lambda\end{aligned} \tag{4.101}$$

since the probabilities $\gamma(z_{nc})$ sum to 1. Inserted back into the original form one obtains

$$\begin{aligned}\sum_{n=1}^{N}\frac{\gamma(z_{nc})}{\pi_c} - N &\stackrel{!}{=} 0\\ \Rightarrow \hat{\pi}_c &= \frac{N_c}{N}\end{aligned} \tag{4.102}$$

as result for the mixing coefficients. In other words, the c-th mixing coefficient is simply the average responsibility taken by the c-th component.

With the most important derivations for the EM algorithm in place, an overall routine can be defined to maximize the likelihood function of a Gaussian mixture model with respect to means μ, covariances $\boldsymbol{\Sigma}$ and mixing coefficients π:

1. Initialize model parameters $\hat{\theta}^0 = \left(\hat{\pi}^0, \hat{\mu}^0, \hat{\boldsymbol{\Sigma}}^0\right)$ with random values.
2. In the E step, use the parameters of the current iteration t to calculate the responsibilities

$$\gamma^t(z_{nc}) = p\left(\mathbf{z}_n \mid \mathbf{m}_n, \theta^t\right) = \frac{\hat{\pi}_c^t \mathcal{N}\left(\mathbf{m}_n; \hat{\mu}_c^t, \hat{\boldsymbol{\Sigma}}_c^t\right)}{\sum_{k=1}^{K}\hat{\pi}_k^t \mathcal{N}\left(\mathbf{m}_n; \hat{\mu}_k^t, \hat{\boldsymbol{\Sigma}}_k^t\right)}. \tag{4.103}$$

3. In the M step, find new parameters $\hat{\theta}^{t+1}$ by solving

$$\hat{\theta}^{t+1} = \underset{\hat{\theta}^t}{\arg\max}\, Q(\hat{\theta}^t), \tag{4.104}$$

   using a maximum likelihood estimator for $\hat{\theta}^{t+1} = \left(\hat{\pi}^{t+1}, \hat{\mu}^{t+1}, \hat{\boldsymbol{\Sigma}}^{t+1}\right)$:

$$\begin{aligned}\hat{\mu}_c^{t+1} &= \frac{1}{N_c^t}\sum_{n=1}^{N}\gamma^t(z_{nc})\mathbf{m}_n\\ \hat{\boldsymbol{\Sigma}}_c^{t+1} &= \frac{1}{N_c^t}\sum_{n=1}^{N}\gamma^t(z_{nc})(\mathbf{m}_n - \hat{\mu}_c^t)(\mathbf{m}_n - \hat{\mu}_c^t)^\top\\ \hat{\pi}_c^{t+1} &= \frac{N_c^t}{N}.\end{aligned} \tag{4.105}$$

4. Evaluate the log-likelihood

$$l\left(\hat{\theta}^{t+1}\right) = \ln p\left(\mathcal{D} \mid \hat{\pi}^{t+1}, \hat{\mu}^{t+1}, \hat{\Sigma}^{t+1}\right) \tag{4.106}$$

and check for convergence by evaluating

$$l\left(\hat{\theta}^{t+1}\right) - l\left(\hat{\theta}^{t}\right) < \varepsilon \tag{4.107}$$

for some threshold ε.
5. If not converged, set $t = t + 1$ and continue with the E step.

This routine allows for the estimation of parameters for a Gaussian mixture model in an iterative fashion. Although it can be challenging to derive explicit update rules for other distributions, it is not restricted to GMMs. More details on utilizing other distributions with the EM algorithm can be obtained from Bishop [2007]. In general, the EM algorithm serves as a robust framework to manage unobserved (latent) data, while guaranteeing to maximize the likelihood of observed data. Due to its adaptivity and its solid probabilistic foundation, the EM framework is employed in various applications, from clustering to the estimation of parameters in hidden Markov models. The *Baum–Welch* algorithm, which will be presented in Section 7.9.2, is also a variation of the EM algorithm.

An example of the clustering abilities of the EM algorithm is shown in Figure 4.4. The 50 one-dimensional observations $\mathcal{D}$ are sampled from a GMM consisting of three components with means of $\mu = [7, 12, 20]$ and covariances of $\Sigma = [1.0, 1.1, 1.3]$ with equal mixing coefficients $\pi_1 = \pi_2 = \pi_3 = \frac{1}{3}$. The objective is to identify, which mixture component generated a given datapoint with all parameters unknown. The EM algorithm converges within 10 iterations and accurately reconstructs the original means.

However, there remain several unresolved issues regarding the practical application of the EM algorithm. First, the EM algorithm is not guaranteed to converge to a global optimum. In the illustration of Figure 4.3, the log-likelihood function may not always be strictly concave. The simplest way to approach this limitation is to run the EM algorithm with different sets of initial parameters to avoid getting stuck in a local maximum with a particular initialization. Additionally, one may use the result of a different parameter estimation algorithm such as *k-means* as initialization for θ. Another challenge is determining the optimal number of components, which is heavily influenced by both the application and the underlying data. Therefore, there is no universal rule, although some heuristics such as the *Bayesian information criterion* exist, which is out of scope for this book.

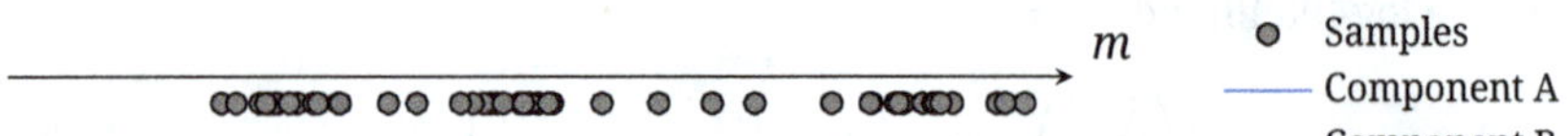

(a) Original data

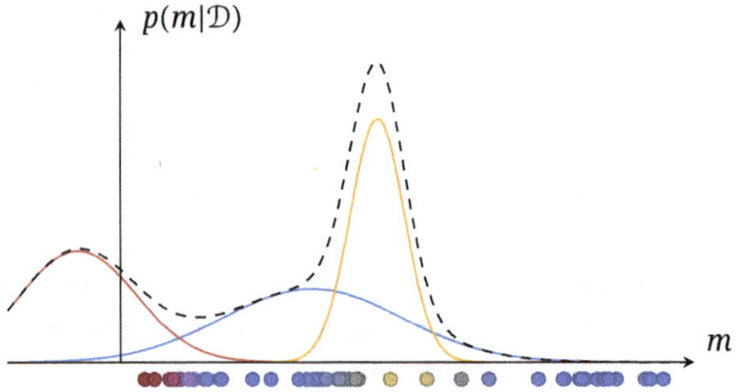

(b) Distribution after random initialization

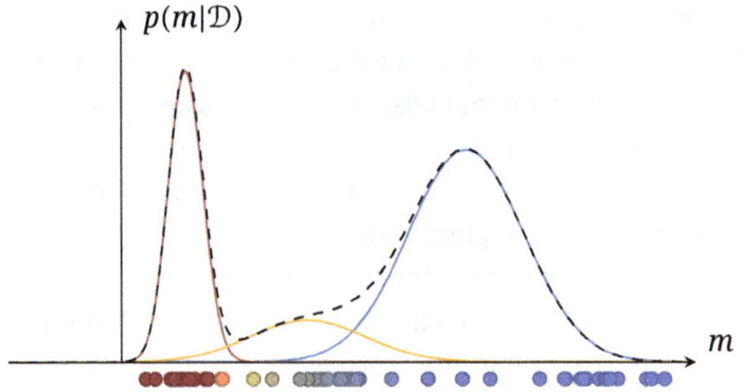

(c) Distribution after 3 iterations of EM algorithm

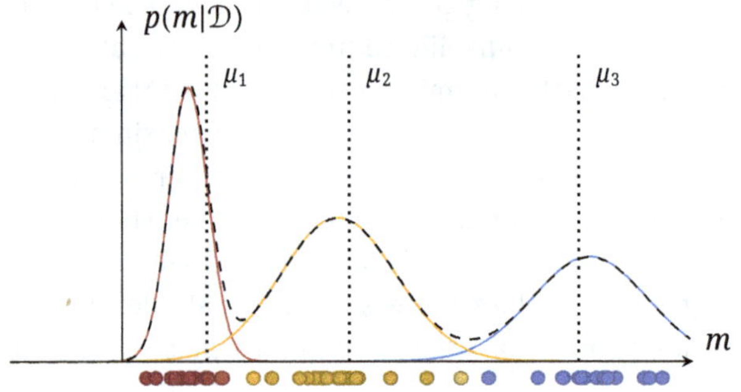

(d) Distribution after 10 iterations of EM algorithm

Fig. 4.4: Different iterations of the EM algorithm on a one-dimensional dataset. The points are colored with gradients of the mixture components to indicate the responsibility of each component for the points. Vertical lines in the last plot show the positions of the means μ which were used to sample the dataset.

4.6 Exercises

(4.1) The weight of a letter m in grams varies between $m = 10$ and $m = 20$. There are two possibilities for estimating the weight of a given letter:

- Estimate $\hat{m}_1 = 15$, independently of the true weight of the letter.
- Estimate $\hat{m}_2 = x$, where x is the display of an inaccurate scale with $\mathrm{E}\{x\} = m$ and $\mathrm{Var}\{m\} = 36$.

How large is the mean squared error (MSE) for each estimator? Which estimator has the smaller MSE?

(4.2) Let $m_1, \ldots, m_N$ be a sample of N i.i.d. elements that follow a Laplace distribution with the density

$$p(m) = \frac{1}{2\sigma} \exp\left(-\frac{|m-\mu|}{\sigma}\right). \tag{4.108}$$

Show that the following estimator is a maximum likelihood estimator for the parameter μ:

$$\hat{\mu} := \arg\min_{\mu} \sum_{i=1}^{N} |m_i - \mu|. \tag{4.109}$$

(4.3) Let the random variable $\underline{x}$ be distributed with the density

$$p(x \mid \theta) = \theta \mathrm{e}^{-\theta x} \tag{4.110}$$

n with unknown parameter $\theta > 0$. An i.i.d. sample $x_1, \ldots, x_N$ of size N is drawn from this distribution. Calculate the maximum likelihood estimator $\hat{\theta}_{\mathrm{ML}}$ of θ.

(4.4) Let $\underline{x}$ be a random variable over a population with expectation μ and variance σ^2. Further, let $x_1, \ldots, x_N$ be an i.i.d. sample of size $N > 4$ over the population. The following estimator of the expected value μ of $\underline{x}$ is proposed:

$$\hat{\mu} := \frac{1}{N-4} \sum_{i=3}^{N-2} x_i, \tag{4.111}$$

i.e., the first and last two elements of the sample are discarded.

1. Show that $\hat{\mu}$ is an unbiased estimator of μ.
2. Is $\hat{\mu}$ a better estimator than the maximum likelihood estimator

$$\hat{\mu}_{\mathrm{ML}} = \frac{1}{N} \sum_{i=1}^{N} x_i\,? \tag{4.112}$$

(4.5) Let $m_1, \ldots, m_N$ be a sample of N i.i.d. elements and consider the following estimator of the sample variance $\sigma^2 = \mathrm{Var}\{m\}$:

$$\widehat{\sigma^2} = \frac{1}{a-N} \sum_{i=1}^{N} (m_i - \mu)^2, \tag{4.113}$$

where $\mu = \mathrm{E}\{m\}$. For which values of a will $\widehat{\sigma^2}$ be an unbiased estimator of σ^2?

(4.6) Let $m_1, \ldots, m_N$ be a sample of N i.i.d. elements and consider the following estimator of the expected value $\mu = \mathrm{E}\{f(m)\}$:

$$\hat{\mu} = \frac{N}{N-a} \sum_{i=1}^{N} f(m_i), \qquad (4.114)$$

for some function $f(\cdot)$. For which values of a will $\hat{\mu}$ be an unbiased estimator of μ?

5 Parameter free methods

At the beginning of this chapter we will first review what has been done so far. The principal goal is to assign a class to an unknown object given its features and a training set of objects with known features and classes. From a more abstract point of view, one wants to learn some kind of rule, given a finite training sample of special cases. The rule will then be applied to a new situation, where one hopes that the proposed rule has general significance.

This is a two-step process: in the first step, the general rule must be found from specific instantiations, the second step is to apply the (hopefully) general rule to a new specific situation. If necessary, the rule found from the first step will need some intermediate formal rewriting into a form that is applicable in the second step. The first step, from the special to the general, is called "induction", the second step, from the general to the special, is called "deduction" (see Figure 5.1). In the context of this textbook, the induction is to find a class-specific feature distribution given a dataset $\mathcal{D}$; the deduction is to apply the a posteriori probability to an unknown feature vector. The necessary formal rewriting is Bayes' law in order to obtain the a posteriori probability.

Instead of following this indirection, it is sometimes possible to directly infer from the given data to the new situation. The term "transduction" was introduced by Vapnik in the context of support vector machines (see Section 7.7) to describe this shortcut. In this chapter, however, we are concerned with the induction step.

The induction step from $\mathcal{D}$ to $p(\mathbf{m}|\omega)$ is an ill-posed inverse problem. For a deeper understanding of inverse problems, see, e.g., the work of Aster et al. [2013] or Rieder [2003]. For our purposes, the following intuition is sufficient: The induction is called "inverse", because the deduction is thought of as the forward model. It is ill-posed if one of the following conditions (going back to Jacques Hadamard) holds:

- The inverse mapping is not well defined,
- the inverse mapping is not unique, or
- the inverse mapping is not continuous.

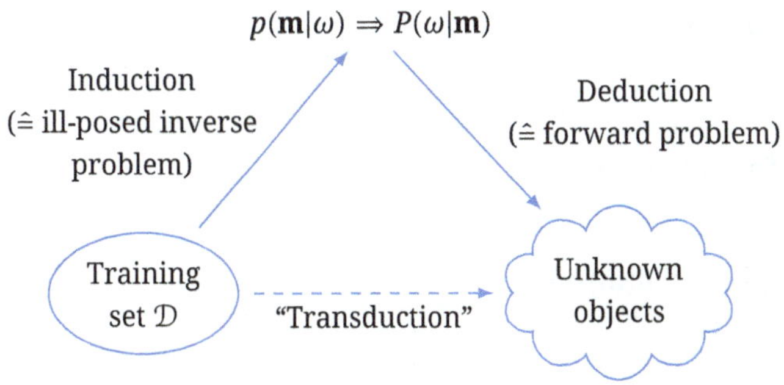

Fig. 5.1: The triangle of inference.

https://doi.org/10.1515/9783111339207-005

In this context, the dataset poses only a finite number of conditions on the infinite-dimensional solution space of all density functions. This means that in general, the data does not suffice to determine a solution. Regularization, i.e., enforcing further restrictions on the space of solutions, can offer a way out. Such additional restrictions might be

- to make additional assumptions, e.g., on the range of parameters,
- to bring in additional prior knowledge, and
- to formulate desirable traits of the solution as auxiliary constraints.

The risk of regularization lies in restricting the space of permissible solutions in such a way that the true solution is unintentionally excluded. In the previous chapter, the class-specific feature distribution was assumed to be structurally known. Hence, the space of all densities was restricted to a finitely parametrized family of densities. Unfortunately, there are only a handful of standard densities that are still analytically and computationally feasible. But it is questionable how well these densities fit the applications. Especially the assumption that a multi-dimensional feature space is governed by a product of simple densities seems bold.

Although this chapter still considers the induction step and tries to find a density $p(\mathbf{m} \mid \omega)$, it follows a totally different approach. The parameter-free methods do not constitute a specific form of the density right from the beginning, but try to look at the samples as a kind of discrete approximation of the true density. While the number of samples increases, a sequence of densities is created that eventually converges to the true density. This textbook will present two such methods: the Parzen window method and the k-nearest neighbor method[1].

Let $\underline{\mathbf{m}}$ denote a random vector, $p(\mathbf{m})$ its density, $\mathbf{m}$ a realization of $\underline{\mathbf{m}}$, and $\mathcal{A}_{\mathbf{m}} \subseteq \mathcal{M}$ a neighborhood around $\mathbf{m}$. Eventually, $p(\mathbf{m})$ will be the unknown, true density we want to approximate. Then

$$P_{\mathbf{m}} = \int_{\mathcal{A}_{\mathbf{m}}} p(\breve{\mathbf{m}})\,\mathrm{d}\breve{\mathbf{m}} \tag{5.1}$$

is the probability that $\underline{\mathbf{m}}$ attains a value in $\mathcal{A}_{\mathbf{m}}$. Let V denote the volume of $\mathcal{A}_{\mathbf{m}}$.

$$\breve{p}(\mathbf{m}) = \frac{P_{\mathbf{m}}}{V} = \frac{1}{V} \int_{\mathcal{A}_{\mathbf{m}}} p(\breve{\mathbf{m}})\,\mathrm{d}\breve{\mathbf{m}} \tag{5.2}$$

defines a new density. In order to check this, let $\mathcal{A}_{\breve{\mathbf{m}}} = \{\mathbf{m} | \mathcal{A}_{\mathbf{m}} \ni \breve{\mathbf{m}}\}$ denote the *inverse* neighborhood, i.e., the set of all vectors $\mathbf{m}$ whose neighborhoods $\mathcal{A}_{\mathbf{m}}$ contain $\breve{\mathbf{m}}$. One can calculate

$$\int_{\mathcal{M}} \breve{p}(\mathbf{m})\,\mathrm{d}\mathbf{m} = \int_{\mathcal{M}} \int_{\mathcal{A}_{\mathbf{m}}} \frac{1}{V} p(\breve{\mathbf{m}})\,\mathrm{d}\breve{\mathbf{m}}\,\mathrm{d}\mathbf{m} = \int_{\mathcal{M}} \int_{\mathcal{A}_{\breve{\mathbf{m}}}} \frac{1}{V} p(\breve{\mathbf{m}})\,\mathrm{d}\mathbf{m}\,\mathrm{d}\breve{\mathbf{m}}$$

1 Not to be confused with the k-nearest neighbor classifier.

$$= \int_{\mathcal{M}} \frac{1}{V} p(\breve{\mathbf{m}}) \underbrace{\int_{\mathcal{A}_{\breve{\mathbf{m}}}} d\mathbf{m}}_{=V} d\breve{\mathbf{m}} = \int_{\mathcal{M}} p(\breve{\mathbf{m}})\, d\breve{\mathbf{m}} = 1. \tag{5.3}$$

Without a proof, we claim that $\breve{p}(\mathbf{m}) \to p(\mathbf{m})$ if $V \to 0$ for all $\mathbf{m}$ where $p(\mathbf{m})$ is continuous.

In summary, $\breve{p}(\mathbf{m})$ is a density that represents a moving average of $p(\mathbf{m})$. The moving average converges to the true density if the neighborhood over which the average is calculated shrinks. Hence, for reasonably small volumes V one can set

$$p(\mathbf{m}) \approx \breve{p}(\mathbf{m}) = \frac{P_{\mathbf{m}}}{V}. \tag{5.4}$$

Consequently, the problem is reduced to finding estimates of the probability $P_{\mathbf{m}}$ for each volume V respectively for each neighborhood $\mathcal{A}_{\mathbf{m}}$.

Assume $\underline{\mathbf{m}}_1, \ldots, \underline{\mathbf{m}}_N \sim p(\mathbf{m})$ are i.i.d. and let $\underline{k}$ be the random variable denoting the number of $\underline{\mathbf{m}}_i$ falling into $\mathcal{A}_{\mathbf{m}}$. Then $\underline{k}$ is binomially distributed with probability mass function

$$P(k \mid P_{\mathbf{m}}) = \binom{N}{k} P_{\mathbf{m}}^k (1 - P_{\mathbf{m}})^{N-k} \tag{5.5}$$

and the unknown parameter $P_{\mathbf{m}}$. Given a realization k of $\underline{k}$, a good estimator (the unbiased ML estimator) is given by

$$\hat{P}_{\mathbf{m}} = \frac{k}{N} \tag{5.6}$$

and hence one can set

$$p(\mathbf{m}) \approx \frac{k/N}{V} = \frac{k}{NV} = \hat{p}(\mathbf{m}). \tag{5.7}$$

Note that the term $\frac{k}{NV}$ on the right hand side actually depends on $\mathbf{m}$, because k is the number of points within the neighborhood $\mathcal{A}_{\mathbf{m}}$ and V is its volume. For the rest of this discussion, the dependence on the center point $\mathbf{m}$ is omitted, but we write $k = k_{N,V}$ to indicate that the number of samples in $\mathcal{A}$ depends on the total number of samples N and on the volume V of the neighborhood.

The overall approximation is based on the nested limiting process

$$p(\mathbf{m}) = \lim_{V \to 0} \frac{P_V}{V} = \lim_{V \to 0} \lim_{N \to \infty} \frac{k_{N,V}}{NV}, \quad \text{where} \tag{5.8}$$

$$P_V := \frac{k_{N,V}}{N}. \tag{5.9}$$

The convergence of the outer limit requires $P_V \to 0$, which means that

$$\lim_{V \to 0} \lim_{N \to \infty} \frac{k}{N} = 0. \tag{5.10}$$

For any fixed V, however, the convergence of the inner limit requires $k \to \infty$ for $N \to \infty$.

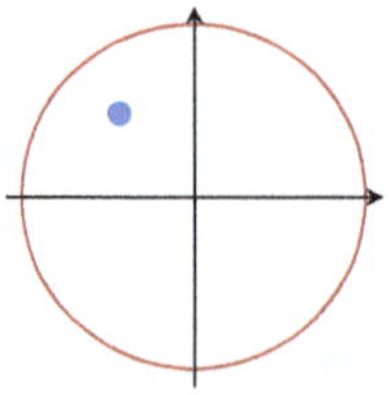

(a) PW, $N = 1, r = 1^{-\frac{1}{4}} = 1 \rightsquigarrow k = 1$

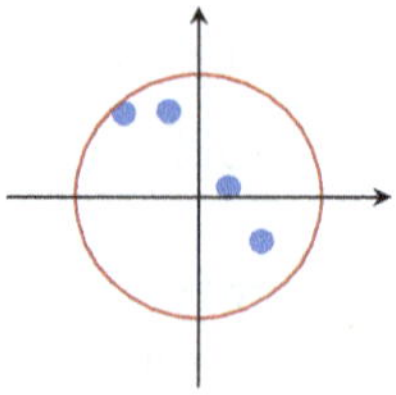

(b) PW, $N = 4, r = 4^{-\frac{1}{4}} \approx 0.71 \rightsquigarrow k = 4$

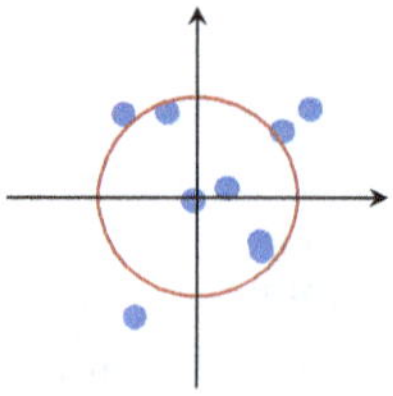

(c) PW, $N = 9, r = 9^{-\frac{1}{4}} \approx 0.58 \rightsquigarrow k = 5$

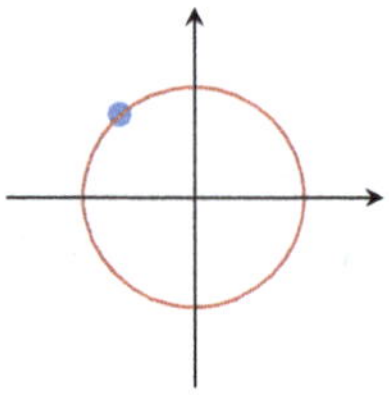

(d) k-NN, $N = 1, k = \sqrt{1} = 1 \rightsquigarrow r \approx 0.64$

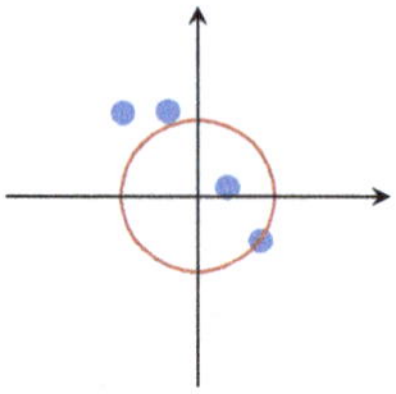

(e) k-NN, $N = 4, k = \sqrt{4} = 2 \rightsquigarrow r \approx 0.44$

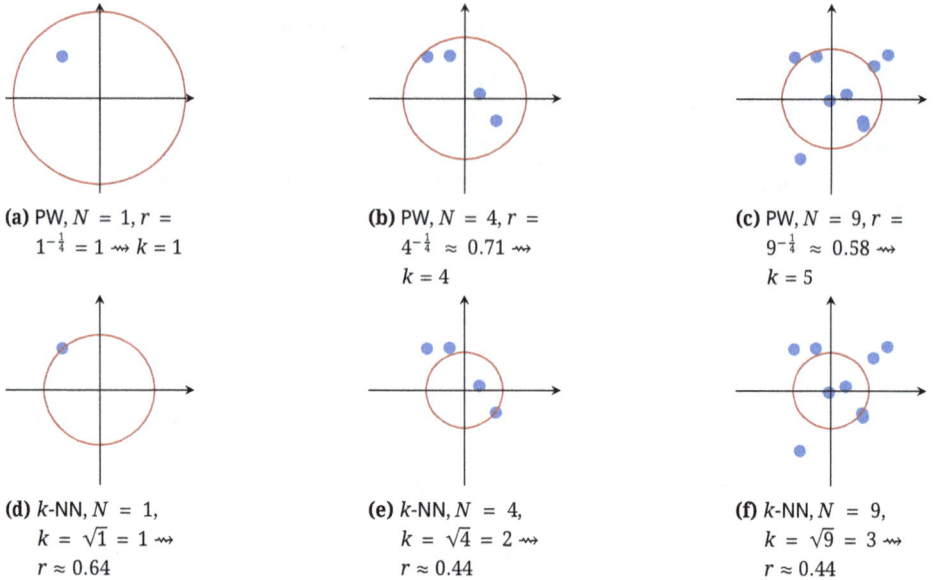

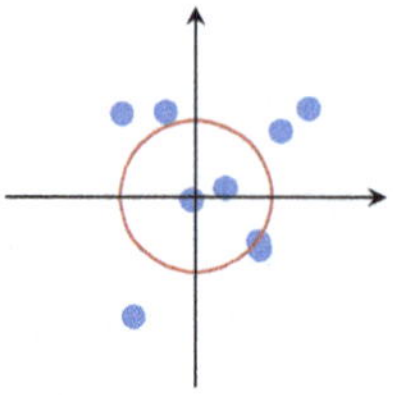

(f) k-NN, $N = 9, k = \sqrt{9} = 3 \rightsquigarrow r \approx 0.44$

Fig. 5.2: Comparison of neighborhood size V and sample size k between Parzen window (PW) and k-nearest neighbor (k-NN) density estimation.

The aforementioned conditions are only necessary but not sufficient conditions to ensure convergence. In particular, the limits must change position, but without any additional assumptions, one has

$$p(\mathbf{m}) = \lim_{V \to 0} \lim_{N \to \infty} \frac{k_{N,V}}{NV} \neq \lim_{N \to \infty} \lim_{V \to 0} \frac{k_{N,V}}{NV} = 0. \tag{5.11}$$

The latter is easy to see: If N is arbitrary but fixed at some bound, the volume V is so small that all the points are located outside and k becomes constantly zero.

In theory this is no problem, because one can first define a sequence of decreasing volumes (outer limit) and then take as many samples as necessary to get a good approximation of $\lim {}^{k}/_{N}$ (inner limit). In practice, the situation is more complicated. Generally, the number of samples N is given in advance or is at least bounded and there is normally no option for getting a fresh sequence of samples for each volume V. Hence the question is: what is a reasonable size for V, given some samples?

Choosing a large V helps to get a reliable approximation for $P_V = \frac{k_{N,V}}{N}$, because there are many samples that fall in $\mathcal{A}$. But unfortunately, the outer approximation $\frac{P_V}{V}$ becomes too coarse. In the extreme case, the neighborhood $\mathcal{A} = \mathcal{M}$ equals the whole support. Then $k = N$, because all the samples must fall in $\mathcal{M}$ and $P_{\mathbf{m}} = 1$ is a perfect approximation, but the moving average degenerates to the uniform distribution. In contrast, choosing a small V is appropriate for getting a good local approximation of the density if $P_V \approx \frac{k(N,V)}{N}$ can be reliably estimated. This becomes more difficult as smaller Vs are chosen, because the event of a sample's falling in $\mathcal{A}$ becomes more unlikely. In the extreme case, there is

no sample and so $P_V = 0$. Then the approximate density is ragged, with areas that are constantly zero and small peaks around each sample.
Informally, the volume V must not diminish too fast with respect to N. In Section 5.1, a proof of convergence is presented if $V^{-1} = \mathcal{O}(\sqrt{N})$. Two approaches are well established in practice. The Parzen window method assigns the volume $V \propto \frac{1}{\sqrt{N}}$ and k is estimated from the sampling. The k-nearest neighbor method assigns $k \propto \sqrt{N}$ and the volume V is estimated from the sampling, i.e., the neighborhood around each point is blown up until exactly k points are included.

Figure 5.2 shows an example of a comparison between the Parzen window method and the k-nearest neighbor method. The samples (blue points) are drawn uniformly within the unit disk with radius $r = 1$. The center point of the neighborhood is $\mathbf{m} = \mathbf{0}$ and $\mathcal{A}_0$ is chosen to be a disc (red line). Note that for the Parzen window method, the radius decreases with $N^{-\frac{1}{4}}$, because the area of the disk is proportional to $N^{-\frac{1}{2}}$.

5.1 The Parzen window method

The Parzen window method assigns the volume of the neighborhood with respect to the sample size. For now, the neighborhood is chosen to be a simple d-dimensional cube with edge length h_N and volume $V_N = h_N^d$. To this end, let

$$\varphi(\mathbf{u}) = \operatorname{rect}(\mathbf{u}) := \begin{cases} 1 & \text{if } |u_j| \le \frac{1}{2} \text{ for all } j = 1, \dots, d \\ 0 & \text{else} \end{cases} \tag{5.12}$$

denote the indicator function of the unit cube centered at the origin. Then $\mathbf{u} \mapsto \varphi(\frac{\mathbf{m}-\mathbf{u}}{h_N})$ denotes the indicator function of a cube centered at $\mathbf{m}$ with edge length h_N. Let $\mathbf{m}_1, \dots, \mathbf{m}_N$ denote the samples. For each $\mathbf{m} \in \mathcal{M}$, the number of samples within its neighborhood can be counted by

$$k_N(\mathbf{m}) = \sum_{i=1}^{N} \varphi\Big(\frac{\mathbf{m} - \mathbf{m}_i}{h_N}\Big). \tag{5.13}$$

Thus, together with Equation (5.7) and setting

$$\delta_N(\mathbf{m}) := \frac{1}{V_N} \varphi\Big(\frac{\mathbf{m}}{h_N}\Big), \tag{5.14}$$

it follows that

$$\hat{p}_N(\mathbf{m}) = \frac{1}{N} \sum_{i=1}^{N} \frac{1}{V_N} \varphi\Big(\frac{\mathbf{m} - \mathbf{m}_i}{h_N}\Big) = \frac{1}{N} \sum_{i=1}^{N} \delta_N\Big(\frac{\mathbf{m} - \mathbf{m}_i}{h_N}\Big) \tag{5.15}$$

is an estimator of the true density. This corresponds to an approximation of the probability density function by an interpolation between the samples by virtue of the function $\delta_N(\cdot)$.

The function φ is called a window function. A window function is any function that decreases to zero sufficiently rapidly; more precisely

$$\lim_{\|\mathbf{u}\|\to\infty} \varphi(\mathbf{u}) \prod_{i=1}^{d} |u_i| = 0. \tag{5.16}$$

The window function φ satisfies this condition as it is constantly zero outside the unit cube.

The symbol δ_N for the sequence of scaled window functions in Equation (5.15) was not chosen arbitrarily, but serves to highlight a connection with Dirac sequences.

Definition 5.1 (Dirac sequence). A sequence of integrable functions δ_N over $\mathcal{M}$ with $\mathcal{M} = \mathbb{R}^d$ or $\mathcal{M} = \mathbb{C}^d$ is called a Dirac sequence iff

1. $\delta_N(\mathbf{m}) \geq 0$ for all $\mathbf{m} \in \mathcal{M}$ and $N \in \mathbb{N}$,
2. $\int_{\mathcal{M}} \delta_N(\mathbf{m})\,d\mathbf{m} = 1$ for all $N \in \mathbb{N}$, and
3. $\lim_{N\to\infty} \int_{\|\mathbf{m}\|>\varepsilon} \delta_N(\mathbf{m})\,d\mathbf{m} = 1$ for all $\varepsilon > 0$.

The first two requirements state that δ_N is a formal density function. The third requirement demands that all the probability is eventually concentrated in an arbitrary small neighborhood around the origin. In other words, the δ_N approach the Dirac distribution δ. Unlike δ, however, all the δ_N are regular distributions. Quite often, δ is even defined as the weak limit of a Dirac sequence. In this case, the convergence holds by definition.

Here, $\delta_N(\mathbf{m}) = \frac{1}{V_N}\varphi(\frac{\mathbf{m}}{h_N})$ was initially chosen to be the uniform distribution over a rectangular neighborhood. A natural generalization is to replace the uniform window function by a Gaussian density. To this end, redefine

$$\varphi(\mathbf{u}) := \exp\{-\tfrac{1}{2}\|\mathbf{u}\|^2\} \tag{5.17}$$

and set

$$\delta_N(\mathbf{m}) = \frac{1}{V_N}\varphi\left(\frac{\mathbf{m}}{h_N}\right) = \frac{1}{(2\pi)^{\frac{d}{2}} h_N^d} \exp\left\{-\frac{1}{2h_N^2}\|\mathbf{m}\|^2\right\}. \tag{5.18}$$

Here, $V_N = (2\pi)^{\frac{d}{2}} h_N^d$ denotes the volume of the (infinite) support of the window function scaled by h_N. Hence, the parameter h_N can be understood as controlling the variance of the Gaussian window function.

The estimated density $\hat{p}_N(\mathbf{m})$ can be regarded as a random quantity $\underline{\hat{p}}_N(\mathbf{m})$ on its own if the samples that support the density are considered as random variables. Hence, it is possible to examine its expectation $\mu_N(\mathbf{m})$ and variance $\sigma_N^2(\mathbf{m})$ point-wise for every $\mathbf{m}$. The estimated density $\underline{\hat{p}}_N(\mathbf{m})$ converges to the true density $p(\mathbf{m})$ in terms of mean squared error if

$$\lim_{N\to\infty} \mu_N(\mathbf{m}) = p(\mathbf{m}), \tag{5.19}$$

$$\lim_{N\to\infty} \sigma_N^2(\mathbf{m}) = 0 \tag{5.20}$$

point-wise. To prove convergence, the following four requirements must be satisfied:

$$\sup_{\mathbf{u}\in\mathcal{M}} \varphi(\mathbf{u}) < \infty, \tag{5.21}$$

$$\lim_{\|\mathbf{u}\|\to\infty} \varphi(\mathbf{u}) \prod_{i=1}^{d} |u_i| = 0, \tag{5.22}$$

$$\lim_{N\to\infty} V_N = 0, \tag{5.23}$$

$$\lim_{N\to\infty} NV_N = \infty. \tag{5.24}$$

Note that the second condition is actually a repetition of the definition of a window function in Equation (5.16). The first condition forces the window function to be modest in the neighborhood of the origin. The last two conditions force the spread of the window function to vanish but not faster than the number of samples increases.

The expectation is calculated in two steps, because the first equality is needed again later. For any $i = 1, \dots, N$,

$$\mathrm{E}\{\delta_N(\mathbf{m} - \underline{\mathbf{m}}_i)\} = \int_{\mathcal{M}} \delta_N(\mathbf{m} - \mathbf{u})p(\mathbf{u})\,\mathrm{d}\mathbf{u} = [\delta_N * p](\mathbf{m}), \tag{5.25}$$

where $*$ denotes the convolution operator. Likewise

$$\mu_N(\mathbf{m}) = \mathrm{E}\{\underline{\hat{p}}_N(\mathbf{m})\} = \frac{1}{N}\sum_{i=1}^{N} \mathrm{E}\{\delta_N(\mathbf{m} - \underline{\mathbf{m}}_i)\} = [\delta_N * p](\mathbf{m}) \tag{5.26}$$

follows. As the Dirac distribution is the neutral element with respect to convolution,

$$\lim_{N\to\infty} \mu_N(\mathbf{m}) = \lim_{N\to\infty} [\delta_N * p](\mathbf{m}) = [\delta * p](\mathbf{m}) = p(\mathbf{m}) \tag{5.27}$$

yields the desired result. Note that in the above line, the limit and the integral were silently swapped. This step used the requirement Equation (5.22).

The calculation of the variance uses Equation (5.26) and exploits the fact that the variance of a sum of independent variables is the sum of the individual variances:

$$\begin{aligned}
\sigma_N^2(\mathbf{m}) &= \mathrm{Var}\{\underline{\hat{p}}_N(\mathbf{m})\} \overset{\text{i.i.d.}}{=} \frac{1}{N^2}\sum_{i=1}^{N} \mathrm{Var}\{\delta_N(\mathbf{m} - \underline{\mathbf{m}}_i)\} \\
&= \frac{1}{N^2}\sum_{i=1}^{N} \mathrm{E}\{\delta_N(\mathbf{m} - \underline{\mathbf{m}}_i)^2\} - \Big(\underbrace{\mathrm{E}\{\delta_N(\mathbf{m} - \underline{\mathbf{m}}_i)\}}_{=\mu_N(\mathbf{m})\geq 0}\Big)^2 \\
&\leq \frac{1}{N}\int_{\mathcal{M}} \delta_N^2(\mathbf{m} - \mathbf{u})p(\mathbf{u})\,\mathrm{d}\mathbf{u} \\
&\overset{(5.14)}{=} \frac{1}{N}\int_{\mathcal{M}} \frac{1}{V_N}\varphi\Big(\frac{\mathbf{m} - \mathbf{u}}{h_N}\Big)\delta_N(\mathbf{m} - \mathbf{u})p(\mathbf{u})\,\mathrm{d}\mathbf{u} \\
&\overset{(5.21)}{\leq} \frac{\sup \varphi(\mathbf{u})}{NV_N}\int_{\mathcal{M}} \delta_N(\mathbf{m} - \mathbf{u})p(\mathbf{u})\,\mathrm{d}\mathbf{u} \overset{(5.26)}{=} \frac{\sup \varphi(\mathbf{u})}{NV_N}\mu_N(\mathbf{m}).
\end{aligned} \tag{5.28}$$

The function $\mu_N(\mathbf{m})$ is a density function and is at least as well-behaved as the true density function $p(\mathbf{m})$, $\sup \varphi(\mathbf{u})$ is bounded by hypothesis and $\lim_{N\to\infty} NV_N = \infty$ holds by assumption, too. Hence, it follows that $\sigma_N^2(\mathbf{m}) \to 0$ and the proof is concluded.

In summary, the estimated density $\hat{p}(\mathbf{m})$ converges to the true density $p(\mathbf{m})$ in terms of the mean square error if—besides other requirements—the volume V_N vanishes more slowly than $\frac{1}{N}$. This is by far the most important point that should be noted from the proof, because this is a design parameter that can be influenced by the application.

This result will now be used to present a complete example using a Gaussian window function. Recalling Equations (5.14), (5.15), (5.17) and (5.18) and combining these with the design decision $V_N \propto \frac{1}{\sqrt{N}}$ to satisfy the condition in Equation (5.24) yields

$$V_N = (2\pi)^{\frac{d}{2}}\, h_N^d \overset{!}{\propto} N^{-\frac{1}{2}} \;\Rightarrow\; h_N = h\, N^{-\frac{1}{2d}}\,, \tag{5.29}$$

where $h \in \mathbb{R}^{\geq 0}$ denotes the initial width of the window function and is a design parameter. Thus, altogether, there follows

$$\begin{aligned}
\hat{p}(\mathbf{m}) &= \frac{1}{N}\sum_{i=1}^{N} \frac{1}{V_N}\,\varphi\Big(\frac{\mathbf{m}-\mathbf{m}_i}{h_N}\Big) = \frac{1}{N\,V_N}\sum_{i=1}^{N} \varphi\Big(\frac{\mathbf{m}-\mathbf{m}_i}{h_N}\Big) \\
&= \frac{1}{N\,(2\pi)^{\frac{d}{2}}\, h_N^d}\sum_{i=1}^{N} \exp\left\{-\frac{1}{2\,h_N^2}\|\mathbf{m}-\mathbf{m}_i\|^2\right\} \\
&= \frac{1}{\sqrt{N}\,(2\pi)^{\frac{d}{2}}\, h^d}\sum_{i=1}^{N} \exp\left\{-\frac{N^{\frac{1}{d}}}{2\,h^2}\|\mathbf{m}-\mathbf{m}_i\|^2\right\}.
\end{aligned} \tag{5.30}$$

The Parzen window method can be used to estimate the class-specific feature distributions $p(\mathbf{m}\,|\,\omega_i)$, $i = 1,\dots,c$, which are in turn used in a MAP classifier. Figure 5.3 shows the decision regions of the ongoing example, where the class-specific densities $p(\mathbf{m}\,|\,\omega_1)$ and $p(\mathbf{m}\,|\,\omega_2)$ were estimated using Parzen windows. One can see that the decision boundary comes very close to the decision boundary of the Bayesian optimal classifier and with 7 %, the asymptotic test error is only slightly larger than the 6.16 % of the optimal classifier. Note, however, that the outcome very much depends on the number and the location of the training samples, as well as the dimensionality of the feature space itself (see Section 6.1).

In summary, the Parzen window method is characterized by three crucial traits.

Universality The Parzen window method does not require any prior knowledge about the probability distribution. Even sophisticated multi-modal distributions can be estimated.

Choice of parameter Although the theoretical convergence to the true density holds in any case, the quality of the result in practice depends heavily on the initial choice of the volume V or the associated spread h.

Data-independent size of neighborhood For a fixed sample size N, any point of the feature space is covered by the same neighborhood, independently of the sample density.

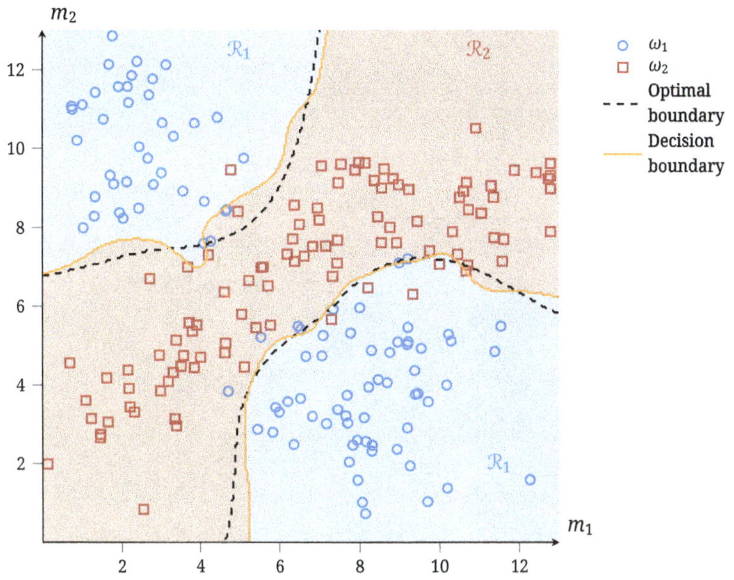

Fig. 5.3: Application to the reference example of Section 3.3.2. Decision regions of a classifier with Parzen window density estimators ($h_N \approx 0.5$, $N = 100$ for either class) with Gaussian window function. The training error is $e_{\text{train}} = 6\,\%$; the testing error is $e_{\text{test}} = 6.5\,\%$, but asymptotically approaches $e_{\text{test}} \approx 7\,\%$. The training set is the same as in Figure 3.8. Test samples are shown with hollow marks.

Figures 5.4 and 5.5 depict the estimation of a normal distribution with a Gaussian window function for different sample sizes N and initial spreads h for $m \in \mathbb{R}$ and $\mathbf{m} \in \mathbb{R}^2$, respectively.

5.2 The k-nearest neighbor method

Similar to the Parzen window method, the k-nearest neighbor method aims to estimate a density in virtue of

$$\hat{p}(\mathbf{m}) = \frac{k/N}{V} = \frac{k}{NV}. \tag{5.31}$$

But in contrast to the Parzen window method, the number of considered samples k only depends on the total number of samples N and instead the volume V is fitted so that exactly k samples fall in the neighborhood of $\mathbf{m}$. Review Figure 5.2 for a graphical illustration of the difference between both methods.

The neighborhood of a point $\mathbf{m}$ can be thought of as a cell centered at $\mathbf{m}$ that is inflated until it contains k_N samples; the cell is small if the neighborhood is dense and large if the neighborhood is sparsely populated. Unfortunately, this intuition is difficult to express in a closed mathematical formula and the proof of convergence to the true density is quite involved. Nonetheless, Loftsgaarden and Quesenberry [1965] have shown

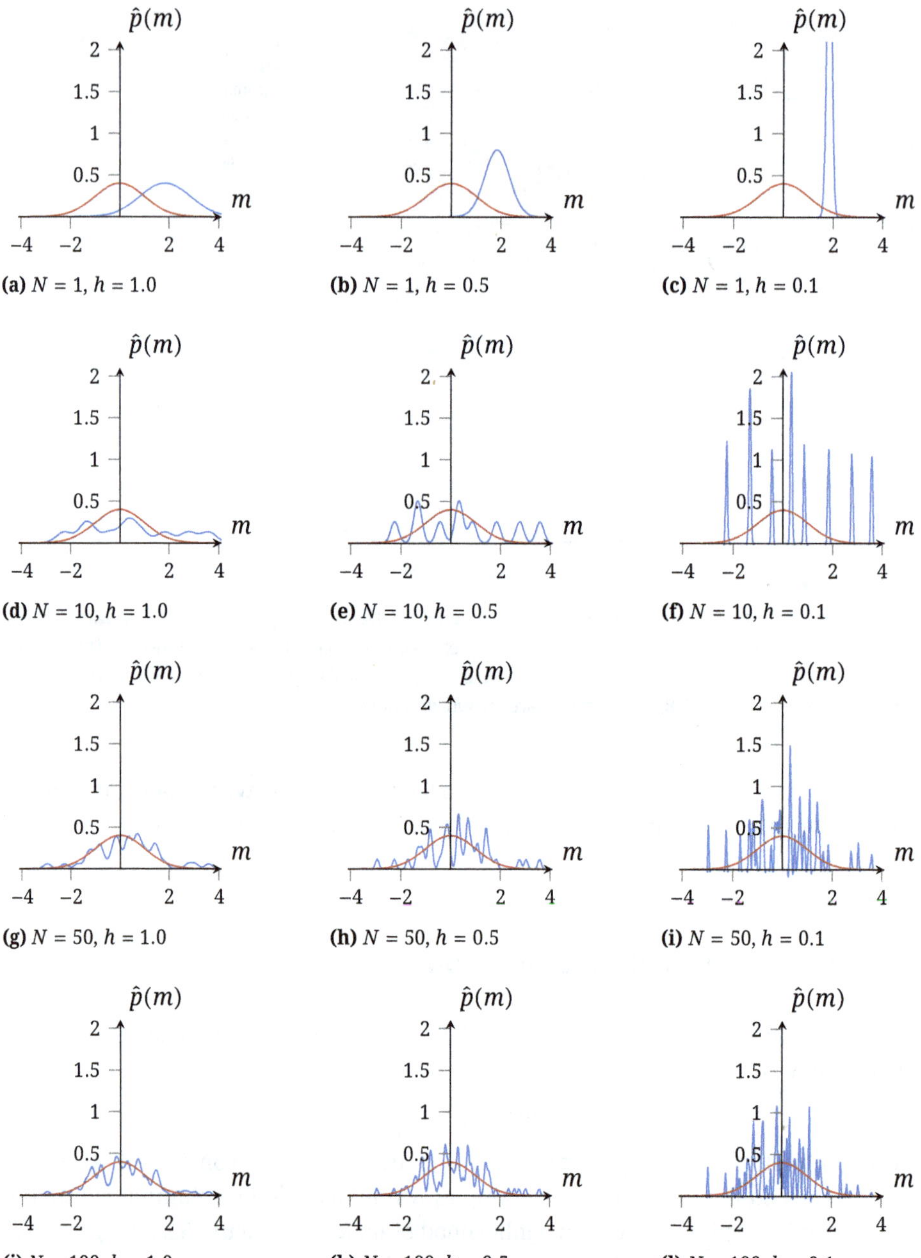

(a) $N = 1, h = 1.0$ (b) $N = 1, h = 0.5$ (c) $N = 1, h = 0.1$

(d) $N = 10, h = 1.0$ (e) $N = 10, h = 0.5$ (f) $N = 10, h = 0.1$

(g) $N = 50, h = 1.0$ (h) $N = 50, h = 0.5$ (i) $N = 50, h = 0.1$

(j) $N = 100, h = 1.0$ (k) $N = 100, h = 0.5$ (l) $N = 100, h = 0.1$

Fig. 5.4: Parzen window density estimation with a Gaussian window function for varying sample sizes N (blue curves) and spreads h with $m \in \mathbb{R}$; the true density is drawn in red.

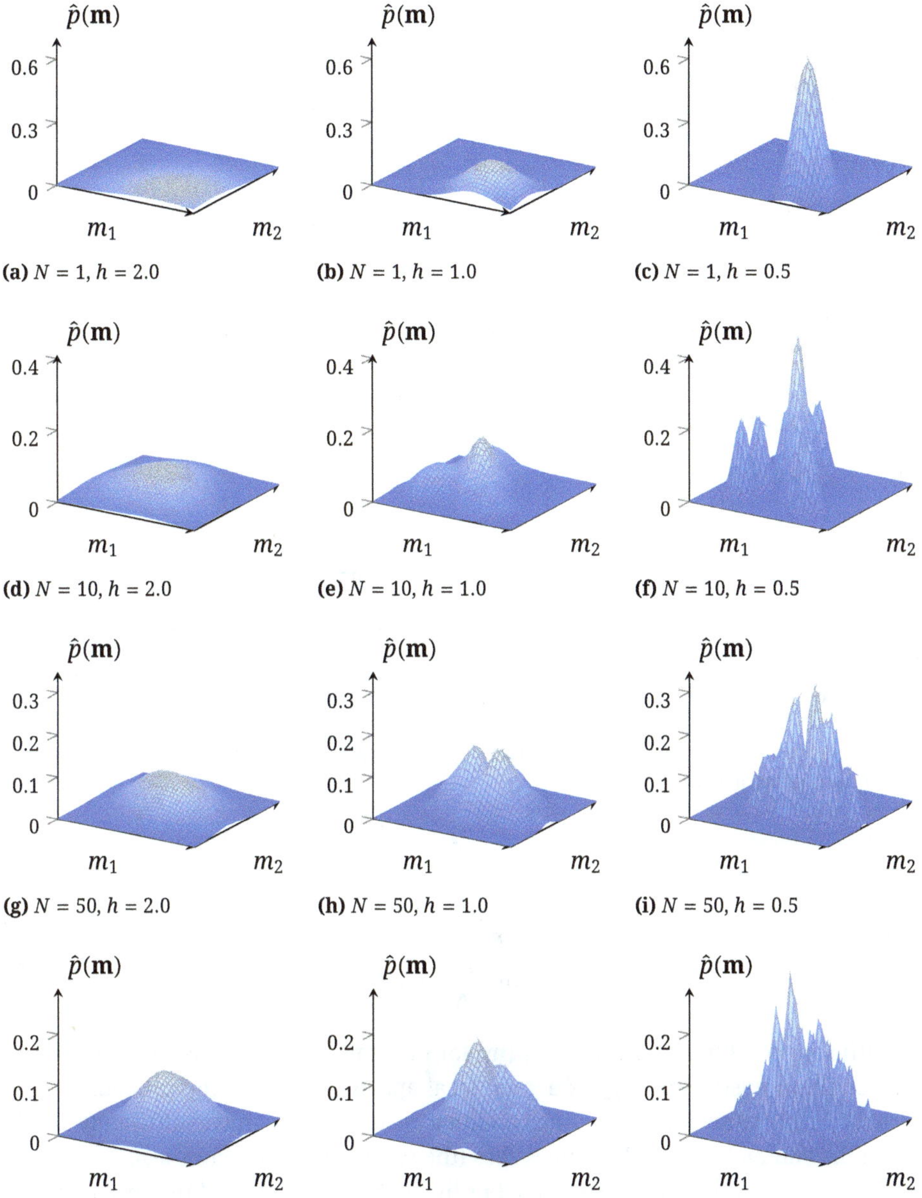

(a) $N = 1, h = 2.0$ (b) $N = 1, h = 1.0$ (c) $N = 1, h = 0.5$

(d) $N = 10, h = 2.0$ (e) $N = 10, h = 1.0$ (f) $N = 10, h = 0.5$

(g) $N = 50, h = 2.0$ (h) $N = 50, h = 1.0$ (i) $N = 50, h = 0.5$

(j) $N = 100, h = 2.0$ (k) $N = 100, h = 1.0$ (l) $N = 100, h = 0.5$

Fig. 5.5: Parzen window density estimation with a Gaussian window function for varying sample sizes N and spreads h with $\mathbf{m} \in \mathbb{R}^2$; the true density is a single Gaussian with $\boldsymbol{\mu} = \mathbf{0}$ and $\boldsymbol{\Sigma} = \mathbf{I}$, $p(\mathbf{m}) = \mathcal{N}(\boldsymbol{\mu},\mathbf{I})$ (not shown). Note that the scale of the applicate is different in each row.

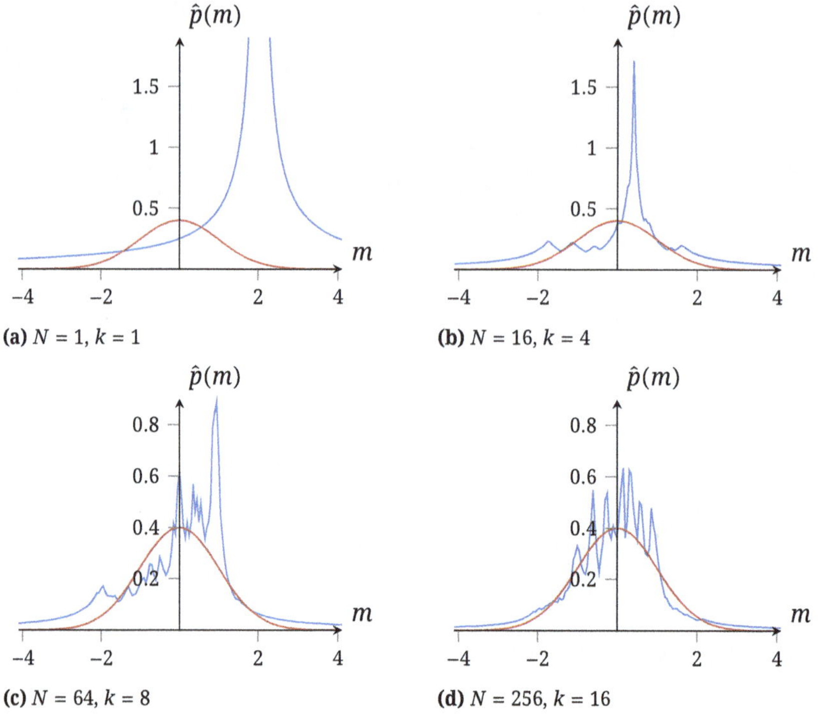

(a) $N = 1$, $k = 1$

(b) $N = 16$, $k = 4$

(c) $N = 64$, $k = 8$

(d) $N = 256$, $k = 16$

Fig. 5.6: k-nearest neighbor density estimation for varying sample sizes N and $k = \sqrt{N}$ (blue curve); the true density is drawn in red.

that

$$\lim_{N\to\infty} k_N = \infty, \tag{5.32}$$

$$\lim_{N\to\infty} \frac{k_N}{N} = 0 \tag{5.33}$$

are sufficient conditions for $\hat{p}(\mathbf{m}) \xrightarrow{\text{P}} p(\mathbf{m})$ for every $\mathbf{m}$ where $p(\mathbf{m})$ is continuous. The first condition ensures that $\frac{k_N}{N}$ is a good local approximation of the probability (the stochastic error vanishes); the second condition ensures that k grows sufficiently slowly so that the volume of the cell becomes zero (the systematic error vanishes).

Instead of reproducing the proof, let us highlight a substantial difference from the Parzen window method: when employing the Parzen window method and a differentiable window function is chosen, each function of the sequence is differentiable, too. Moreover, every approximation fulfills the formal requirements of a density.

Densities estimated by the k-nearest neighbor method are usually continuous, but not differentiable for $k > 1$, even if the true density is. In the case of $k = 1$, the estimated density even has poles (see Figure 5.6a). In this case, the neighborhood of $\mathbf{m}$ has zero volume if $\mathbf{m}$ is one of the samples $\mathbf{m}_i$. Hence the denominator of Equation (5.31) vanishes.

In the simplest scenario, i.e., for $N = k = 1$, sample $\mathbf{m}_1$ and a spherical neighborhood, the density is

$$\hat{p}(\mathbf{m}) = \frac{1}{a\|\mathbf{m} - \mathbf{m}_1\|} \tag{5.34}$$

with one pole at $\mathbf{m}_1$ (here, a denotes the volume of the d-dimensional unit sphere). For $k > 1$, the volume of the neighborhood of $\mathbf{m}$ changes smoothly with respect to $\mathbf{m}$, as long as the k samples that define the neighborhood stay the same. If $\mathbf{m}$ moves into the area of influence of a different sample, the volume of the neighborhood changes in a non-differentiable way. Generally, the points of nondifferentiability of the estimated function do not match the samples, but are placed in between.

Moreover, in the 1-dimensional case, i.e., $m \in \mathbb{R}$, the integral value of each approximated density is not equal to one but diverges to infinity even for $k > 1$. Outside of the finite number of samples N, every approximation asymptotically behaves like $m \mapsto \frac{1}{m}$ and therefore the integral becomes infinite. This means that although the density estimate converges to the true density function point-wise in probability, the approximation is not a density on its own.

5.3 k-nearest neighbor classification

The technique discussed in the previous Section aimed at estimating an approximation to the class-specific feature distribution $p(\mathbf{m}|\omega)$, but ultimately we are interested in the classification of an unknown sample. Estimating $p(\mathbf{m}|\omega)$ is only a step on this way. This section introduces a direct classifier that adopts the previous idea, does not need any distribution assumption, and whose decision is only based on the neighboring samples.

As always, $\mathcal{D} = \{\mathbf{m}_1, \ldots, \mathbf{m}_N\}$ denotes the set of all training samples and $\mathbf{m}$ a new unknown sample. Let $\mathcal{A}$ denote a reasonably small neighborhood of $\mathbf{m}$, k the total number of samples that fall in $\mathcal{A}$, and k_i the number of samples that belong to class ω_i. Then due to the same line of argument as before,

$$\hat{P}(\mathbf{m}, \omega_i) = \frac{k_i/N}{V} = \frac{k_i}{NV} \tag{5.35}$$

is an estimator of the joint probability. Applying Bayes' law leads to

$$\hat{P}(\omega_i|\mathbf{m}) = \frac{\hat{P}(\mathbf{m}, \omega_i)}{\sum_{j=1}^{c} \hat{P}(\mathbf{m}, \omega_j)} = \frac{k_i}{k}. \tag{5.36}$$

The appealing point of the last line is that it not only provides the a posteriori probability directly, but that it does not suffer from the technical difficulties of the previous result. Because there are only finitely many classes, $\hat{P}(\omega|\mathbf{m})$ is not a probability density function, but a probability mass function and sums to one. Moreover, the volume V and the number of samples N that caused those difficulties cancel out.

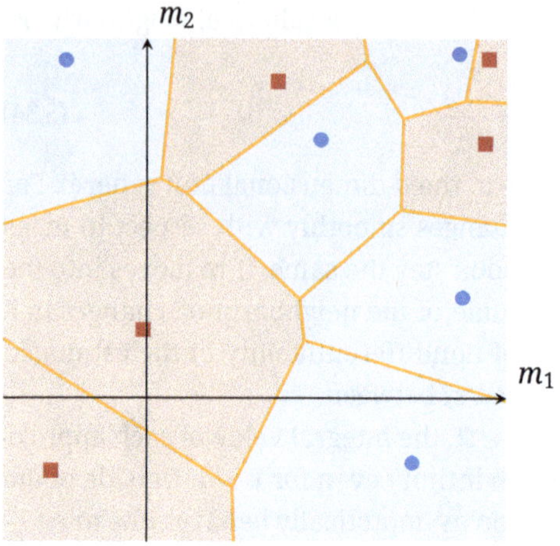

Fig. 5.7: Example Voronoi tessellation of a two-dimensional feature space.

Together with the usual maximum a posteriori rule, the estimated class is the class of the most frequently represented samples in the neighborhood. For $k = 1$, this classifier is called the nearest neighbor classifier. The formal decision rule is

$$\hat{\omega}(\mathbf{m}) = \omega_i \qquad \Leftrightarrow \qquad \arg\min_{\mathbf{m}_j \in \mathcal{D}} \|\mathbf{m} - \mathbf{m}_j\| \in \mathcal{D}_i, \tag{5.37}$$

where $\mathcal{D}_i$ is the set of training samples of class ω_i. The decision function is

$$k_i(\mathbf{m}) = \alpha \cdot \frac{1}{\min_{\widetilde{\mathbf{m}} \in \mathcal{D}_i} \|\mathbf{m} - \widetilde{\mathbf{m}}\|} = \alpha \cdot \max_{\widetilde{\mathbf{m}} \in \mathcal{D}_i} \frac{1}{\|\mathbf{m} - \widetilde{\mathbf{m}}\|}, \tag{5.38}$$

where α is a normalization constant that ensures that the $k_i(\mathbf{m})$ sum to one:

$$\alpha := \frac{1}{\sum_{j=1}^{c} k_j(\mathbf{m})}. \tag{5.39}$$

Plotting the decision boundaries in the feature space yields a Voronoi tessellation (see Figure 5.7). The only free design parameter is the scaling of the individual components or, equivalently, the choice of the metric. This parameter heavily influences which neighbor is considered to be "near". Figure 5.8 illustrates this effect for a fixed standard Euclidean metric but different scales of the first axis.

A natural extension is to base the classification decision not only on the nearest neighbor but on the majority class of several neighbors, i.e., assign the most frequent class among the $k > 1$ neighbors. Let $\mathcal{A}_k(\mathbf{m})$ denote a neighborhood of $\mathbf{m}$ that includes exactly k samples. The decision function can then be written as

$$k_i(\mathbf{m}) = \frac{1}{k}\left|\{\widetilde{\mathbf{m}} \,|\, \widetilde{\mathbf{m}} \in \mathcal{A}_k(\mathbf{m}) \cap \mathcal{D}_i\}\right|. \tag{5.40}$$

A sketch of this classification method is shown in Figure 5.9.

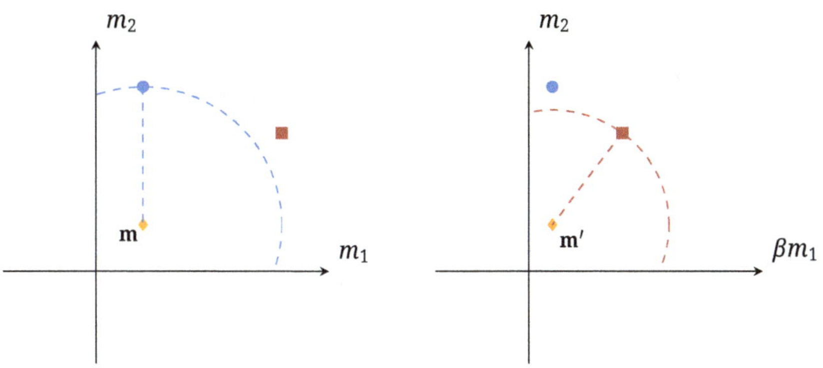

(a) Euclidean distance

(b) Transforming the feature space, e.g., by scaling m_1, can change the nearest neighbor. The transformation also applies to the vector itself, e.g., $\mathbf{m}' = \begin{pmatrix} \beta & 0 \\ 0 & 1 \end{pmatrix} \mathbf{m}$.

Fig. 5.8: Dependence of the nearest neighbor classifier on the metric

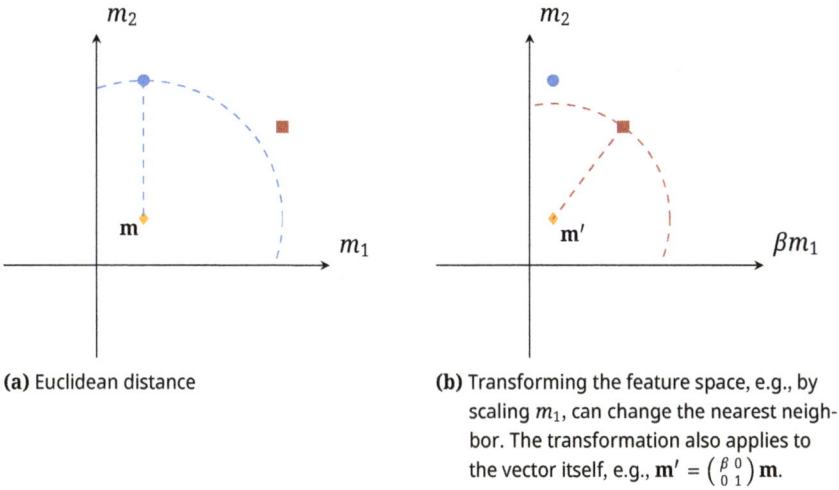

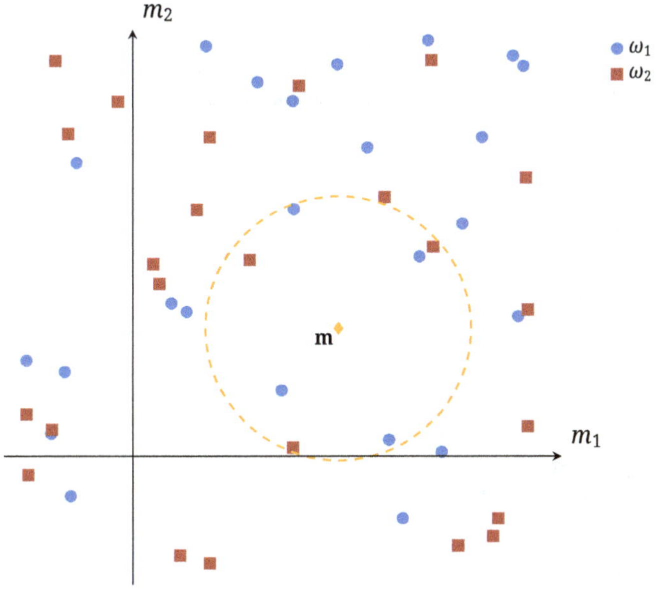

Fig. 5.9: k-nearest neighbor classifier for $c = 2$, $d = 2$ and $k = 7$; $\mathbf{m}$ is assigned to ω_1 (blue class) since 4 out of its 7 neighbors are in this class.

Figures 5.10 to 5.12 show the decision boundaries of a k-nearest neighbor classifier with varying values of k for the reference example of Section 3.3.2. With $k = 1$, the decision region is complicated and rugged, while higher values of k smooth out the details. As with the Parzen window method, the optimal choice of k depends on the number of training samples and the dimensionality of the feature space, but also on the distance measure used.

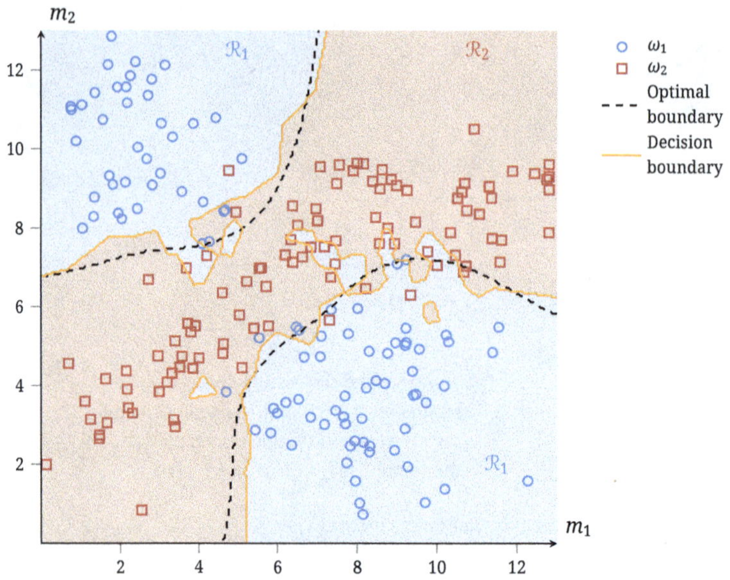

Fig. 5.10: Decision regions of a nearest neighbor classifier with $k = 1$. The training error is $e_{\text{train}} = 0\,\%$ by definition of the classifier. The testing error is $e_{\text{test}} = 10\,\%$ and asymptotically approaches $e_{\text{test}} \approx 9.4\,\%$. The training set is the same as in Figure 3.8. Test samples are shown with hollow marks.

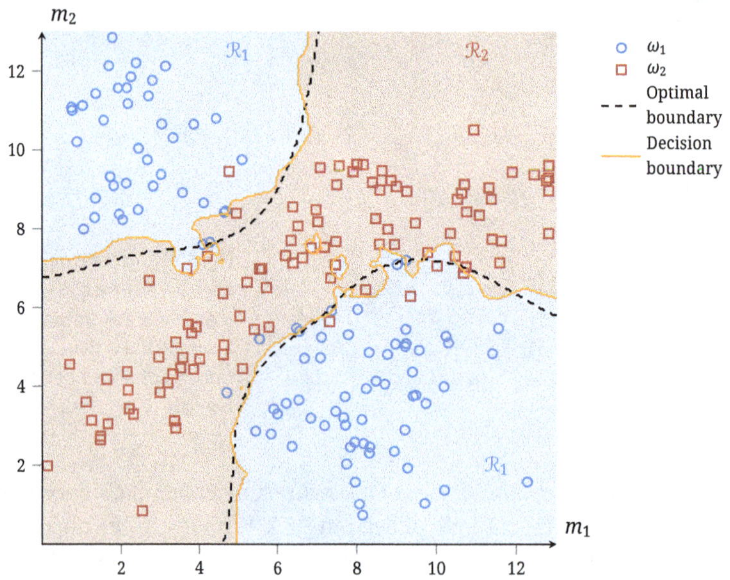

Fig. 5.11: Decision regions of a nearest neighbor classifier with $k = 3$. The training error is $e_{\text{train}} = 4.5\,\%$. The testing error is $e_{\text{test}} = 8\,\%$ and asymptotically approaches $e_{\text{test}} \approx 7.5\,\%$. The training set is the same as in Figure 3.8. Test samples are shown with hollow marks.

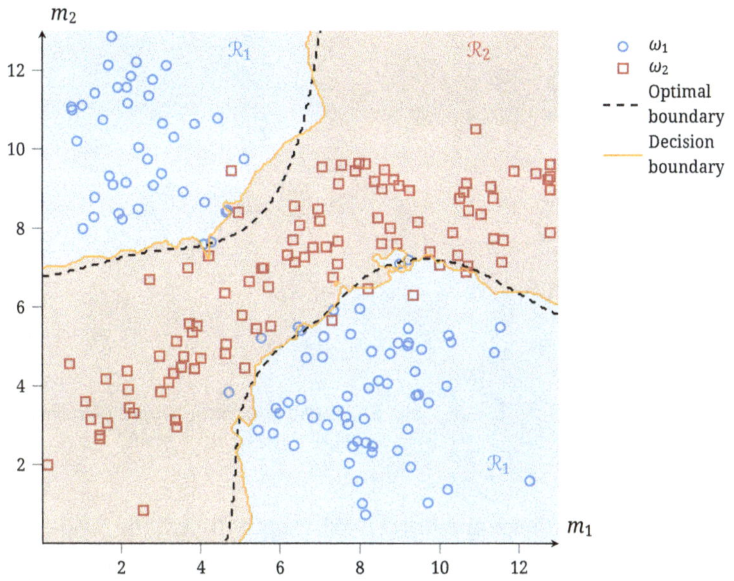

Fig. 5.12: Decision regions of a nearest neighbor classifier with $k = 5$. The training error is $e_{\text{train}} = 7\,\%$. The testing error is $e_{\text{test}} = 6\,\%$ and asymptotically approaches $e_{\text{test}} \approx 7.1\,\%$. The training set is the same as in Figure 3.8. Test samples are shown with hollow marks.

An interesting question is how well the nearest neighbor classifier performs in comparison to the MAP classifier. In the latter case, the error probability P^* equals the overall risk and we recall from Equation (3.38) that

$$P^* = R = 1 - \int_{\mathcal{M}} P(\hat{\omega}(\mathbf{m}), \mathbf{m})\,\mathrm{d}\mathbf{m}. \tag{5.41}$$

In the case of the nearest neighbor classifier, the error probability depends on the number of samples being drawn. The classification is erroneous if the sample $\mathbf{m}_i$ nearest to the test sample $\mathbf{m}$ has a different class than the true class of $\mathbf{m}$. Let P_N denote the error probability of the nearest neighbor classifier with N samples. We state without proof that $P = \lim_{N\to\infty} P_N$ exists and call P the asymptotic error probability of the nearest neighbor classifier. With c classes, both error probabilities are at most $\frac{c-1}{c}$, because this is the probability of being wrong if one merely guesses. The optimal Bayes error probability P^* is a lower bound for P. But it is also possible to bound the error probability of the nearest neighbor classifier from above in terms of P^*. Cover and Hart [1967] have shown that

$$P^* \le P \le P^*\left(2 - \frac{c}{c-1}P^*\right). \tag{5.42}$$

This means the error probability of the nearest neighbor classifier is somewhere in the blue shaded area of 5.13.

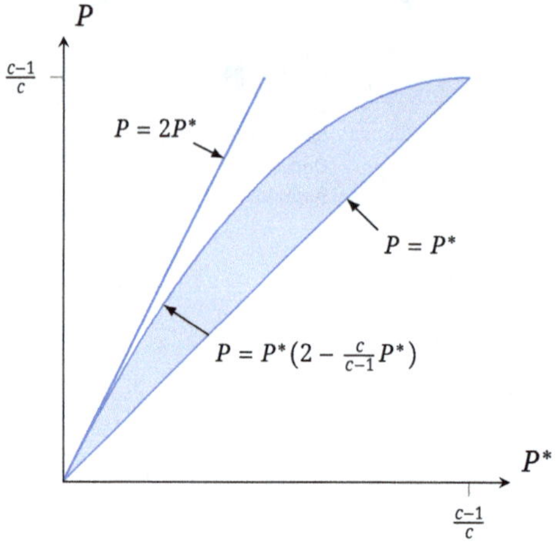

Fig. 5.13: Asymptotic error bounds of the nearest neighbor classifier.

Instead of reproducing the proof in Cover and Hart [1967], we just discuss the result. The first observation is that the upper and lower bounds coincide for the boundary values $P^* \in \{0, \frac{c-1}{c}\}$. This is a reasonable result. If $P^* = 0$ for the optimal error probability, then the supports of the class-specific feature distributions must be disjoint. Hence, for infinite many samples, the nearest sample asymptotically has the correct class and the nearest neighbor classifier always decides correctly. In contrast, if $P^* = \frac{c-1}{c}$, then the Bayes classifier is not better than guessing and the nearest neighbor classifier is not worse.

By dropping the last term in Equation (5.42), the weaker upper bound

$$P \leq 2P^* \tag{5.43}$$

follows. In real-world applications, where the true distribution is typically unknown, any classifier—as sophisticated as it might possibly be—needs to base its decision rule somehow on the given samples. On the one hand, no matter how this rule might look, the classifier cannot be better than P^*, because this is the optimal lower bound. On the other hand, the nearest neighbor classifier does asymptotically not perform worse than $2P^*$. In consequence, roughly speaking, half of the information that can be somehow extracted from the samples is already included in the nearest sample.

The last paragraph sounds promising, as a very easy classification rule has a guaranteed not so bad upper bound. Unfortunately, this bound only holds asymptotically and no such bound exists for the case of a finite sample. There is not much that can be done about this situation, because without any additional assumptions, there are few starting points to control the quality of the approximation. Even worse, in the most general case, examples can be constructed such that convergence is arbitrarily slow and not even monotonic.

5.4 Exercises

(5.1) Given the mapping $y = A(x) = x^2$, why is the inference from y to x an ill-posed inverse problem?

(5.2) One wishes to establish a functional relation between two scalar measurements x and y. Considering the underlying physics, it is known that the relation must be linear. Suppose that there are $N > 2$ noisy measurements (x_i, y_i), $i = 1, \ldots, N$. The task is formulated as follows:

> Find the parameters $a, b \in \mathbb{R}$ of a straight line that interpolates the data points, i.e., $y_i = a\,x_i + b$ for all $i = 1, \ldots, N$.

Why is this (inverse) problem ill-posed? How can the task be reformulated so that the inverse problem is well-posed?

(5.3) Given the following sample of one-dimensional features m_i

$$\mathcal{D} = \{8.1, 8.9, 7.6, 9.7, 12.2, 7.1, 10.4, 9.3, 14.9, 10.1\},$$

estimate the sample density $\hat{p}(m)$ at the locations $m = 6$, $m = 8$, $m = 10$, $m = 12$ and $m = 14$ using the Parzen window method with a rectangular window as in Equation (5.12) and a window size of $h_N = 1$.

(5.4) Given the following sample of one-dimensional features m_i

$$\mathcal{D} = \{8.0, 8.5, 7.6, 9.7, 12.2, 7.1, 10.5, 9.3, 14.9, 10.0\},$$

estimate the sample density $\hat{p}(m)$ at the locations $m = 6$, $m = 8$, $m = 10$, $m = 12$ and $m = 14$ using the k-nearest neighbor method with $k = 3$ neighbors.

(5.5) Use the following sample $\mathcal{D} = \{\mathbf{m}_1, \ldots, \mathbf{m}_6\}$ to graphically classify the points $\mathbf{m}'_1 = (-2,2)^\mathsf{T}$, $\mathbf{m}'_2 = (2,0)^\mathsf{T}$ and $\mathbf{m}'_3 = (-1,-5)^\mathsf{T}$ using the nearest neighbor method:

$$\mathbf{m}_1 = \begin{pmatrix} 0 \\ 0 \end{pmatrix}, \mathbf{m}_2 = \begin{pmatrix} 4 \\ 2 \end{pmatrix}, \mathbf{m}_3 = \begin{pmatrix} 2 \\ 6 \end{pmatrix}$$

with $\omega(\mathbf{m}_1) = \omega(\mathbf{m}_2) = \omega(\mathbf{m}_3) = \omega_1$ and

$$\mathbf{m}_4 = \begin{pmatrix} 4 \\ -4 \end{pmatrix}, \mathbf{m}_5 = \begin{pmatrix} -6 \\ -6 \end{pmatrix}, \mathbf{m}_6 = \begin{pmatrix} -4 \\ 2 \end{pmatrix}$$

with $\omega(\mathbf{m}_4) = \omega(\mathbf{m}_5) = \omega(\mathbf{m}_6) = \omega_2$.
Use the Euclidean distance $d(\mathbf{x},\mathbf{y}) = \|\mathbf{x} - \mathbf{y}\|_2$ as distance measure. Sketch the decision boundaries in the two-dimensional feature space.

6 General considerations

This chapter explicitly discusses two issues that are scattered around several chapters in many other books about pattern recognition. These issues would fit equally well into any other chapter of this book and would emerge in slightly different flavors depending on the surrounding context. Hence, one could discuss each of the issues on demand partially as required, but this approach would miss the point that both issues pervade the entire scope of pattern recognition.

Actually both matters are rather short and therefore this chapter will be, too. But we decided to spent an entire chapter to accommodate their fundamental importance.

The first topic addresses the question of dimension and clarifies the term the *curse of dimensionality*. The second topic is *overfitting.*

6.1 Dimensionality of the feature space

Section 2.7 introduced techniques to reduce the dimension of a feature space but lacked an explanation of the reasons why a small number of dimensions is favorable. The introduction of this book established some design principles for how to select features and referred to the "curse of dimensionality" (see Section 1.4, but did not give an explanation of the term. This section will fill this gap.

The beginning will be the exact opposite and first give an example that seems to support the commonsense (but false) belief that a large number of features should lead to better classification. After that, this belief is disproved and it will be shown why the example is misleading.

Recall an example from Section 3.3.5. The number of classes will be $c = 2$ and both class-specific feature distributions will be $p(\mathbf{m}|\omega_i) = \mathcal{N}(\mathbf{m}; \boldsymbol{\mu}_i, \boldsymbol{\Sigma})$ with shared covariance matrix $\boldsymbol{\Sigma}$. Moreover, the a priori distribution $P(\omega_1) = P(\omega_2) = \frac{1}{2}$ is assumed. The log-likelihood for the decision between ω_1 and ω_2 is

$$\ln\left(\frac{P(\omega_2|m)}{P(\omega_1|m)}\right) = \ln\left(\frac{p(m|\omega_2)}{p(m|\omega_1)}\right) = \Lambda(\mathbf{m}) = (\boldsymbol{\Sigma}^{-1}(\boldsymbol{\mu}_2 - \boldsymbol{\mu}_1))^{\top}(\mathbf{m} - \frac{1}{2}(\boldsymbol{\mu}_1 + \boldsymbol{\mu}_2)) \tag{6.1}$$

and the decision rule is

$$\hat{\omega}(\mathbf{m}) = \begin{cases} \omega_1 & \Lambda(\mathbf{m}) < 0 \\ \omega_2 & \text{else} \end{cases} . \tag{6.2}$$

Note that $\Lambda(\mathbf{m}) = 0$ corresponds with he decision boundary according to Equation (3.59). Putting the random variable $\underline{\mathbf{m}}$ into $\Lambda(\underline{\mathbf{m}})$ makes this a Gaussian distributed random variable itself, because it is a linear transformation of $\underline{\mathbf{m}}$. Thus, it is possible to calculate

https://doi.org/10.1515/9783111339207-006

the conditional expectation and variance with respect to the true class. There follows

$$
\begin{aligned}
\mathrm{E}\{\Lambda(\underline{\mathbf{m}})\,|\,\underline{\omega} = \omega_1\} &= \mathrm{E}\left\{(\boldsymbol{\Sigma}^{-1}(\boldsymbol{\mu}_2 - \boldsymbol{\mu}_1))^{\mathsf{T}}(\underline{\mathbf{m}} - \frac{1}{2}(\boldsymbol{\mu}_1 + \boldsymbol{\mu}_2))\,\big|\,\underline{\omega} = \omega_1\right\} \\
&= (\boldsymbol{\mu}_2 - \boldsymbol{\mu}_1)^{\mathsf{T}}\boldsymbol{\Sigma}^{-1}\Big(\mathrm{E}\{\underline{\mathbf{m}}\,|\,\underline{\omega} = \omega_1\} - \frac{1}{2}(\boldsymbol{\mu}_1 + \boldsymbol{\mu}_2)\Big) \\
&= -\frac{1}{2}(\boldsymbol{\mu}_2 - \boldsymbol{\mu}_1)^{\mathsf{T}}\boldsymbol{\Sigma}^{-1}(\boldsymbol{\mu}_2 - \boldsymbol{\mu}_1)
\end{aligned} \tag{6.3}
$$

and likewise $\underline{\omega} = \omega_2$ yields

$$
\mathrm{E}\{\Lambda(\underline{\mathbf{m}})\,|\,\underline{\omega} = \omega_2\} = \frac{1}{2}(\boldsymbol{\mu}_1 - \boldsymbol{\mu}_2)^{\mathsf{T}}\boldsymbol{\Sigma}^{-1}(\boldsymbol{\mu}_1 - \boldsymbol{\mu}_2). \tag{6.4}
$$

The calculation of the variance can be combined for both cases $\underline{\omega} = \omega_i$ $(i = 1, 2)$:

$$
\begin{aligned}
\mathrm{Var}\{\Lambda(\underline{\mathbf{m}})\,|\,\underline{\omega} = \omega_i\} &= \mathrm{Var}\left\{(\boldsymbol{\Sigma}^{-1}(\boldsymbol{\mu}_2 - \boldsymbol{\mu}_1))^{\mathsf{T}}(\underline{\mathbf{m}} - \frac{1}{2}(\boldsymbol{\mu}_1 + \boldsymbol{\mu}_2))\,\big|\,\underline{\omega} = \omega_i\right\} \\
&= (\boldsymbol{\mu}_2 - \boldsymbol{\mu}_1)^{\mathsf{T}}\boldsymbol{\Sigma}^{-1}\,\mathrm{Cov}\{\underline{\mathbf{m}}\,|\,\underline{\omega} = \omega_i\}\,\boldsymbol{\Sigma}^{-1}(\boldsymbol{\mu}_2 - \boldsymbol{\mu}_1) \\
&= (\boldsymbol{\mu}_2 - \boldsymbol{\mu}_1)^{\mathsf{T}}\boldsymbol{\Sigma}^{-1}(\boldsymbol{\mu}_2 - \boldsymbol{\mu}_1).
\end{aligned} \tag{6.5}
$$

To simplify the syntax, define

$$
s := \|\boldsymbol{\mu}_2 - \boldsymbol{\mu}_1\|_{\mathrm{m}} = \sqrt{(\boldsymbol{\mu}_2 - \boldsymbol{\mu}_1)^{\mathsf{T}}\boldsymbol{\Sigma}^{-1}(\boldsymbol{\mu}_2 - \boldsymbol{\mu}_1)} \tag{6.6}
$$

as the Mahalanobis distance w.r.t. $\boldsymbol{\Sigma}^{-1}$ between the expectation values $\boldsymbol{\mu}_1$ and $\boldsymbol{\mu}_2$. In summary, this leads to

$$
p(\mathbf{m}|\omega_1) = \mathcal{N}(\Lambda(\mathbf{m}); -\tfrac{1}{2}s^2, s^2), \tag{6.7}
$$

$$
p(\mathbf{m}|\omega_2) = \mathcal{N}(\Lambda(\mathbf{m}); \tfrac{1}{2}s^2, s^2). \tag{6.8}
$$

An object is incorrectly classified if $\Lambda(\mathbf{m}) \geq 0$ but the true class equals ω_1, or if $\Lambda(\mathbf{m}) < 0$ but the true class equals ω_2. Hence, the error probability is

$$
\begin{aligned}
R &= \frac{1}{2}P(\Lambda(\underline{\mathbf{m}}) \geq 0\,|\,\underline{\omega} = \omega_1) + \frac{1}{2}P(\Lambda(\underline{\mathbf{m}}) < 0\,|\,\underline{\omega} = \omega_2) \\
&= P(\Lambda(\underline{\mathbf{m}}) \geq 0\,|\,\underline{\omega} = \omega_1) \\
&= \frac{1}{\sqrt{2\pi}s}\int_0^{\infty} \exp\Big(-\frac{(\Lambda + \frac{s^2}{2})^2}{2s^2}\Big)\,\mathrm{d}\Lambda \\
&= \frac{1}{\sqrt{2\pi}}\int_{\frac{s}{2}}^{\infty} \exp(-\frac{u^2}{2})\,\mathrm{d}u.
\end{aligned} \tag{6.9}
$$

The last line shows that the error probability vanishes ($R \to 0$) if the Mahalanobis distance of the expectation values increases ($s \to \infty$). Until now, no assumption about

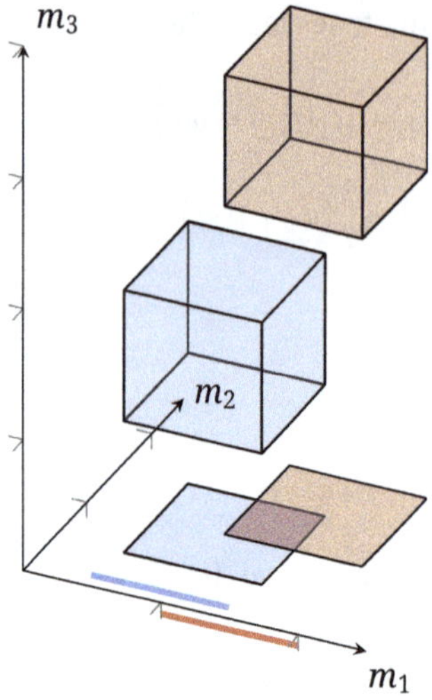

Fig. 6.1: Illustration: Increasing dimension vs. overlapping densities. The densities do not overlap in three dimensions, yet in one dimension the overlap is nearly 50 %.

the dimension d of the feature space has been made. The mutual positions of the expectation values cannot be easily changed, because they are given by the nature of the application, but one could try to put more features into the feature vector and thereby increase its dimension. As long as these additional features contain new information about the problem, the Mahalanobis distance is increased. In mathematical terms, $d \to \infty \rightsquigarrow s \to \infty \Rightarrow R \to 0$. Note that simply duplicating components does not increase the Mahalanobis distance: the increase depends on the correlation between the existing and the additional features. In consequence, one could argue the more features the better, or, at greater length, the higher the dimension of the feature space, the greater the distance, the lower the error. The illustration in Figure 6.1 seems to support this statement. The plot indicates the support of two uniform distributions in different dimensions. Under a projection onto the first dimension, both supports overlap each other by one half. This overlapping region is responsible for a false classification. In two dimensions, the overlapping region of the rectangles only counts for one quarter of the area. So the proportion of possible false classifications declines. In three dimensions, the cubes can be perfectly separated.

Although this seems to support the belief that a larger number of dimensions improves the classification, the statement is generally wrong in real-world applications. The conclusion is only true if the class-specific feature distributions are perfectly known or if there were infinitely many training samples.

Consider another, this time factual, example that illustrates the real effect of increasing the number of dimensions (Beyerer [1994]). The task was to automatically assess the quality of honed surfaces of cylinders given a catalog of $N = 33$ pictures of such a surface with manually assigned grades on an ordinal scale between 1 and 10. To solve the problem, 25 heuristic and model-driven features were defined. The idea for classification was

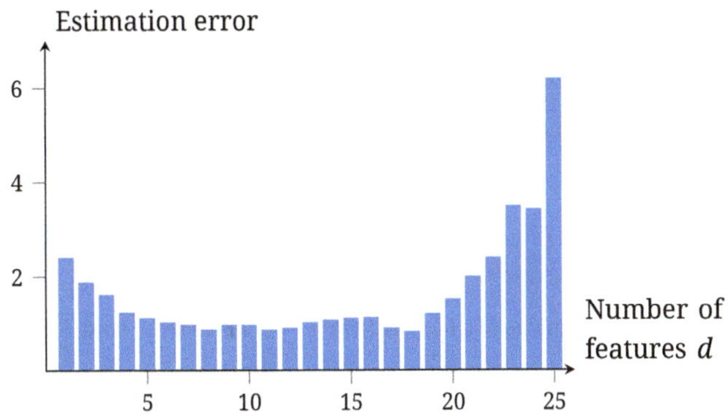

Fig. 6.2: Dependence of error rate on the dimension of the feature space in Beyerer [1994].

to estimate the grade n by a linear regression $\hat{n} = \mathbf{A}\mathbf{m}$ with the smallest mean squared error. A feature selection according to Section 2.7.7 showed that increasing the number of features improved the classification result at the beginning but beyond a certain point, additional features increased the error again (see Figure 6.2).

This phenomenon can be best understood by the following allegory. Each feature bears some net payload and some irrelevant payload that counts as a disturbance. As long as a new feature adds more important information into the system than disturbance, the classification performance increases. But if all the net payload that a feature potentially could add is already included by the existing features, then only the disturbance goes on top and the performance degrades. To rectify the misleading perception, the concept of *interval probability* will be introduced with the help of an example. Let $\underline{\mathbf{m}} \sim \mathcal{N}(\boldsymbol{\mu}, \sigma^2\mathbf{I})$ with dimension $d \in \mathbb{N}$. As all components are independent, the probability that a sample falls in the d-dimensional cube with edge length 4σ around its expectation value equals

$$\Pr(|m_1 - \mu_1| < 2\sigma, \dots, |m_d - \mu_d| < 2\sigma) \approx 0.95^d. \tag{6.10}$$

This is a strictly decreasing function with respect to d. For $d = 1$, the bulk of the samples lies within the interval $[\mu_1 - 2\sigma, \mu_1 + 2\sigma]$. But for $d = 100$, it follows that $P \approx 0.95^{100} \approx 0.0059$ and only a small fraction of the samples lies within the cube.

Another example of counter-intuitive interval probability can be found when regarding a d-dimensional unit hypersphere with radius $r = 1$. Let $r' < r$ be the radius of a smaller sphere such that $\varepsilon = r - r'$ is the width of the sphere shell. The situation is illustrated in Figure 6.3. If $\underline{\mathbf{m}}$ is equally distributed within the unit hypersphere, one can show that $1 - (1 - \varepsilon)^d$ of the probability mass lies in the shell of the unit sphere. In a 3-dimensional sphere with a small shell width of $\varepsilon = 0.01$, only $1 - 0.99^3 \approx 0.03$ of the probability mass lies in the sphere shell. However, for $d = 1000$ nearly the whole probability is distributed within the shell (≈ 0.99996). Note that this observation underlines the claim in Section 2.4.5 that the cosine distance is a good distance measure in high-dimensional feature spaces.

Generally, the higher the dimension, the more sparsely the samples are scattered. This leads to the notion of *sample density*, to conceptualize this statement more precisely.

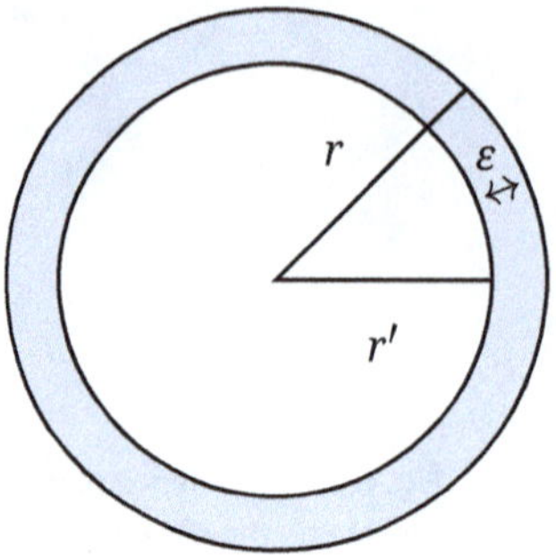

Fig. 6.3: Scheme of a hypersphere ($d = 2$) with a small shell.

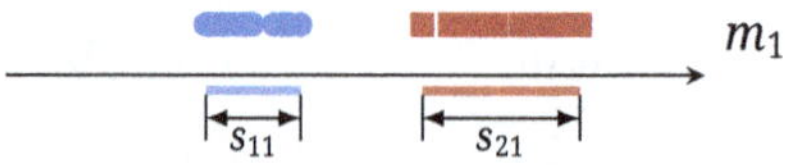

(a) $d = 1$

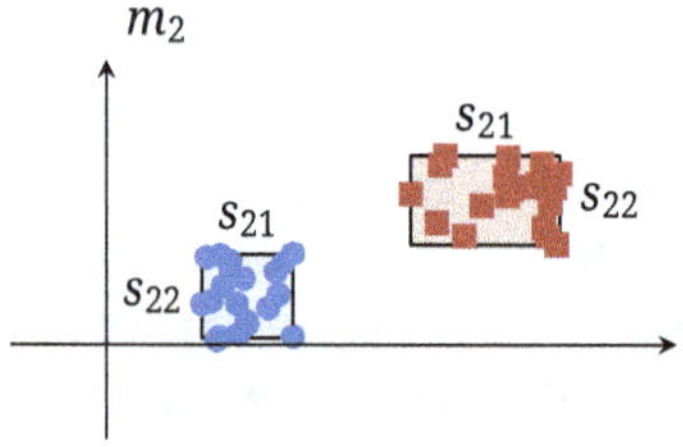

(b) $d = 2$

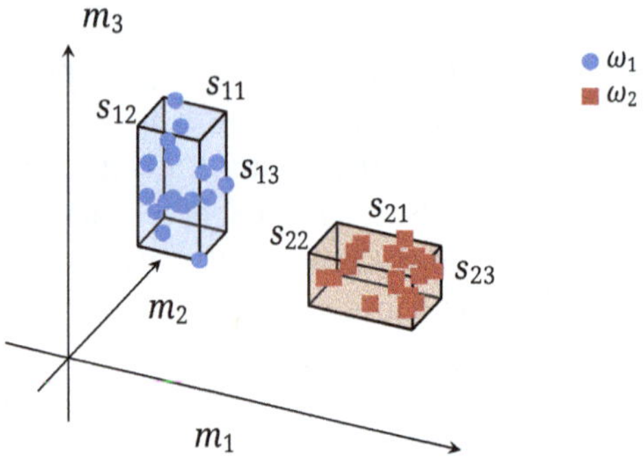

(c) $d = 3$

Fig. 6.4: Density of a sample for feature spaces of increasing dimensionality. In each plot, the number of samples per class is the same.

Given a finite number of samples N_i per class, the smallest axis-aligned enclosing cuboid is considered (see Figure 6.4). The sample density will be defined as the number of samples per unit volume: $\gamma_i := \frac{N_i}{V_i}$. So as to better compare the cuboids and abstract from the different ratios of the edges, the geometric mean of the edge lengths, $\overline{s}_i = \sqrt[d]{\prod_{j=1}^{d} s_{ij}}$, is used. Then $\gamma_i = \frac{N_i}{V_i} = \frac{N_i}{(\overline{s}_i)^d}$. In order to keep the density constant for different dimensions,

$$\gamma_i = \frac{N_i}{(\overline{s}_i)^d} = \text{const.} \qquad \Leftrightarrow \qquad N_i \propto (\overline{s}_i)^d \tag{6.11}$$

must be fulfilled. This means that the number of samples has to increase exponentially with the number of dimensions.

The decision regions in the feature space are bounded by $(d-1)$-dimensional hypersurfaces. These decision boundaries are determined by pairwise equalities of the decision functions $k_l(\mathbf{m}) = k_j(\mathbf{m})$. A parametric description of the decision boundaries requires a mathematical model whose parameter space generally has a higher dimension $q \geq d$ than the dimension of the feature space (see Figure 6.5). The simplest $(d-1)$-dimensional manifold is a hyperplane that is defined by $q = d$ parameters (see Figures 6.5a and 6.5c).

The parameter vector $\theta \in \Theta$ with $\dim \Theta = q$ needs to be estimated from the samples. The quality of the decision boundary depends on the estimation error involved. Hence, it is natural to ask how the dimension q of the parameter space influences the magnitude of the error. To this end, we will reconsider the multi-dimensional CRB (see Equation (4.14)) and apply it to an example with normally distributed, independent features. Let $\underline{\mathbf{m}} \sim \mathcal{N}(\mu, \sigma^2\mathbf{I})$ with unknown expectation μ but known covariance $\sigma^2\mathbf{I}$. Hence, the parameter vector is $\theta = \mu$ and $q = d$. Then

$$
\begin{aligned}
\ln p(\mathbf{m}) &= -\frac{1}{2\sigma^2}(\mathbf{m}-\mu)^{\mathsf{T}}(\mathbf{m}-\mu) - \frac{d}{2}\ln 2\pi - \ln\sigma \\
\Rightarrow \nabla_\mu \ln p(\mathbf{m}) &= \frac{1}{\sigma^2}(\mathbf{m}-\mu) \\
\Rightarrow (\nabla_\mu \ln p(\mathbf{m}))(\nabla_\mu \ln p(\mathbf{m}))^{\mathsf{T}} &= \frac{1}{\sigma^4}(\mathbf{m}-\mu)(\mathbf{m}-\mu)^{\mathsf{T}} \\
\Rightarrow \mathbf{J}(\mu) &= \mathrm{E}\left\{(\nabla_\mu \ln p(\underline{\mathbf{m}}))(\nabla_\mu \ln p(\underline{\mathbf{m}}))^{\mathsf{T}}\right\} \\
&= \frac{1}{\sigma^4}\mathrm{E}\left\{(\underline{\mathbf{m}}-\mu)(\underline{\mathbf{m}}-\mu)^{\mathsf{T}}\right\} = \frac{1}{\sigma^2}\mathbf{I}.
\end{aligned}
\tag{6.12}
$$

So the inverse of the Fisher information matrix is $\mathbf{J}^{-1}(\mu) = \sigma^2\mathbf{I}$ and it follows that

$$
\operatorname{tr}\operatorname{Cov}\left\{\underline{\hat{\mu}}\right\} \geq \frac{1}{N}\operatorname{tr}\mathbf{J}^{-1}(\mu) = \frac{q\sigma^2}{N} \xrightarrow{d\to\infty \Rightarrow q\to\infty} \infty. \tag{6.13}
$$

The trace of the covariance matrix is the sum of the squared variances and grows linearly in the number of parameters. Unfortunately, the trace has no direct geometric interpretation. But since in this example the features are all pairwise independent, the only nonzero entries are on the diagonal: hence one also has $\det\operatorname{Cov}\left\{\underline{\hat{\mu}}\right\} \geq \det\mathbf{J}^{-1}(\mu)$ and therefore

$$
\sqrt{\det\operatorname{Cov}\left\{\underline{\hat{\mu}}\right\}} \geq \frac{1}{\sqrt{N}}\sqrt{\det\mathbf{J}^{-1}(\mu)} = \frac{\sigma^q}{\sqrt{N}} \xrightarrow{d\to\infty \Rightarrow q\to\infty} \infty. \tag{6.14}
$$

$|\operatorname{Cov}\{\underline{\hat{\mu}}\}|$ equals the volume of the parallelepiped spanned by the columns of the matrix $\operatorname{Cov}\{\underline{\hat{\mu}}\}$. One could say that the volume of the estimation error grows exponentially with the dimension of the parameter space.

In summary, if the feature space dimension tends to infinity $(d \to \infty)$, then the parameter space dimension tends to infinity as well $(q \to \infty)$. Thus, also the overall

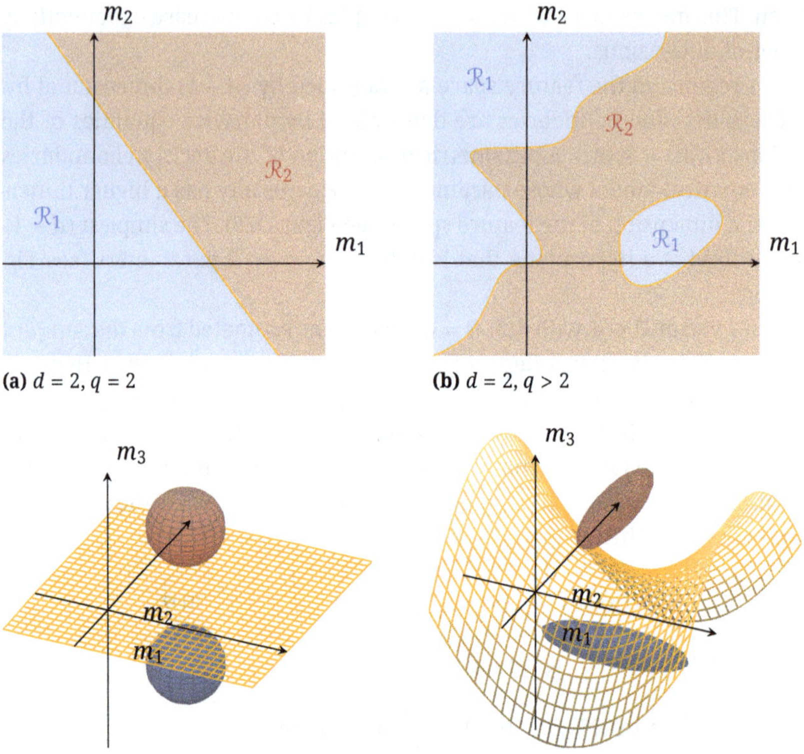

(a) $d = 2, q = 2$ **(b)** $d = 2, q > 2$

(c) $d = 3, q = 3$ **(d)** $d = 3, q > 3$

Fig. 6.5: Examples for the interrelation of the feature space dimension d and the parameter space dimension q according to the parametrized decision boundaries. The simplest decision boundaries are hyperplanes with $q = d$. To describe the orientation, $q - 1$ parameters are necessary and one parameter determines the location (distance to origin) within the feature space (see examples (a) and (c)). Additional degrees of freedom that allow deformation of the separating hyperspace need additional parameters and thus $q > d$ (see examples (b) and (d)).

estimation errors of the parameters of the decision boundary (Equations (6.13) and (6.14)) grow infinitely.

As the last point in this section, the computational complexity of both tasks, estimation and classification, will be considered. To this end, we examine the complexity of the decision function k_i on the basis of the MAP classifier for the Gaussian case:

Learning (Estimation)

$$k_i(\mathbf{m}) = -\frac{1}{2}\underbrace{(\mathbf{m}-\hat{\boldsymbol{\mu}}_i)^{\mathsf{T}}}_{\mathcal{O}(dN)}\underbrace{\hat{\boldsymbol{\Sigma}}_i^{-1}}_{\mathcal{O}(d^2N)}\underbrace{(\mathbf{m}-\hat{\boldsymbol{\mu}}_i)}_{\mathcal{O}(dN)}-\underbrace{\frac{d}{2}\ln 2\pi}_{\mathcal{O}(1)}-\frac{1}{2}\ln\det\underbrace{\hat{\boldsymbol{\Sigma}}_i}_{\mathcal{O}(d^2)}+\ln\underbrace{P(\omega_i)}_{\mathcal{O}(N)}. \tag{6.15}$$

The covariance matrix $\hat{\mathbf{\Sigma}}$ has $\frac{d(d+1)}{2}$ distinct entries, hence the estimation is asymptotically dominated by $\mathcal{O}(d^2N)$. The necessary matrix inversion $\hat{\mathbf{\Sigma}}^{-1}$ requires $\mathcal{O}(d^{2.4})$ operations, but $d^{2.4} < d^2N$ due to the fact that $d < N$. Therefore, the overall complexity to determine k_i is $\mathcal{O}(d^2N)$. As there are $i = 1, \ldots, c$ such decision functions, the total cost is $\mathcal{O}(cd^2N)$.

Classification

$$k_i(\mathbf{m}) = \underbrace{-\frac{1}{2}(\mathbf{m} - \hat{\mu}_i)^\mathsf{T}\hat{\mathbf{\Sigma}}_i^{-1}(\mathbf{m} - \hat{\mu}_i)}_{\mathcal{O}(d^2)} \underbrace{- \frac{d}{2}\ln 2\pi - \frac{1}{2}\ln\det\hat{\mathbf{\Sigma}}_i + \ln P(\omega_i)}_{\mathcal{O}(1)} . \tag{6.16}$$

Evaluating a single decision function needs, asymptotically, $\mathcal{O}(d^2)$ operations. Again, as there are c such functions that need to be compared, the total complexity is $\mathcal{O}(cd^2)$.

This short analysis reveals that the complexity grows quadratically with the dimension d. Slower growth would be even better, but this is actually a good result and far from posing a serious problem. Usually, for currently available systems and reasonable applications, the computational complexity is not an issue. If a classification task fails due to the high dimensionality of the feature space, then the previously discussed points are much more likely to bear the blame than is the computational complexity.

To summarize, if the dimension d of the feature space increases, …

- …the interval probability $\Pr(|m_1 - \mu_1| < 2\sigma, \ldots, |m_d - \mu_d| < 2\sigma)$ decreases;
- …the dimension of the parameter space q increases;
- …the total parameter estimation error increases;
- …the density of the samples y decreases and;
- …the complexity increases.

6.2 Overfitting

The term *overfitting* denotes a phenomenon that generally occurs when a model with a large number of parameters is fit to a set with too few samples. After the model is chosen and the number of parameters is fixed, the remaining objective is to minimize the classification error over the dataset. If the model is flexible enough, the error with respect to the specifically given dataset can be reduced to zero: the model learns the data by heart (see Figure 6.6b). However, this usually does not coincide with a good general solution and the classification error on new and unseen samples will be large. An overly simple model, on the other hand, is not able to sufficiently reduce the error at all (see Figure 6.6a), because it lacks the necessary flexibility to match the data. The ability to achieve a low error rate on both the training data and the testing data is called *generalization*.

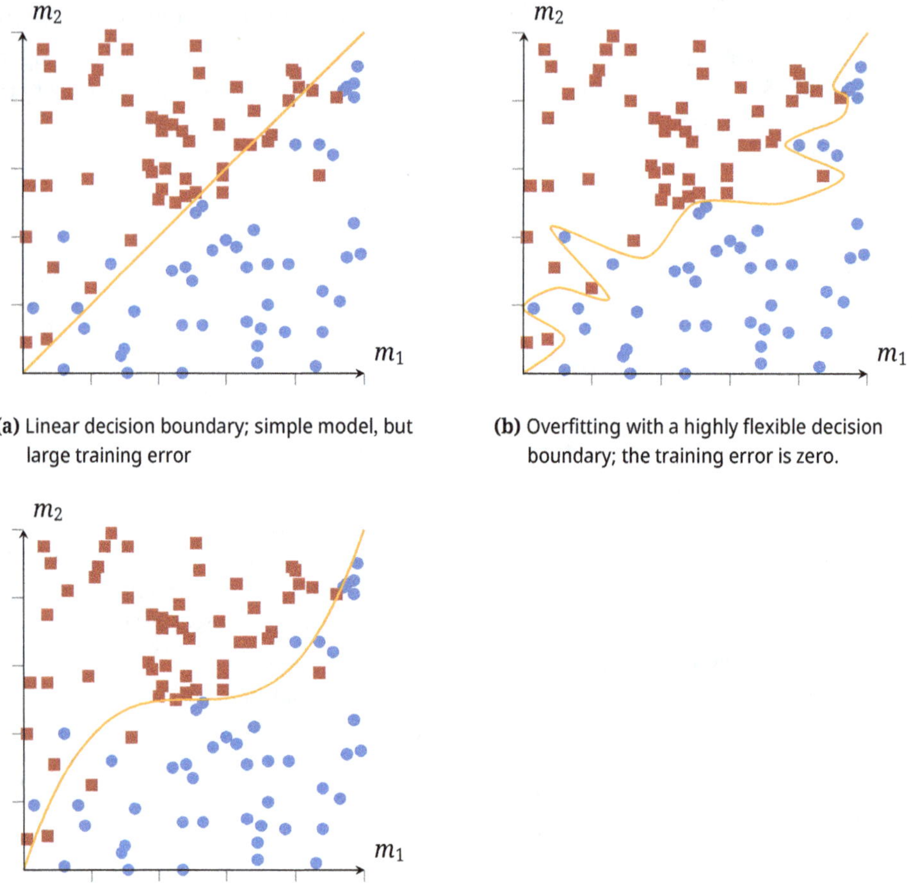

(a) Linear decision boundary; simple model, but large training error

(b) Overfitting with a highly flexible decision boundary; the training error is zero.

(c) Optimal decision boundary (Bayesian classifier); note that the training error is greater than zero.

Fig. 6.6: Trade-off between generalization and training error. The classifier should neither be too simple to represent the underlying classes, nor too complex to not generalize from the training data.

In order to check whether a chosen model fits the problem, the dataset can be divided into a training set $\mathcal{D}$ and a test set $\mathcal{T}$ (see Figure 1.5). The training set is used to estimate the parameters of the model, the test set is used to assess the model's performance and ability to generalize. Nonetheless, such a check can never be a strict proof that the model is the correct one, but only a test for plausibility. Hence, the question remains how to find the right model. In Figure 6.6c, the optimal decision boundary (of a Bayesian classifier) can be given, because the example was artificially created and the underlying model from which the dataset was generated was known. In reality, this is hardly ever true. Hence, the viable approach is to employ *Occam's razor*. This principle states that among different

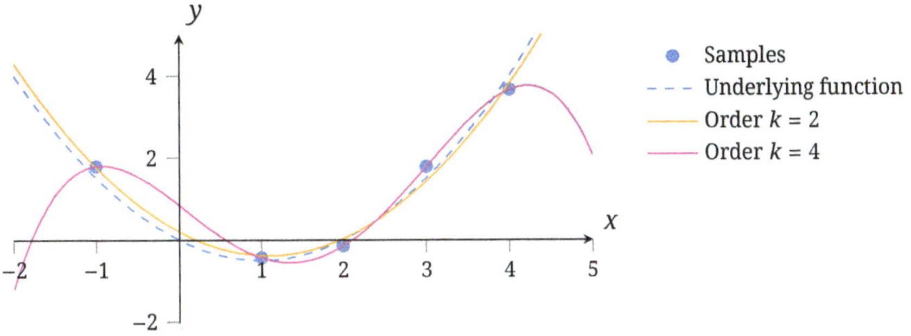

Fig. 6.7: Overfitting in a regression scenario. Two polynomials with order $k = 2$ and $k = 4$ are fitted to data that was generated from a polynomial with order $k = 2$.

competing hypotheses that are equally consistent with the given data, the hypothesis with the fewest assumptions should be selected.

Figure 6.6 illustrates the effect of overfitting by means of an example. The decision boundary tries to optimally separate the classes, within the limits of its ability.

The next example of overfitting is in the context of regression analysis. Let $f(x) = \frac{1}{2}x^2 - x$, $\underline{y} = f(x) + \underline{r}$ where $\underline{r}$ is some Gaussian distributed noise. Five samples (x_1, y_1), ..., (x_5, y_5) are given and it is only known that x and y are governed by some polynomial rule. The task is to find the best estimation $\hat{f}$ with $\hat{f}(x) = \sum_{i=0}^{k} a_i x^i$. For order $k = 4$, the regression $\hat{f}$ is able to perfectly fit the given samples, but overall the regression with order $k = 2$ resembles the true function much better although it exhibits a small training error (see Figure 6.7). If there had been a sixth sample (x_6, y_6) and both regression functions had been kept fixed, the quadratic polynomial would very likely have been a better fit.

But if there had been many more samples (possible infinitely many), both regression functions would eventually converge to the true function. Hence, the model of order $k = 4$ is not generally worse than the model of order $k = 2$, it is only worse for a small number of samples. This leads to the following rules of thumb, which are in accordance with Occam's razor:

- The smaller the dataset, the simpler the model should be and
- the higher the number of parameters of the model (or classifier), the more samples are required.

6.3 Exercises

(6.1) Given a sample $\mathcal{D}$ of five-dimensional feature vectors $\mathbf{m} \in \mathcal{D} \subset \mathbb{R}^5$, how many parameters need to be estimated for the feature density $p(\mathbf{m})$ to be represented by a Gaussian mixture model with 3 components, i.e.,

$$\hat{p}(\mathbf{m}) = \sum_{k=1}^{3} a_k g_k(\mathbf{m}), \tag{6.17}$$

where $g_k(\mathbf{m})$ denotes a Gaussian density? How many parameters need to be estimated when using a Parzen window method instead?

(6.2) Given two classes ω_1 and ω_2 that are to be classified using four-dimensional features $\mathbf{m} \in \mathbb{R}^4$, how many parameters must be estimated when using a linear classifier? How many parameters must be estimated for a maximum a posteriori classifier, under the assumption that the features are class-conditionally normally distributed, i.e., $p(\mathbf{m} \mid \omega_c) = \mathcal{N}(\boldsymbol{\mu}_c, \boldsymbol{\Sigma}_c)$, $c = 1,2$?

(6.3) A micro-controller for Internet of Things applications can only save up to 256 parameters. You are tasked to use this micro-controller for classification in a six-dimensional feature space.

1. How many linear classifiers can be realized using this micro-controller? How many parameters will remain unused?
2. How many classes can be separated using a maximum a posteriori classifier with multivariate Gaussian distributions as class-dependent feature distributions? How many parameters will remain unused?

(6.4) Suppose given a d-dimensional feature vector $\mathbf{m} = (m_1, \ldots, m_d)^\mathsf{T}$ of d stochastically independent features $m_i, i = 1, \ldots, d$. Each of the m_i is uniformly distributed over the interval $[-10,11]$. What is the smallest d that fulfills the following inequality:

$$\Pr\left(\mathbf{m} \notin [-2,5]^d\right) > \frac{9}{10}? \tag{6.18}$$

In other words, what is the smallest dimension d in which more than 90% of the probability mass will be outside of the hypercube $[-2,5]^d$?

7 Special classifiers

The remaining chapters of this book collect some further topics of pattern recognition. Except for Section 9.5, these chapters deal with certain important classifier methods. In contrast to the techniques of Chapters 3 to 5, these classifiers do not estimate a distribution first, but try to find a classification rule directly from the given data instead. This means that these classifiers follow the transduction path from Figure 5.1.

7.1 Linear discriminants

A linear discriminant is a linear function that operates on the feature space. With two classes ($c = 2$), $\mathbf{w} \in \mathbb{R}^d$ and $b \in \mathbb{R}$, it is given by

$$k(\mathbf{m}) = \mathbf{w}^\mathsf{T}\mathbf{m} + b \tag{7.1}$$

and the decision rule becomes

$$\hat{\omega} = \hat{\omega}(\mathbf{m}) = \begin{cases} \omega_1 & \text{if } k(\mathbf{m}) > 0 \\ \omega_2 & \text{if } k(\mathbf{m}) < 0\;. \\ \text{whatever} & \text{if } k(\mathbf{m}) = 0 \end{cases} \tag{7.2}$$

This means that the decision boundary is given by the $(d-1)$-dimensional hyperplane $\mathcal{H} = \{\mathbf{m} \in \mathcal{M} | k(\mathbf{m}) = 0\}$. Here, $\mathbf{w}$ is a vector perpendicular to the hyperplane and b determines the distance to the origin. The oriented distance of a point to the plane is given by $D(\mathbf{m}, \mathcal{H}) = k(\mathbf{m})$, provided that $\mathbf{m}$ is normalized ($\|\mathbf{m}\|_2 = 1$).

The resemblance to the decision function of Chapter 3 is not an accident. In case of a Bayesian classifier for normally distributed features with identical covariances, the decision function is linear. However, here and in the following sections, the approach is reversed: instead of inspecting the decision boundary of a Bayesian classifier, the decision boundary is explicitly stated. The parameters of the plane are directly determined from the training samples in such a way that the classification error is minimized. How to carry out such a minimization will be discussed in the following sections.

7.1.1 More than two classes

Equation (7.2) shows the decision rule for the case of $c = 2$ classes, but the approach can be extended to more classes as well. In the upcoming discussion, it is implicitly assumed that the classes are linearly separable.

The most straightforward solution is to determine one hyperplane per class. Each hyperplane divides the space into two half-spaces so that the samples that belong to the class fall in one half-space, while the samples that belong to the other classes fall in the

https://doi.org/10.1515/9783111339207-007

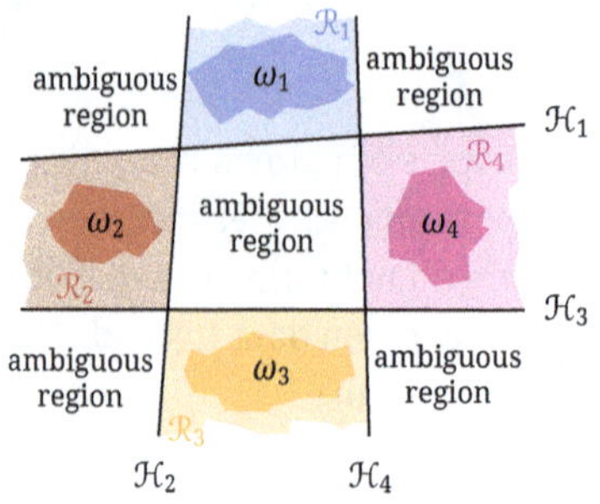

(a) One linear discriminant function per class

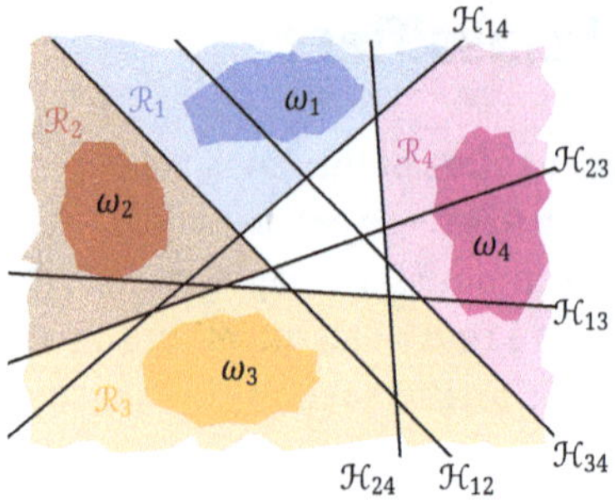

(b) One linear discriminant for each pair of classes

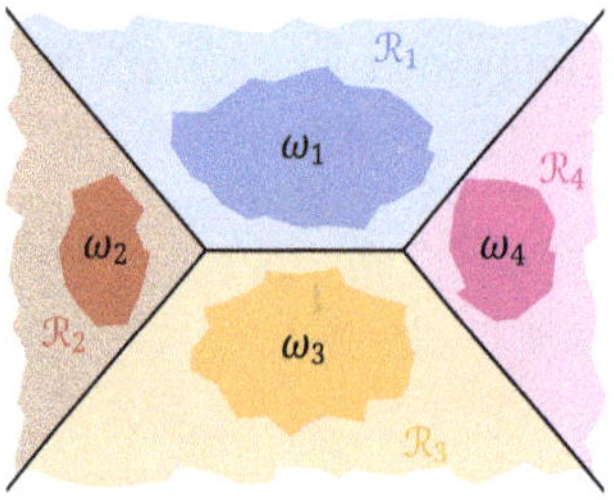

(c) Linear machine (no ambiguous regions)

Fig. 7.1: Different techniques for extending linear discriminants to more than two classes. All these methods except for the linear machine introduce ambiguous regions.

other half-space. An unseen sample is classified as belonging to class ω_i if it falls in the corresponding half-space, but not in a half-space that corresponds to the other classes, i.e., as before

$$\hat{\omega}(\mathbf{m}) := \omega_i \Leftrightarrow k_i(\mathbf{m}) > 0 \quad \text{and} \quad k_j(\mathbf{m}) < 0 \text{ for } i \neq j. \tag{7.3}$$

The resulting decision regions are depicted in Figure 7.1a. A major drawback of this approach is that a large volume of the feature space belongs to an ambiguous region, where no classification is possible.

Another approach is to determine one hyperplane $\mathcal{H}_{ij}$ with $i, j = 1, \ldots, c$, $i \neq j$ for each pair of classes, resulting in $\frac{c(c-1)}{2}$ linear discriminants. A sample is classified as being in class ω_i if its feature vector lies on the correct side of all hyperplanes that separate ω_i from the other classes ω_j,

$$\hat{\omega}(\mathbf{m}) := \omega_i \Leftrightarrow k_{ij}(\mathbf{m}) > 0 \text{ for all } j \neq i. \tag{7.4}$$

This approach is depicted in Figure 7.1b and usually leads to smaller ambiguous regions, at the cost of a more complicated classifier.

A third approach is given by the *linear machine*. As with the first approach, there is only one linear discriminant k_i for each class, but the decision rule is different: a point is assigned to a class if the corresponding linear discriminant is larger than any other:

$$\hat{\omega}(\mathbf{m}) := \omega_i \Leftrightarrow k_i(\mathbf{m}) > k_j(\mathbf{m}) \text{ for all } j \neq i. \tag{7.5}$$

As there is always a largest linear discriminant, there are no ambiguous regions. This leads to the Voronoi tessellation shown in Figure 7.1c. Note that the depicted hyperplanes are not the hyperplanes $\mathcal{H}_i$ given by $k_i(\mathbf{m}) = 0$. Instead, the final decision boundaries are given by

$$\begin{aligned} 0 &= k_i(\mathbf{m}) - k_j(\mathbf{m}) = \mathbf{w}_i^\mathsf{T}\mathbf{m} + b_i - \mathbf{w}_j^\mathsf{T}\mathbf{m} + b_j \\ &= (\mathbf{w}_i - \mathbf{w}_j)^\mathsf{T}\mathbf{m} + (b_i - b_j). \end{aligned} \tag{7.6}$$

As with the second technique, at most $\frac{c(c-1)}{2}$ decision boundaries are possible.

In summary, all three techniques have in common that the decision regions are convex and connected, because they are the intersection of half-spaces. Moreover the decision boundary consists of at most $\frac{c(c-1)}{2}$ segments of hyperplanes. Usually the number is smaller, because not all regions are mutually adjacent.

7.1.2 Nonlinear separation

Going back to the two-class cases $c = 2$, there is another possibility of extending the linear discriminants. Explicitly writing out the vectorized term,

$$k(\mathbf{m}) = \mathbf{w}^\mathsf{T}\mathbf{m} + b = \sum_{i=1}^{d} w_i m_i + b, \tag{7.7}$$

suggests that it can be extended by higher order combinations, e.g., quadratic terms,

$$k(\mathbf{m}) = \sum_{j=1}^{d}\sum_{i=1}^{d} w_{ij} m_i m_j + \sum_{i=1}^{d} w_i m_i + b = \mathbf{m}^\mathsf{T}\mathbf{W}\mathbf{m} + \mathbf{w}^\mathsf{T}\mathbf{m} + b. \tag{7.8}$$

As discussed in Equation (3.60), the decision boundary $k(\mathbf{m}) = 0$ is a hyperquadric that coincides with the decision boundary of a Bayesian classifier with class-specific features that are normally distributed. Adding higher order terms (cubic, quartic, quintic, etc.) allows deriving polynomial decision functions of any degree.

Indeed, the monomials can be replaced by arbitrary functions y_i, where by convention $y_0(\mathbf{m}) = 1$. This leads to an extended linear discriminant:

$$k(\mathbf{m}) = \sum_{i=0}^{d^*} a_i y_i(\mathbf{m}) = \underbrace{\begin{pmatrix} a_0 & \dots & a_{d^*} \end{pmatrix}}_{=\mathbf{a}^\mathsf{T}} \begin{pmatrix} 1 \\ y_1(\mathbf{m}) \\ \vdots \\ y_{d^*}(\mathbf{m}) \end{pmatrix} = \mathbf{a}^\mathsf{T}\mathbf{y}. \tag{7.9}$$

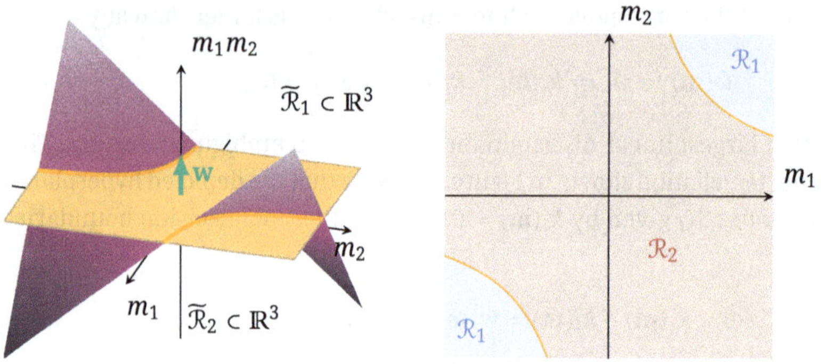

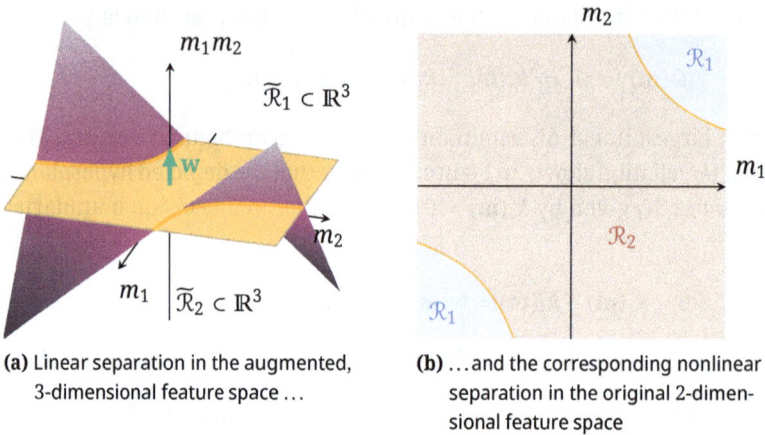

(a) Linear separation in the augmented, 3-dimensional feature space ...

(b) ...and the corresponding nonlinear separation in the original 2-dimensional feature space

Fig. 7.2: Nonlinear separation by augmentation of the feature space. The purple surface in (a) shows the embedding of the original feature space, the orange plane is the decision boundary of the linear discriminant in $\mathbb{R}^3$. The augmentation feature vector is defined as $\mathbf{y} := (1, m_1, m_2, m_1m_2)^\top$ and the parameters of the linear discriminant are $\mathbf{a} = (1,0,0,1)^\top$.

In the special case $\mathbf{y} := (1, m_1, \ldots, m_d)^\top$, Equation (7.9) corresponds to the ordinary linear discriminant in Equation (7.1). Although in the general case $k(\mathbf{m})$ is not linear in $\mathbf{m}$, it is linear in $\mathbf{y}$. In general, separability can be ensured by artificially increasing the dimension of the feature space by augmenting the feature space with nonlinear combinations of the existing features. Note that linear separation in the augmented space usually results in nonlinear decision boundaries in the original feature space (see Figure 7.2). In other words: augmenting the feature space makes it possible to use linear methods, even if the samples in the original feature space were not linearly separable.

Nonetheless, there are two pitfalls:

1. Although the coefficient vector $\mathbf{a}$ can be easily estimated, there is no standard technique to find appropriate nonlinear functions $y_i(\mathbf{m})$.
2. Increasing the dimensionality means that more parameters $\mathbf{a}$ have to be determined. All the effects discussed in Section 6.1 apply.

Figure 7.3 shows the decision regions of a linear regression classifier with an augmented feature vector (see description). By design of the feature vector, the decision boundary is a conic section and is visually very similar to the decision boundary of the Gaussian classifier in Figure 3.14. The linear regression classifier does not make any explicit assumption about the density of the features. However, such assumptions are implicit in the choice of feature augmentation.

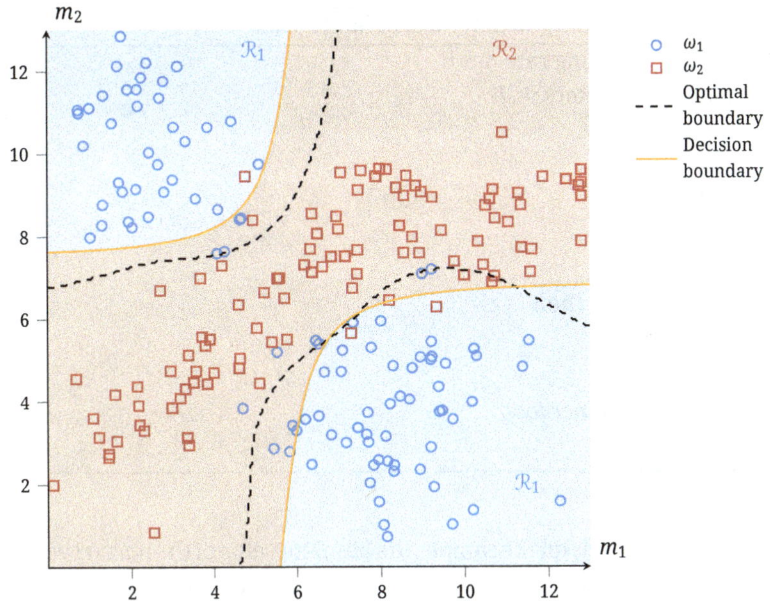

Fig. 7.3: Application to the reference example of Section 3.3.2. Decision regions of a linear regression classifier with augmented feature vector $\mathbf{y} := \left(1, m_1, m_2, m_1 m_2, m_1^2, m_2^2\right)^\mathsf{T}$. The training and testing errors are $e_{\text{train}} = 8.5\,\%$ and $e_{\text{test}} = 8.5\,\%$. The testing error asymptotically approaches $e_{\text{test}} \approx 8.7\,\%$. The training set is the same as in Figure 3.8. Test samples are shown with hollow marks.

7.2 The perceptron

What remains are techniques to determine a separating hyperplane from the given samples. The perceptron algorithm introduced byRosenblatt [1957, 1962] serves as the first example. A perceptron is a binary classifier ($c = 2$) and requires that the training set $\mathcal{D}$ is linearly separable. The pseudocode to learn the classifier is shown in Algorithm 7.1. For each sample $\mathbf{m}_i$, an indicator variable

$$z_i := \begin{cases} 1 & \text{if } \omega(\mathbf{m}) = \omega_1 \\ -1 & \text{if } \omega(\mathbf{m}) = \omega_2 \end{cases} \tag{7.10}$$

is introduced so that both correct and false classifications can be covered with a single statement (see line 4 of Algorithm 7.1). This indicator variable is often found in binary classifiers and will reappear in Section 7.7.

The perceptron algorithm starts with an arbitrary but fixed hyperplane and iteratively constructs a sequence of hyperplanes until the training error is 0. This can only work if the training data is linearly separable, but if it is, then the algorithm is guaranteed to converge.

Algorithm 7.1: The perceptron algorithm (Rosenblatt [1962]).

Data: Training set $\mathcal{D}$, Learning rate $\eta > 0$
Result: Hyperplane parameters $\mathbf{w}$, b
$\mathbf{w} \leftarrow \mathbf{0}$
$b \leftarrow 0$
$R \leftarrow \max_{1 \le i \le N} \|\mathbf{m}_i\|$
repeat
    forall $\mathbf{m}_i \in \mathcal{D}$ **do**
        if $z_i \left(\mathbf{w}^\top \mathbf{m}_i + b\right) \le 0$ **then**
            $\mathbf{w} \leftarrow \mathbf{w} + \eta \, z_i \, \mathbf{m}_i$
            $b \leftarrow b + \eta \, z_i \, R^2$
until *no training error in inner loop*
return $\mathbf{w}$, b

The speed of convergence depends on the sample, the absolute value of the normal vector $\mathbf{m}$, and the learning rate η. Novikoff [1962] showed that the influence of a single update eventually vanishes and that the sequence of hyperplanes converges.

Theorem 7.1 (Novikoff's perceptron theorem). Let $\mathcal{D}$ be linearly separable with a margin $\gamma > 0$. This means there exists a hyperplane given by $\mathbf{w}$ and b with $\|\mathbf{w}\| = 1$ and for all $\mathbf{m}_i \in \mathcal{D}$ one has

$$z_i(\mathbf{w}^\top \mathbf{m}_i + b) \ge \gamma. \tag{7.11}$$

Then the perceptron algorithm makes at most

$$k \le \left(\frac{2R}{\gamma}\right)^2 \tag{7.12}$$

errors on the training set.

Though the perceptron algorithm is guaranteed to find a separating hyperplane with zero training error if such a hyperplane exists, the solution is often not good in an intuitive sense. Figure 7.4 shows three intermediate hyperplanes and the final hyperplane found by the perceptron algorithm. Clearly, a human would have drawn a different decision boundary. In other words, the hyperplane found by the perceptron does not necessarily generalize well to unseen data. Furthermore, the final results depend on the initial hyperplane $\mathbf{w}_0$, b_0, the learning rate η, and the order in which the samples in $\mathcal{D}$ are processed. Theorem 7.1 already required the existence of a hyperplane with margin γ. Surely, it would be desirable to find such a hyperplane with γ as large as possible. This idea is picked up in Section 7.7.

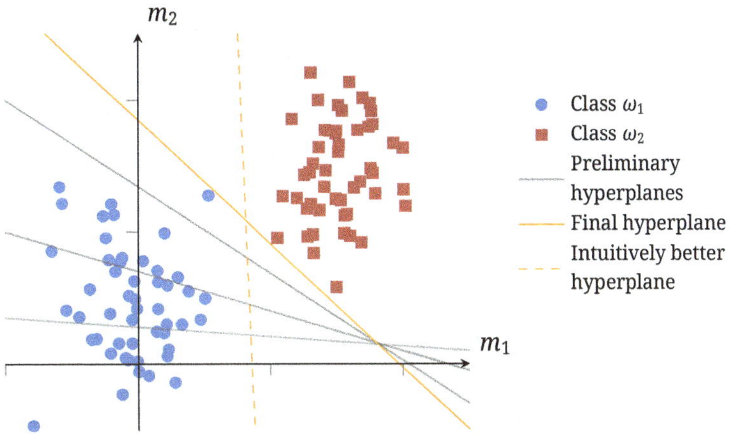

Fig. 7.4: Four steps of the perceptron algorithm. The algorithm converged to a separating, but suboptimal hyperplane.

7.3 Linear regression

The aim of linear regression is to find a linear function (see Equation (7.1)) that best maps a set of input vectors to their corresponding output. Note that "best" is only loosely defined and can, for example, mean minimal squared error, minimal absolute error, or any other loss function.

In the context of pattern recognition, linear regression can be applied to learn a linear decision function for each class ω_i. The input is given by the dataset $\mathcal{D}$ and the corresponding output is the (perfect) decision function, that is, the objective is to find a decision function $k_i(\mathbf{m}) = \mathbf{a}_i^\top \mathbf{m}$, $\mathbf{a}_i \in \mathbb{R}^d$ such that

$$k_i(\mathbf{m}_k) = \mathbf{a}_i^\top \mathbf{m} \stackrel{!}{=} \begin{cases} 1 & \text{if } \omega(\mathbf{m}) = \omega_i \\ 0 & \text{otherwise} \end{cases} \tag{7.13}$$

for every $i = 1, \ldots, c$. Given the training sample $\mathcal{D}$, the optimization goals become

$$\mathbf{m}_k^\top \mathbf{a}_i \stackrel{!}{=} z_{ki} := \begin{cases} 1 & \text{if } \omega(\mathbf{m}_k) = \omega_i \\ 0 & \text{otherwise} \end{cases}, \tag{7.14}$$

where $k = 1, \ldots, N$. These conditions can be conveniently expressed in matrix form:

$$\underbrace{\begin{pmatrix} \mathbf{m}_1^\top \\ \vdots \\ \mathbf{m}_1^\top \end{pmatrix}}_{:=\mathbf{M}} \mathbf{a}_i \stackrel{!}{=} \underbrace{\begin{pmatrix} z_{1i} \\ \vdots \\ z_{Ni} \end{pmatrix}}_{:=\mathbf{z}_i}, \tag{7.15}$$

where $\mathbf{z}_i$ is the membership vector to class ω_i and known for $\mathcal{D}$.

If $N > d$, Equation (7.14) is an overdetermined system of linear equations, hence an approximate solution can be obtained using the method of least squares:

$$e := \|\mathbf{M}\mathbf{a}_i - \mathbf{z}_i\|^2 \stackrel{!}{\rightarrow} \text{minimal}. \tag{7.16}$$

Factoring out Equation (7.16) and taking the gradient with respect to $\mathbf{a}_i$ yields

$$e = \mathbf{a}_i^\mathsf{T}\mathbf{M}^\mathsf{T}\mathbf{M}\mathbf{a}_i - \mathbf{z}_i^\mathsf{T}\mathbf{M}\mathbf{a}_i - \mathbf{a}_i^\mathsf{T}\mathbf{M}\mathbf{z}_i + \mathbf{z}_i^\mathsf{T}\mathbf{z}_i, \tag{7.17}$$

$$\nabla_{\mathbf{a}_i} e = 2\underbrace{\mathbf{M}^\mathsf{T}\mathbf{M}}_{\text{symmetric and invertible}}\mathbf{a}_i - 2\mathbf{M}^\mathsf{T}\mathbf{z}_i. \tag{7.18}$$

Setting $\nabla_{\mathbf{a}_i} = 0$ and solving for $\mathbf{a}_i$ yields the optimal (in the sense of minimal squared error) solution

$$\hat{\mathbf{a}}_i = \left(\mathbf{M}^\mathsf{T}\mathbf{M}\right)^{-1}\mathbf{M}^\mathsf{T}\mathbf{z}_i, \quad i = 1, \ldots, c. \tag{7.19}$$

The term $\left(\mathbf{M}^\mathsf{T}\mathbf{M}\right)^{-1}\mathbf{M}^\mathsf{T}$ is called a *pseudo-inverse* of the matrix $\mathbf{M}$. Substituting this result into Equation (7.13) gives the decision functions

$$k_i(\mathbf{m}) = \mathbf{z}_i^\mathsf{T}\mathbf{M}\left(\mathbf{M}^\mathsf{T}\mathbf{M}\right)^{-1}\mathbf{m}, \quad i = 1, \ldots, c. \tag{7.20}$$

As the pseudo-inverse and the feature vector $\mathbf{m}$ do not depend on ω_i, the entire decision vector can be written as

$$\mathbf{k}(\mathbf{m}) = \begin{pmatrix} \mathbf{z}_1^\mathsf{T} \\ \vdots \\ \mathbf{z}_c^\mathsf{T} \end{pmatrix} \mathbf{M}\left(\mathbf{M}^\mathsf{T}\mathbf{M}\right)^{-1}\mathbf{m}. \tag{7.21}$$

Note that the decision function does not take into account an offset b. However, an offset, as well as nonlinearities, can be included using the techniques discussed in Section 7.1.2. The structure of the classifier remains the same.

7.4 Artificial neural networks

artificial neural networks (ANNs) are biologically inspired networks of artificial neurons and synapses. The neurons sum the input from all incoming synapses, apply a nonlinear function, and output the result of the computation to all outgoing synapses. Artificial neural networks are modelled as directed graphs, where the nodes correspond to neurons and the edges correspond to synapses. A special type of neural network is the *feed-forward network*. These networks are directed acyclic graphs that are organized into layers such that the neurons of one layer have outgoing edges only to the neurons of the next layer, but not to neurons in the same or other layers. The first layer of a feed-forward network is called the *input layer* and the last layer is called the *output layer*. The layers in between are called *hidden layers*. The processing flow goes from the input to the output layer, but not the other way around.

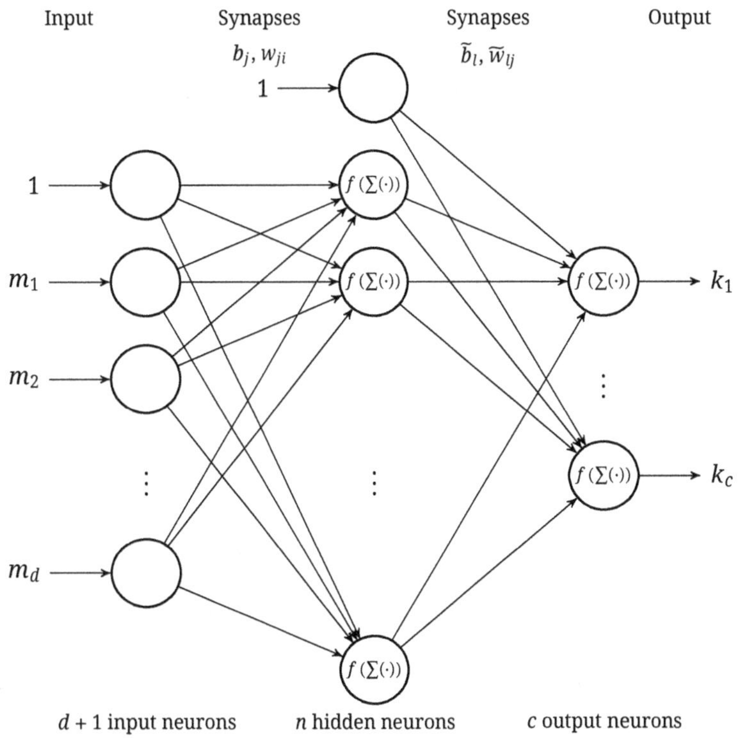

Fig. 7.5: Feed-forward neural network with one hidden layer.

Figure 7.5 shows a feed-forward neural network with one hidden layer. The input layer consists of $(d + 1)$ neurons, where d neurons distribute the features $m_1, \ldots, m_d$ to the neurons of the hidden layer and one neuron outputs a constant level. Such neurons are called *bias neurons*. The hidden layer consists of n neurons, where one neuron is, again, a bias neuron and the other $(n - 1)$ neurons compute the features from the input layer. Finally, the output layer consists of c neurons, where in this example each output computes the decision function k_l for the corresponding class. Each synapse is endowed with a weight. Here, w_{ji} denotes the weight from the i-th input neuron to the j-th hidden neuron, $\tilde{w}_{lj}$ denotes the weight from the j-th hidden neuron to the l-th output neuron, and b_j and $\tilde{b}_l$ denote the weights from the bias neurons.

Overall, this network computes the discriminant functions

$$k_l(\mathbf{m}) = f\left(\sum_{j=1}^{n} \tilde{w}_{lj} f\left(\sum_{i=1}^{d} w_{ji} m_i + b_j\right) + \tilde{b}_l\right) \qquad l = 1, \ldots, c$$

$$= f\left(\tilde{\mathbf{w}}_l^\mathsf{T} \mathbf{h} + \tilde{b}_l\right) \quad \text{where } \mathbf{h} := \begin{pmatrix} f\left(\mathbf{w}_1^\mathsf{T} \mathbf{m} + b_1\right) \\ \vdots \\ f\left(\mathbf{w}_n^\mathsf{T} \mathbf{m} + b_n\right) \end{pmatrix}. \tag{7.22}$$

The *activation function* $f(\cdot)$ is not further specified, but a typical choice is the *Fermi function* $f(\xi) := \frac{1}{1+e^{-\xi}}$, which approaches 0 as ξ goes to $-\infty$, approaches 1 as ξ goes to ∞, and is $\frac{1}{2}$ for $\xi = 0$ (a sigmoid activation). Surprisingly, feed-forward neural networks with one hidden layer can represent *any* continuous function $\mathbf{m} \mapsto \mathbf{k}$ that maps features to decision vectors. This result can be proven using Kolmogorov's general representation theorem (Kolmogorov [1963]):

Theorem 7.2 (General representation theorem). Every real-valued continuous function $g(\mathbf{m})$ defined on the cube $[0,1]^d$, $d > 1$ can be represented by

$$g(\mathbf{m}) = \sum_{j=1}^{2d+1} \Xi_j \left(\sum_{i=1}^{d} \psi_{ij}(m_i) \right), \tag{7.23}$$

where Ξ_j and ψ_{ij} are continuous functions of one variable.

In general, Ξ_j and ψ_{ij} are nonlinear functions. Because of this result, neural networks are sometimes also called universal function approximators.

A feed-forward neural network is commonly trained through *backpropagation*. Backpropagation iteratively minimizes the training error by propagating the error from the output layer to the input layer, all while adjusting the weights along the way. This process will now be examined in greater detail.

For the sake of convenience, the biases b_j and weights w_{ij} for each layer of the network are combined in a single matrix

$$\widetilde{\mathbf{W}} = \begin{bmatrix} w_{10} = b_1 & w_{11} & w_{12} & \cdots & w_{1d} \\ w_{20} = b_2 & w_{21} & w_{22} & \cdots & w_{2d} \\ w_{30} = b_3 & w_{31} & w_{32} & \cdots & w_{3d} \\ \vdots & \vdots & \vdots & \ddots & \vdots \\ w_{\tilde{d}0} = b_{\tilde{d}} & w_{\tilde{d}1} & w_{\tilde{d}1} & \cdots & w_{\tilde{d}d} \end{bmatrix} \tag{7.24}$$

and the vector $\widetilde{\mathbf{m}} = \left[1, \mathbf{m}^{\mathsf{T}}\right]^{\mathsf{T}}$ is augmented accordingly. Input and output dimension of that layer are d and $\tilde{d}$, respectively. The calculation of a single neuron layer can then be expressed by

$$\mathbf{k} = f(\mathbf{W}\mathbf{m} + \mathbf{b}) = f(\widetilde{\mathbf{W}}\,\widetilde{\mathbf{m}}) \tag{7.25}$$

with the activation function $f(\cdot)$ applied component-wise. This process is depicted in Figure 7.6.

For a network with H layers excluding the input and corresponding weight matrices $\widetilde{\mathbf{W}}^{\nu}, \nu = 1, \ldots, H$, the propagation of an input $\widetilde{\mathbf{m}}$ through the network yields

$$\mathbf{k}^H(\widetilde{\mathbf{m}}) = f\left(\widetilde{\mathbf{W}}^H f\left(\cdots \widetilde{\mathbf{W}}^2 f\left(\widetilde{\mathbf{W}}^1 \widetilde{\mathbf{m}}\right)\right)\right) \tag{7.26}$$

which requires heavy application of the chain rule to calculate the derivative. During the training process, a gradient descent on a squared error term

$$e(\widetilde{\mathbf{W}}) := \mathrm{E}\left\{\left[\|\mathbf{k} - \omega\|^2\right]\right\} \tag{7.27}$$

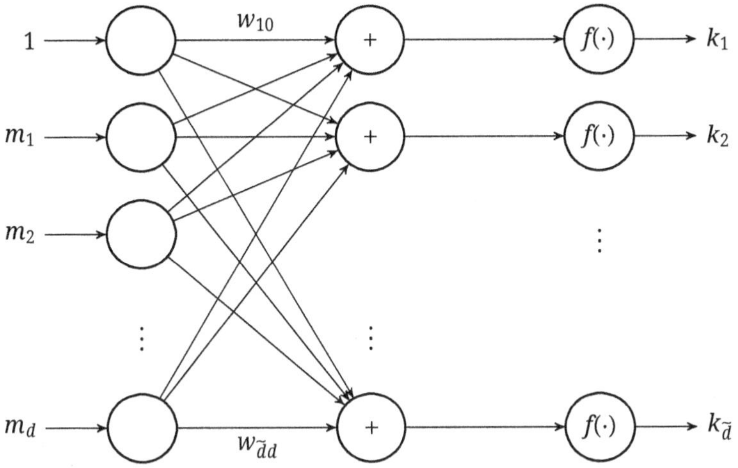

Fig. 7.6: Single layer of a feed-forward neural network.

is performed by adjusting the weights $\widetilde{\mathbf{W}}$ in each iteration. The objective is to find a global minimum. A learning rate η is introduced, which can be adjusted to avoid local minima. The iterative adaptation of $\widetilde{\mathbf{W}}$ expands to

$$\Delta\widetilde{\mathbf{W}} = \eta\nabla e \tag{7.28}$$

and requires the calculation of the gradient $\nabla e = \frac{\partial e}{\partial \widetilde{\mathbf{W}}}$ in each iteration. As a precondition, the activation function must be differentiable over the activation region, which is the case for the Fermi function:

$$\frac{df}{d\xi} = \frac{1}{(1+\mathrm{e}^{-\xi})^2}\mathrm{e}^{(-\xi)}(-1) = \frac{1}{1+\mathrm{e}^{-\xi}}\left(1 - \frac{1}{1+\mathrm{e}^{-\xi}}\right) = f(\xi)(1-f(\xi)). \tag{7.29}$$

Since the expectation of the squared error term in Equation (7.27) is not known, it is estimated with the empirical mean over the training set $\mathcal{D}$ to evaluate the gradient. Therefore, the gradient expands to

$$\nabla e \approx \frac{1}{N}\sum_{n=1}^{N}\nabla e_n \tag{7.30}$$

with $e_n = \frac{1}{2}\|\mathbf{k}(\mathbf{m}_n) - \boldsymbol{\omega}(\mathbf{m}_n)\|^2$. The impact of particular neurons on other layers can be computed to obtain an update rule for the weights. As illustrated in Figure 7.7, we denote the i-th input of layer v as k_i^{v-1} and the j-th output as k_j^v. Applying the weights $\widetilde{\mathbf{W}}^v$ of layer v to its input yields a vector $\mathbf{s}^v := \widetilde{\mathbf{W}}^v\mathbf{k}^{v-1}$.

Using this *weighted input* $\mathbf{s}$, the output of a layer v can be reformulated as

$$k_j^v = f(s_j^v) = f\left(\sum_{i=0}^{d^v} w_{ji}^v k_i^{v-1}\right). \tag{7.31}$$

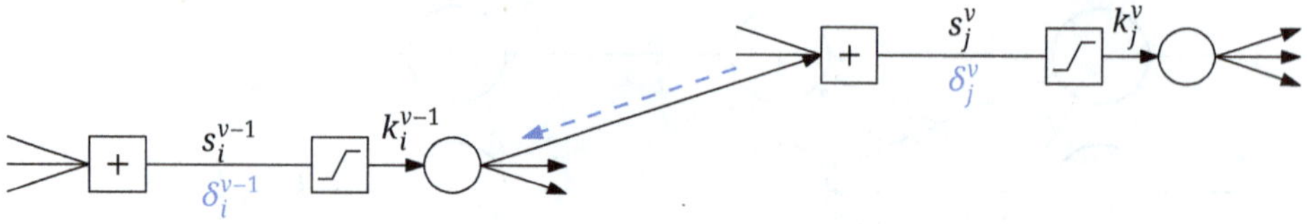

Fig. 7.7: Updates of weights in a feed-forward neural network.

This notation enables the specification of a term

$$\delta_j^v = \frac{\partial e}{\partial s_j^v} \tag{7.32}$$

for the j-th neuron in layer v. The main purpose of this notation is to relate the neuron-specific error back to the weight update $\frac{\partial e}{\partial \mathbf{W}}$ in an algebraically convenient way, even though the direct relation to the output k_j^v would be possible as well.

The above formulation helps to retrieve the partial derivatives

$$\begin{aligned} \frac{\partial e}{\partial w_{ji}^v} &= \frac{\partial e}{\partial s_j^v} \frac{\partial s_j^v}{\partial w_{ji}^v} \\ &= \delta_j^v \frac{\partial s_j^v}{\partial w_{ji}^v} \\ &= \delta_j^v k_i^{v-1} \end{aligned} \tag{7.33}$$

which measure the rate of change of the error term induced by the output k_i^{v-1} of the previous layer. Deriving the neuron specific error term is slightly more complicated. For quadratic error terms, applying the chain rule to the errors in the output layer yields an easily tractable term

$$\begin{aligned} \delta_j^H &= \frac{\partial e}{\partial s_j^H} \\ &= \sum_l \frac{\partial e}{\partial k_l^H} \frac{\partial k_l^H}{\partial s_j^H} \\ &\quad (k_l \text{ only depends on input } s_j \text{ for } j = l) \\ &= \frac{\partial e}{\partial k_j^H} \frac{\partial k_j^H}{\partial s_j^H} \\ &= \frac{\partial \frac{1}{2} \sum_{l=1}^{c} (k_l - \omega_l)^2}{\partial k_j^H} f'\left(s_j^H\right) \\ &= (k_j - \omega_j) f'\left(s_j^H\right). \end{aligned} \tag{7.34}$$

Here, ω_l denotes the l-th entry of $\boldsymbol{\omega}$ and f' the derivative of the activation function f. Putting it all together, the weight update for the last layer is given by:

$$\begin{aligned}\Delta w_{ji}^{H} &= -\eta \frac{\partial e}{\partial w_{ji}^{H}}\\ &= -\eta \delta_j^H k_i^{H-1}\\ &= -\eta\left(k_j - \omega_j\right) f'\left(s_j^H\right) k_i^{H-1}. \end{aligned} \tag{7.35}$$

For all other layers $v < H$, the effect of all inputs s_j^{v-1} on the weightened inputs of the subsequent layer s_k^v must be taken into account. Therefore, the simplification from Equation (7.34) cannot be applied. Using the chain rule and applying the definitions yields

$$\begin{aligned}\delta_j^{v-1} &= \frac{\partial e}{\partial s_j^{v-1}}\\ &= \sum_l \frac{\partial e}{\partial s_l^v}\frac{\partial s_l^v}{\partial s_j^{v-1}}\\ &= \sum_l \delta_l^v \frac{\partial s_l^v}{\partial s_j^{v-1}}\\ &\text{(using the definition of } s_l^v)\\ &= \sum_l \delta_l^v \frac{\partial\left(\sum_i w_{li}^v k_i^{v-1}\right)}{\partial s_j^{v-1}}\\ &\text{(using the definition of } k_i^{v-1} = f\left(s_i^{v-1}\right))\\ &= \sum_l \delta_l^v w_{lj}^v f'\left(s_j^{v-1}\right)\\ &= f'\left(s_j^{v-1}\right)\left[\sum_l w_{lj}^v \delta_l^v\right] \end{aligned} \tag{7.36}$$

which can now be utilized to compute each error within the network and update the weights according to Equation (7.35). Note that as compared to the derivation for the last layer, the errors in each layer δ_j^{v-1} are now directly related to the errors δ_j^v in the subsequent layer. In this sense, the error is being *backpropagated* from layer v to the preceding layer $v-1$.

In summary, the **backpropagation algorithm** consists of the following steps:

Initialization Initialize the weight matrices $\widetilde{\mathbf{W}}^1, \ldots, \widetilde{\mathbf{W}}^H$ of all layers. Then, for each training sample $(\mathbf{m}_i, \boldsymbol{\omega}_i)$, do:

Forward pass Using the weights, calculate the outputs k_i^v for each layer.

Error calculation Calculate the error $\delta_j^H = (k_j - \omega_j) f'\left(s_j^H\right)$ on the neurons in the last layer (Equation (7.34)).

Error backpropagation Propagate the error from the last layer to the preceding ones by calculating $f'\left(s_j^{v-1}\right)\left[\sum_l w_{lj}^v \delta_l^v\right]$ for $v = H-1, H-2, \ldots, 1$ (Equation (7.36)).

Weight update Update the weights according to Equation (7.35): $\Delta w_{ji}^{\nu} = -\eta \frac{\partial e}{\partial w_{ji}^{\nu}} = -\eta \delta_j^{\nu} k_i^{\nu-1}$ for all $i = 1,2,\ldots,d^{\nu-1}$; $j = 1,2,\ldots,d^{\nu}$ and layers $\nu = 1,2,\ldots,H$.

This process is computationally efficient as it only requires one forward-pass and one backward-pass to update all weights simultaneously.

Feed-forward neural networks have the advantage of requiring no prior knowledge and being relatively easy to configure. At the same time, they allow the approximation of arbitrary decision functions and provide a very fast classification.

On the other hand, large neural networks require the estimation of a substantial number of parameters, which comes with all the associated problems (see Section 6.1). Although backpropagation is an efficient algorithm, it is still computationally expensive to train networks compared to other techniques, particulary when the training set is large.

There are no established criteria for constructing neural networks for a given task and it is challenging to interpret the impact of individual weights on the underlying computation. Neural networks are prone to overfitting the training data and there is no assurance that the learned parameters constitute a global optimum, as the loss function in Equation (7.27) is non-convex. Nonetheless, neural networks achieve remarkable results in numerous application domains.

Figure 7.8 shows the decision regions of a feed-forward neural network with two hidden layers, each of which contains ten neurons. The decision boundary is complicated and does not seem to match the optimal decision boundary well, yet with 7.6 %, the asymptotic testing error is only 1.5 percentage points larger than the Bayes error rate. However, the decision regions can look vastly different when the architecture is changed, e.g., when using a different number of hidden layers, or if the layers contain a different number of hidden neurons. Even a different choice of initial weights can have a significant impact on the decision boundary. For these reasons, a thorough evaluation of the parameter space is paramount when using this highly flexible classifier.

7.5 Autoencoders

An artificial neural network can also be used to compress data by letting it learn to reproduce its input (i.e., to learn the identity function). If the hidden layer consists of fewer neurons than the input layer, the network is forced to learn some lower-dimensional encoding of the data. Such networks are called *autoencoders* and have already been introduced in Section 2.7.6. As an example (which is due to Ritter et al. [1990]), consider a dataset of eight training vectors $\mathcal{D} = \{\mathbf{m}_1, \ldots, \mathbf{m}_8\}$ with $\mathbf{m}_i \in \mathbb{R}^8$ for $i = 1,\ldots,8$. Let

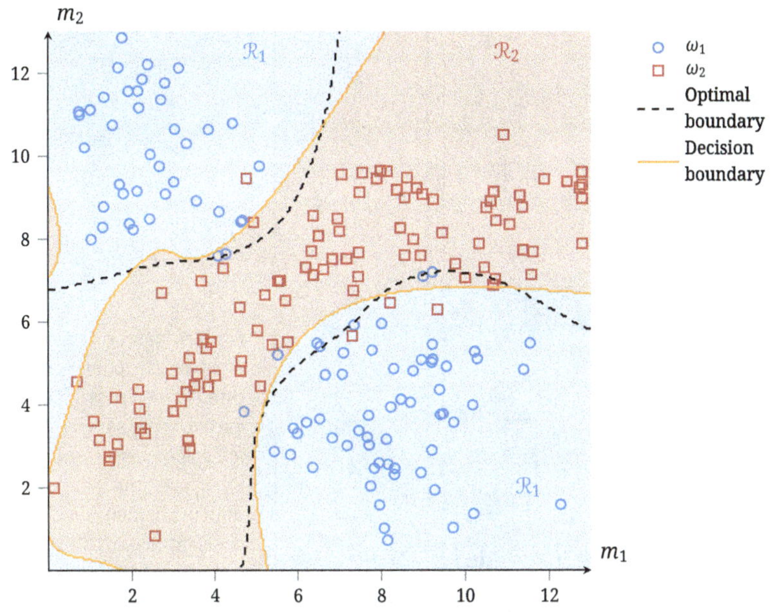

Fig. 7.8: Application to the reference example of Section 3.3.2. Decision regions of a feed-forward neural network with two hidden layers with 10 neurons each. Note that a different initialization usually results in vastly different decision boundaries. The training and testing errors are $e_{\text{train}} = 8\,\%$ and $e_{\text{test}} = 5.5\,\%$. The testing error asymptotically approaches $e_{\text{test}} \approx 7.6\,\%$. The training set is the same as in Figure 3.8. Test samples are shown with hollow marks.

the i-th entry of $\mathbf{m}_i$ be $\frac{9}{10}$ and every other entry be $\frac{1}{10}$:

$$\mathbf{m}_i := \left(\frac{1}{10}, \ldots, \underbrace{\frac{9}{10}}_{i\text{-th position}}, \ldots, \frac{1}{10}\right)^{\mathsf{T}}. \tag{7.37}$$

Figure 7.9 shows the activation of each neuron (except the bias neurons) of an autoencoder that was trained on this dataset. The net was trained for 5000 iterations with a learning rate of $\eta = 0.25$ and the final mean squared training error (see Equation (7.27)) was $\bar{e} = 0.047$. It can be seen that while the input is reconstructed almost perfectly, the network is forced to form some kind of binary-code with 3 bits in the hidden layer, since 8 different inputs are to be compressed to 3 numbers.

Note that the compression is only valid for the seen data: unseen training samples may not be compressed with a low reconstruction error. For example, this network produces a squared error of $e = 6.51$ when reconstructing the vector $\mathbf{m} = \mathbf{1} = (1, \ldots, 1)^{\mathsf{T}}$.

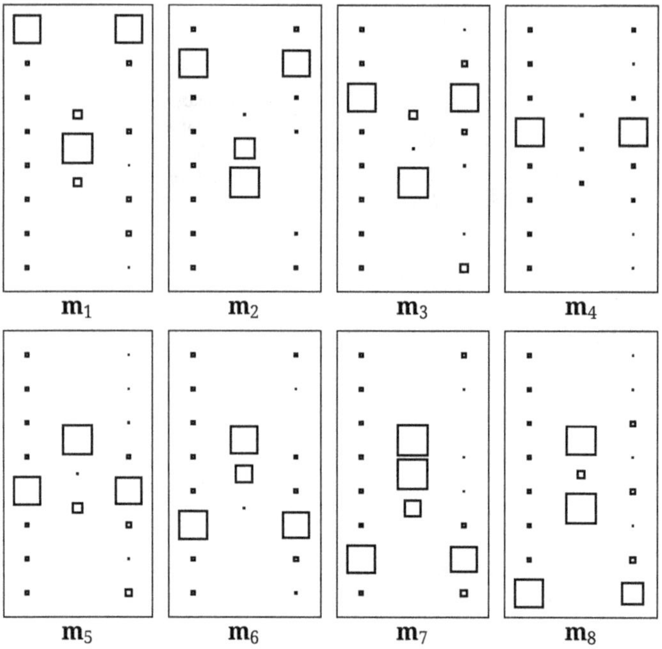

Fig. 7.9: Neuron activation of an autoencoder with three hidden neurons for each training sample. The left column represents the input layer, the middle column the hidden layer, and the right column the output layer. The size of a box indicates the magnitude of its activation.

7.6 Deep learning

Theorem 7.2 implies that a neural network with one hidden layer is sufficient to represent any function and therefore derive arbitrarily complicated decision boundaries, provided that it contains enough neurons. Yet there are still reasons to prefer networks with many layers with fewer neurons: deeper networks have the same approximation capabilities as shallow networks, but generally require fewer parameters (see Schmidhuber [2015]). For example, there are functions that require a polynomial number of parameters (w.r.t. the number of inputs) in a network with n hidden layers, but require an exponential number of parameters with $n-1$ layers (Schmidhuber [2015]). As discussed in Section 6.1, having fewer parameters typically leads to better generalization properties and reduces the risk of overfitting the training data.

Multiple layers can also be used to model hierarchical part–object relationships, where the first few layers model the individual parts and the following layers model the composition of those parts. A typical example of such a model is a car, where the first few layers could model car parts, such as wheels, doors, windows, headlights, etc., and the following layers model the car's frame and body. Lastly, the human visual cortex is also organized in many hierarchical layers, that fulfill increasingly complicated tasks.

Unfortunately, there is no clear definition of what constitutes a "deep" network. Generally, a neural network with one hidden layer is considered shallow, whereas a network with ten hidden layers is already considered deep. One of the pioneering works in deep learning, LeNet-5 by LeCun et al. [1998] had seven layers, but other architectures can have more than 1000 layers (He et al. [2016]).

7.6.1 Historical difficulties and successful approaches

Artificial neural networks with multiple layers are typically trained using the backpropagation algorithm, which, as mentioned above, performs a gradient descent to minimize the squared prediction error on the training set. However, deep networks cause two major issues with backpropagation: First, the gradient may vanish or explode during the backward pass due to exponential changes from one layer to another. In effect, gradient descent may require an unfeasibly large number of iterations to converge. Second, since gradient descent only considers local information about the first derivative, backpropagation often becomes stuck at saddle points or in local optima. This means that even if the gradient descent converges, there is no guarantee that the solution found is also a good solution.

Deep networks also tend to have large parameter spaces and therefore need more training data than models with fewer parameters. Such training data requires storage and computation time, neither of which were always as abundant as they are today. Furthermore, there were no clear concepts for representing practical problems or encoding prior knowledge in deep neural networks. At the same time, alternatives such as the SVM achieved a similar classification performance, but had a much more solid theoretical foundation.

In recent years, these issues have largely been solved. There are huge datasets, such as ImageNet (Deng et al. [2009]), that contain millions of labeled training samples. Graphical processing units allow significantly accelerating the computation of the gradient and therefore allow many more iterations with small learning rates. This means that even an almost vanishingly small gradient can be sufficient to escape a saddle point, provided that it is followed for long enough.

But there have also been significant theoretical advances to improve the training algorithm. Unsupervised pre-training allows using unlabeled training data to initialize a network with a good solution, before the supervised gradient descent is performed. Stochastic gradient descent, momentum, and weight decay speed up the training and avoid falling into local optima. Rectified linear units instead of sigmoid activation avoid vanishing gradients with large activations.

Similarly, specialized architectures have led to breakthroughs in certain areas. The long short-term memory (LSTM) approach (see Section 7.9.3) is well suited to handle sequential data, such as audio, text, or time series. Since LSTMs "remember" training errors during backpropagation, they also solve the problem of vanishing gradients. Convolutional neural networks convolve the input data with banks of trainable filters and are especially well suited for multidimensional data that contain repeating structures, e.g., image data. Residual networks (ResNets) are the most popular kind of CNNs that counteracts the problem of vanishing gradients and allows successful training of deep networks for image recognition with more than 100 layers. Variational autoencoders (VAE) are a special type of autoencoders which learn the underlying distribution of some data and use this knowledge to synthetically generate new data. Mixture density networks (MDNs) can be used to infer the parameters of a Gaussian mixture as direct output of a neural network, which has various practical applications.

A detailed discussion of these techniques is outside the scope of this book, but we will briefly explore the fundamental ideas and motivations in the following.

7.6.2 Unsupervised pre-training

In unsupervised pre-training, the weights are initialized by treating each layer individually. Here, the goal is to find a good initialization for the supervised backpropagation. The intuition is that the pre-training will move the parameter vector near a local optimum, which can then be quickly found by gradient descent. Unsupervised pre-training also does not require annotated training data, but only unlabeled data, which is much easier to obtain in large quantities. This reduces the need for labeled data, since the supervised training will only run for a short time. The only requirement is that the unlabeled data must be from the same domain as the labeled data, e.g., images of cars, if the goal is to classify car models, etc.

One particular approach to pre-training is the use of stacked autoencoders, where each layer of the deep network is treated as a coder of its input (Bengio et al. [2007], see Figure 7.10).

The layers are trained one after the other: the first layer is trained to reconstruct the input of the network with minimal error. When the training of this layer is complete, the weights are fixed and the output of the neurons of the first layer are treated as the input for training the second layer, etc. In this way, each layer performs a feature extraction for the next layer. This approach avoids problems caused by local optima, because training the next layer starts from a reasonable initialization. At the same time, layer-wise training significantly reduces the dimension of the parameter space compared to training the entire network at once. However, pre-training is unnecessary when a large amount of training data is available.

7.6.3 Stochastic gradient descent

Stochastic gradient descent (SGD) approximates the gradient using random subsets of the training data. In particular, in the t-th iteration, only the *batch* $\mathcal{B}_t \subset \mathcal{D}$ is used to approximate the gradient,

$$\frac{\partial e}{\partial \mathbf{w}} = \sum_{\mathbf{v} \in \mathcal{D}} \frac{\partial e_{\mathbf{v}}}{\partial \mathbf{w}_t} \approx \sum_{\mathbf{v} \in \mathcal{B}_t} \frac{\partial e_{\mathbf{v}}}{\partial \mathbf{w}_t}. \tag{7.38}$$

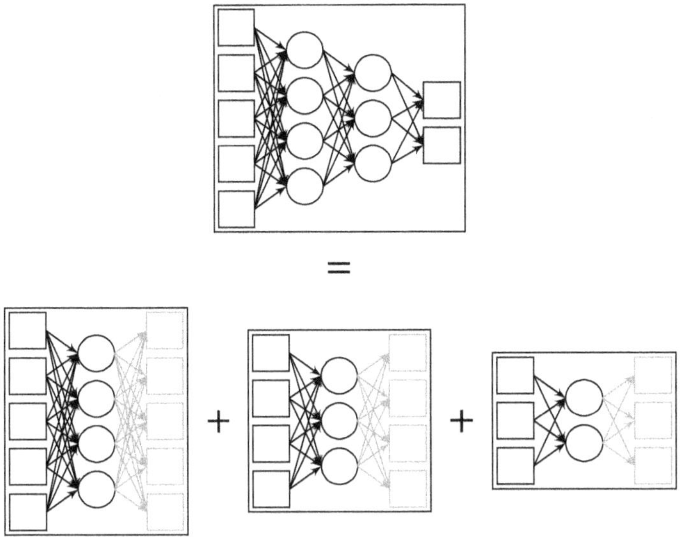

Fig. 7.10: Pre-training with stacked autoencoders decomposes training a deep network layer by layer. Each layer is trained as an autoencoder of its input, where the input is the output of the previous layer.

Doing so essentially randomizes the direction of the parameter update, which helps to escape from saddle points and shallow local optima. Note that stochastic gradient descent is theoretically justified only when the gradient of the error function can be written as the sum of gradients, i.e., if the first part of Equation (7.38) holds.

Another technique to escape from saddle points and local optima is the introduction of *momentum*. The idea is to carry a fraction of the last parameter update over to the current update,

$$\Delta \mathbf{w}_t = \frac{\partial e}{\partial \mathbf{w}} + M \cdot \Delta \mathbf{w}_{t-1}, \tag{7.39}$$

where $M \neq 0$ is the momentum factor. From a mechanical physics perspective, the above Equation (7.39) contains the second derivative of the location, the acceleration, and the factor M corresponds to the inertia of the system. As momentum builds up over time, this technique also enables the gradient descent to escape saddle points and local optima and to traverse long valleys that have only a slight inclination towards the optimum. In such situations, gradient descent without momentum tends to oscillate between the walls of the valley, especially when the direction of the gradient is randomized with stochastic gradient descent (Equation (7.38)).

Lastly, *weight decay* causes the weights to exponentially decay to zero if no other weight update (i.e., $\Delta \mathbf{w}_t = \mathbf{0}$) is performed. In practice, this is a regularization that penalizes large weights and therefore reduces overfitting. Formally, weight decay adds another term $D \cdot \mathbf{w}_t$ to the weight update, where $0 \leq D < 1$ is the weight decay factor. Putting the above together and with η denoting the learning rate, the overall parameter

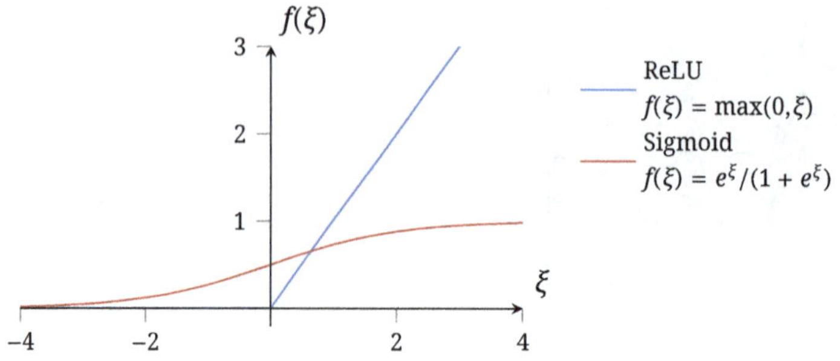

Fig. 7.11: Comparison of ReLU and sigmoid activation functions.

update is calculated as

$$\mathbf{w}_{t+1} = \mathbf{w}_t - \eta\, \Delta\mathbf{w}_t, \quad \text{where} \tag{7.40}$$

$$\Delta\mathbf{w}_t = D \cdot \mathbf{w}_t + M \cdot \Delta\mathbf{w}_{t-1} + \sum_{\mathbf{v}\in\mathcal{B}_t} \frac{\partial e_{\mathbf{v}}}{\partial \mathbf{w}_t}. \tag{7.41}$$

7.6.4 Rectified linear units

A rectified linear unit (ReLU) denotes an activation function that avoids the problem of vanishing gradients. In particular, the ReLU activation function

$$f(\xi) = \max(0,\xi) \tag{7.42}$$

is linear if $\xi \geq 0$ and 0 (rectified) otherwise. As with the sigmoid activation function, ReLU activation is nonlinear, but unlike the sigmoid, the gradient is constant over the range $(0,\infty)$ instead of vanishing for large ξ. Figure 7.11 compares both activation functions.

7.6.5 Convolutional neural networks

Convolutional neural networks (CNNs) are a specialized neural network architecture suitable for multidimensional data such as images. Such data contains repeating local structures, e.g., edges and corners, that are arranged to form higher level concepts. Convolutional neural networks exploit this to achieve a highly accurate classification, but are less suitable for data that are organized in a different manner.

A CNN takes the entire pattern as input. The first Q layers of a CNN essentially perform a hierarchical feature extraction by convolving the input with learned filters. These layers can be seen as a series of convolution blocks, where each block is composed of several layers (see Figure 7.12). Each layer in a convolution block plays a special role. The weights in the first layer are restricted, so that a neuron in the target layer is only

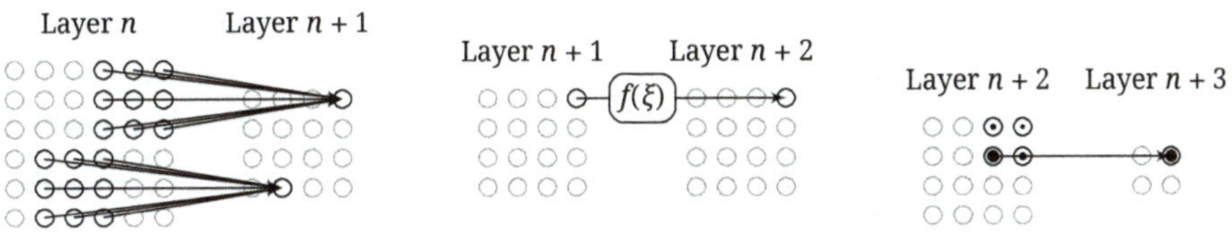

(a) Convolution **(b)** ReLU activation **(c)** Max pooling

Fig. 7.12: A single convolution block in a convolutional neural network.

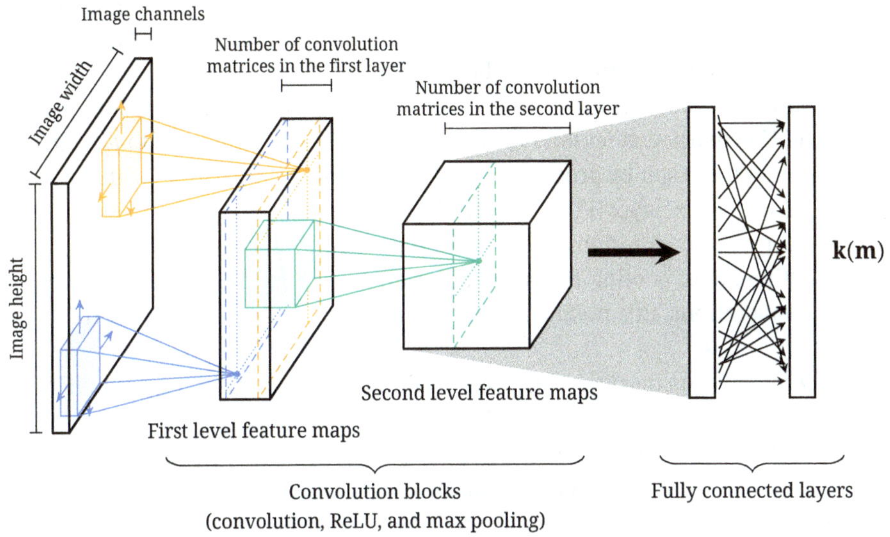

Fig. 7.13: High level structure of a toy example convolutional neural network with $Q = 2$ convolution blocks (convolution layer, ReLU, max pooling) and $R = 2$ fully connected layers. Stride and padding are not shown in the figure. All nodes in the last feature maps are connected without restriction to the nodes in the following (non-convolutional) hidden layer. In the figure this is indicated by the gray shading between these layers. Note: Real networks, e.g., in Krizhevsky et al. [2012], are usually much larger.

connected to a small neighborhood in the source layer (see Figure 7.12a). Furthermore, the weights that map a local neighborhood to a target neuron are shared across all neighborhoods. In effect, the weights can be seen as the entries of a convolution matrix. The convolution layer can be interpreted as performing a convolution of the input with that matrix, where the convolution matrix slides over the input.

Although Figure 7.12a shows only one such convolution, in practice several of these convolution matrices are learned in parallel. This is outlined in Figure 7.13, where the successive blocks in the convolution stage become deeper and deeper.

To reduce the computational effort and the number of parameters to learn, the convolution is often combined with a downscaling operation. A stride of n computes the convolution only at every n-th position in both spatial directions and therefore shrinks the output by a factor of n^2—a factor of n in both directions. For example, the output of a convolution layer with stride two is one fourth the size of the input, because the convolution is computed only on every odd row and column of the input. Stride is also often used to replace the pooling layer, as both have a similar effect, but strides significantly reduce the computation time.

Another reduction due to convolution is that positions on the boundary are omitted, because the convolution can only be computed at positions where the convolution matrix fully fits into the image. As this reduction is typically undesired, the input can be padded by a certain number of pixels to allow convolution at these positions. There is no clear guideline on how to fill the padded area, but common approaches are to tile the input periodically or to reflect the input at the border pixels.

The next layer performs a nonlinear mapping of the convolution result, usually using the ReLU activation function (see Figure 7.12b). Lastly, a so called *max pooling* layer subsamples the result by propagating only the maximum activation within a small neighborhood to the next layer (Figure 7.12c). These max pooling layers cause the features to become tolerant to translation and improve the tolerance for noise in the input data. At the same time, max pooling leads to data reduction (Figure 7.13). Combined with successive convolutions, this means that later stages see a larger portion of the input image.

These Q convolution blocks are followed by R fully connected layers (the final layers in Figure 7.13), where typically $Q > R$. The fully connected layers play the role of a regular multilayer feed-forward neural network that classifies the features extracted by the convolutional layers. However, the parameters of both the feature extraction and the classification are learned at the same time. In effect, the features are very well tuned to the classifier and vice versa.

CNNs are usually trained to minimize the error on the training set using (stochastic) gradient descent and not using pre-training. This way, it is ensured that the convolution matrices in the same layer learn different weights and that the convolution matrices of consecutive layers are geared to each other.

To avoid overfitting in the fully connected layer, one can use so called *dropout*, where a random fraction of neurons are *deactivated* in each training iteration (Srivastava et al. [2014]). The idea is that dropout forces the network to introduce redundancies and therefore reduces the risk of overfitting.

CNNs have been shown to be remarkably successful at difficult image recognition tasks. For example, prior to the advent of CNNs, the state of the art in image categorization (the task of classifying an image into one of many categories) achieved an error rate of 45.7 % on the ImageNet dataset. The convolutional neural network approach by Krizhevsky et al. [2012] reduced this error rate to 37.5 %! Their network consisted of five convolution blocks followed by two fully connected layers. Here, and with CNNs in general, the different convolution blocks produce different types of features: the first block essentially detects edges and corners; the second block responds to primitive textures; the third block detects parts of objects, etc. This can be seen in Figure 7.14. The figure

(a) 1st convolution layer

(b) 2nd convolution layer

(c) 3rd convolution layer

(d) 4th convolution layer

(e) 5th convolution layer

Fig. 7.14: Image patches that produce large filter responses in a convolutional neural network with five convolution blocks. For each block, six convolution matrices are shown. The first layer responds mainly to color, edges, and corners. The following layers capture increasingly complicated structures and even whole concepts, e.g., "group of humans", in the fifth layer. The images were generated as described in Zeiler and Fergus [2014].

shows image patches that produce large filter responses for six distinct convolution matrices for each layer in a convolutional neural network with five convolution blocks. Although the details depend on the input data and the architecture of the network, it is common to all CNNs that the deeper the layer, the higher the level of abstraction in the corresponding features. Peculiarly, a similar structure is found in the human visual cortex.

Example: CNN as feature extraction method

CNNs are not always used only for classification, but sometimes also as a means to learn efficient features for the underlying pattern. Sommer et al. [2017], for example, used CNNs to detect and classify different kinds of vehicles—cars and trucks—in aerial images. To this end, a CNN was trained on small scale images to classify cars from non-cars. The convolutional layers were then enlarged to accommodate the large scale aerial images. Note that this does not require any re-training of the CNN, as the weights of the convolution layers are shared for every convolution matrix as described above.

Of course, the fully connected part no longer matches the size of the feature maps, but in this case, what was desired was that the fully connected part be used in a moving window (see Section 7.8) over the output of the last convolution block to create a list of object detection candidates. Since this list does not contain information about the type of vehicle, a second network was used to classify the detection candidates. This network

Fig. 7.15: Detection and classification of vehicles in aerial images with CNNs. Red and blue boxes show detection of cars and trucks, green and turquoise boxes show the corresponding ground truth. Results of Sommer et al. [2017], images from the DLR3K dataset (Liu and Mattyus [2015]).

used the same convolution blocks as the detection network, but different weights in the fully connected layers. In effect, this network can be seen as an ordinary feed-forward network that uses the intermediate result of the convolution parts of the first object detection network as input features. Some detection and classification results of this method are shown in Figure 7.15.

However, feature representation does not always need to be classified using a feed-forward network. In the work of Herrmann et al. [2016], the goal was to search a large database for faces that are similar to the face shown in a low resolution query video. Since a CNN classifier would have to be retrained every time a new person is added to the database and because videos of different lengths are difficult to process using convolutional neural networks, a CNN was used specifically as a tool for feature extraction only. The structure of the CNN is shown in Figure 7.16. The CNN was trained to produce a 128 dimensional descriptor from images of 32×32 pixels in size with the goal that descriptors of the same person should have a small Euclidean distance. At the same time, the descriptors of faces of different persons should have a large Euclidean distance. To

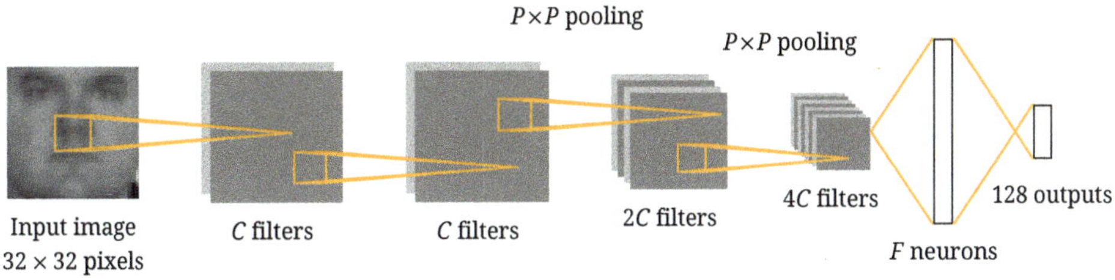

Fig. 7.16: Structure of the CNN used in Herrmann et al. [2016].

this end, two CNNs with the same structure were trained to minimize the loss

$$l = \sum_{i,j} \max\left(0, 1 - y_{ij}\left(b - \|\mathbf{m}_i - \mathbf{m}_j\|^2\right)\right) \tag{7.43}$$

on a training set, where $\mathbf{m}_i$ and $\mathbf{m}_j$ denote the output of the network for the i-th and j-th training sample and $y_{ij} \in \{-1,1\}$ is an indicator variable that is 1 if the training samples show the same person and -1 otherwise. $b > 0$ is the classification threshold, i.e., $\mathbf{m}_i$ and $\mathbf{m}_j$ are said to belong to the same person iff $\|\mathbf{m}_i - \mathbf{m}_j\|^2 \leq b$. Once trained, only one of the CNNs is used in classification and the other one is discarded. Such a training procedure with two or more networks with the same structure (but different parameters) is also known as a *siamese* setup (Bromley et al. [1993]).

7.6.6 Residual networks

A very popular CNN architecture was proposed in He et al. [2016] and is termed *residual network* or short *ResNet*. ResNets learn a *residual mapping* $r(\mathbf{x}) = h(\mathbf{x}) - \mathbf{x}$ of a network layer input $\mathbf{x}$ instead of directly learning the desired mapping $h(\mathbf{x})$ as shown in Figure 7.17. The motivation behind such a *residual learning* is described in the following.

It has been experimentally shown that CNNs with a higher number of layers can have a larger training error than CNNs with fewer layers. Since neural networks are universal approximators, it is possible to build the identity function with a stack of non-linear layers. Thus, the training error of a larger network should be less compared to the error of a smaller network. The authors of ResNet hypothesize that the counter-intuitive degradation problem when adding more layers is caused by the difficulty of learning identity mappings by stacking non-linear layers in non-residual CNN architectures. For this reason, the ResNet architecture contains *shortcut connections* that make it easy for the network to learn identity mappings ($h(\mathbf{x}) = \mathbf{x}$) by simply pushing the layer weights of the convolutions within the residual function to zero such that $r(\mathbf{x}) = \mathbf{0}$ holds.

In practical cases, the identity function is hardly the optimal mapping for $h(\mathbf{x})$. However, the residual learning is beneficial as long as the difference of the optimal mapping to the identity mapping is smaller than to a zero mapping.

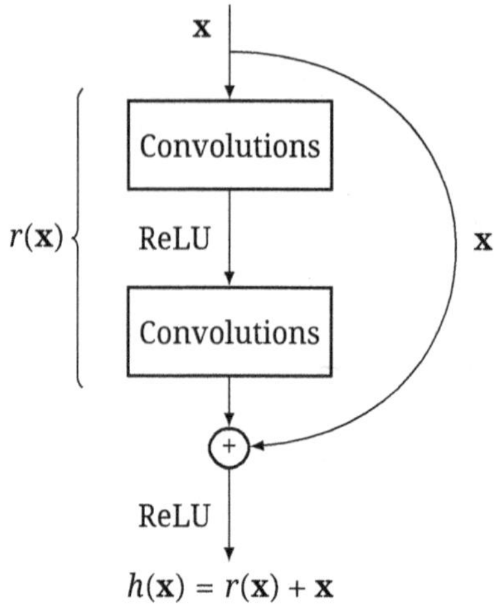

Fig. 7.17: Building block of ResNet.

The large ResNet variant with 152 layers (ResNet-152) achieved an error rate of 19.4 % on ImageNet. Note that the first CNN that has been applied to that dataset in Krizhevsky et al. [2012] had only 7 layers and had a much higher error rate of 37.5 %.

7.6.7 Variational autoencoders

Another important architecture of artificial neural networks is termed *variational autoencoder* (VAE). As the name suggests, it is a variant of the classical autoencoder introduced in Sections 2.7.6 and 7.5. However, besides architectural similarities, the goal of a VAE is not to compress data but to synthetically generate *new* data.

Given a set of data $\mathcal{D} = \{\mathbf{x}_1, \ldots, \mathbf{x}_N\}$, $\mathbf{x} \in \mathbb{R}^d$, the encoder, i.e., an ANN, learns a latent probability distribution $p(\mathbf{z}|\mathbf{x})$ that can be thought of as a low-dimensional representation (encoding) of the distribution $p(\mathbf{x})$. The decoder, i.e., a second ANN, is trained to reconstruct an original data point $\mathbf{x}$ given its latent representation $\mathbf{z}$. The probability distribution of the latent variable $\mathbf{z}$ is mostly chosen to be a Gaussian distribution: $\mathbf{z} \sim p(\mathbf{z}|\mathbf{x}) = \mathcal{N}(\mu_\mathbf{z}, \Sigma_\mathbf{z})$. The basic structure of a VAE is depicted in Figure 7.18.

One goal in the training process of VAEs is to minimize the reconstruction error $\varepsilon = \hat{\mathbf{x}} - \mathbf{x}$ as in traditional AEs. However, the latter ones are not suited for data generation because the learned encodings do not follow a *regular* structure. In contrast, VAEs are trained in a way that the encodings are regularized. Regularization means in this context that two points $\mathbf{z}_1$ and $\mathbf{z}_2$ which are close to each other in the latent space result in similar reconstructed data points $\hat{\mathbf{x}}_1$ and $\hat{\mathbf{x}}_2$ after decoding.

To get a regularized latent space, one can demand that the probability distribution $p(\mathbf{z}|\mathbf{x}) = \mathcal{N}(\mu_\mathbf{z}, \Sigma_\mathbf{z})$ is close to a standard Gaussian distribution $p'(\mathbf{z}|\mathbf{x}) := \mathcal{N}(\mathbf{0}, \mathbf{1})$, which can be achieved by minimizing the Kullback–Leibler divergence of the two probability functions $D(p||p')$.

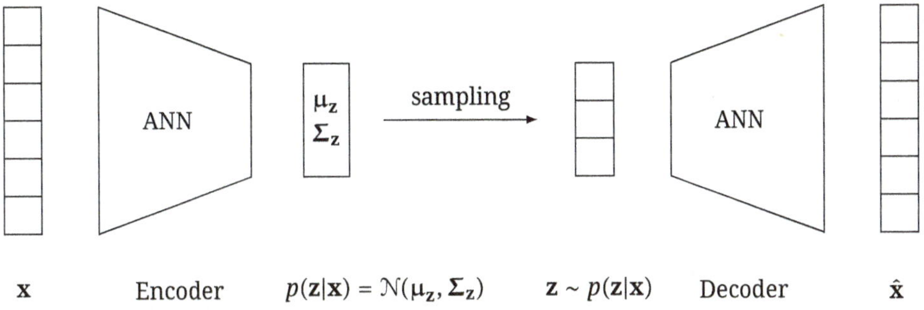

Fig. 7.18: Scheme of a varational autoencoder.

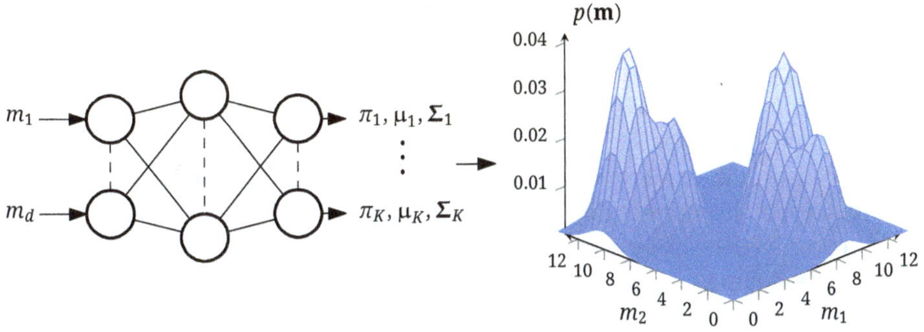

Fig. 7.19: Scheme of a mixture density network

After training, a new datum $\mathbf{x}'$ can be generated by sampling a random variable $\mathbf{z}' \sim p(\mathbf{z}|\mathbf{x})$ which then is used during inference as input to the decoder network.

7.6.8 Mixture density networks

In combination with neural networks, Gaussians can not only be used as form of regularization in the latent space. Instead, a network can also be utilized to estimate the parameters of a Gaussian as an output. This being said, in a number of instances a single distribution is not sufficient to fit the underlying data, particularly when the target density exhibits a multi-modal structure. Since Gaussian mixture models (cf. Section 3.4) serve as universal approximators, the neural network can learn to approximate its parameters and therefore fit arbitrary distributions. Bishop [1994] introduced these models as *mixture density networks* (MDNs). The structure of this idea is depicted in Figure 7.19, where the network's outputs attempt to reassemble the parameters of a GMM that was used to create the reference example.

More formally, a MDN aims to learn the parameters $\theta = (\boldsymbol{\pi}, \boldsymbol{\mu}, \boldsymbol{\Sigma})$ of a GMM that models the conditional probability distribution $p(\mathbf{t} \mid \mathbf{m})$. The objective is to identify the set of functions which map samples $\mathbf{m}$ on their respective outputs $\mathbf{t}$. This conditional can be directly expressed in the shape of a Gaussian mixture

$$p(\mathbf{t} \mid \mathbf{m}) = \sum_{k=1}^{K} \pi_k \mathcal{N}(\mathbf{t} \mid \boldsymbol{\mu}_k(\mathbf{m}), \boldsymbol{\Sigma}_k(\mathbf{m})) \tag{7.44}$$

with K components and corresponding mixture parameters as introduced in Section 3.4. The components are often constrained to be isotropic because this greatly reduces the number of constraints which have to be implemented in the network. Therefore, the simplification $\boldsymbol{\Sigma}_k = \mathbf{I}\sigma_k^2$ applies. Since the reference example requires the same variance $\sigma_k = 1$ for all components (see Section 3.3.2), an MDN with this constraint could still recover the mixture parameters. Note that all parameters are functions of $\mathbf{m}$ as they are the output of the MDN, given a particular sample $\mathbf{m}$ as input.

The sketch depicted in Figure 7.19 is not accurate in the sense that the parameters θ are not estimated directly. Rather, the following auxiliary variables are estimated first and then constrained to meet the requirements of the actual parameters θ:

- k_k^{π} represent the mixture coefficients of the components $\pi_k(\mathbf{m})$.
- k_k^{Σ} represent the variances of the components $\boldsymbol{\Sigma}_k$.
- k_{kj}^{μ} represent the $\tilde{d}$ entries $\mu_{kj}(\mathbf{m})$ of the means of the components $\boldsymbol{\mu}_j(\mathbf{m})$, where $\tilde{d}$ is the dimensionality of the output dimension $\mathbf{t}$.

The total number of parameters to be inferred by the network is the sum of one mixing factor, one variance and L entries for each of the K components, resulting in a total of $K + K + K\tilde{d} = K(\tilde{d} + 2)$ parameters to be learned. Bishop [1994] suggests different regularization techniques for the parameters to comply with the requirements of GMMs:

- The mixing parameters must fulfill the requirements $\sum_{k=1}^{K} \pi_k = 1, \pi_k \geq 0$ which can be accomplished through the use of a softmax activation

$$\pi_k(\mathbf{m}) = \frac{\exp\left(k_k^{\pi}\right)}{\sum_{l=1}^{K} \exp\left(k_l^{\pi}\right)} \tag{7.45}$$

  in the respective nodes.
- The variances simply represent scales that must adhere to $\sigma_k^2(\mathbf{m}) \geq 0$. This is achieved by defining them as exponentials

$$\sigma_k^2(\mathbf{m}) = \exp\left(k_k^{\Sigma}\right) \tag{7.46}$$

  of the network outputs.
- The mean vectors indicate the location and remain unrestricted, hence

$$\mu_{kj}(\mathbf{m}) = k_{kj}^{\mu} \tag{7.47}$$

  is employed directly.

A standard maximum likelihood approach yields the negative log-likelihood

$$e(\underline{\mathbf{W}}) = -\sum_{n=1}^{N} \ln\left(\sum_{k=1}^{K} \pi_k\left(\mathbf{m}_n,\underline{\mathbf{W}}\right), \mathcal{N}\left(\mathbf{t}_n \mid \boldsymbol{\mu}_k\left(\mathbf{m}_n,\underline{\mathbf{W}}\right), \sigma_k^2\left(\mathbf{m}_n,\underline{\mathbf{W}}\right)\right)\right) \tag{7.48}$$

as error term and minimization objective. The weights $\widetilde{\mathbf{W}}$ can be trained through the standard backpropagation algorithm as presented in Section 7.4.

7.7 Support vector machines

The support vector machine (SVM) is one of the most versatile classifiers. It is relatively simple, yet extremely powerful, and provides good generalization even with a small number of training samples. The SVM is a linear discriminant classifier for two classes ($c = 2$), but can be extended to multiple classes using the techniques discussed in Section 7.1.1. The SVM can be explained based on five fundamental ideas:

1. Linear separation with maximum distance of the separating hyperplane to the nearest training samples (the *support vectors*).
2. Dual formulation of the linear classifier to reduce the number of parameters to estimate.
3. Nonlinear mapping of the features to a high-dimensional feature space Φ.
4. Implicit use of the (possibly ∞-dimensional) space of eigenfunctions of a so-called *kernel function* K as the transformed feature space Φ. The transformed features do not have to be explicitly computed and the classifier has a small number of free parameters even though $\dim(\Phi)$ is large (*kernel trick*).
5. Relaxation of the linear separability requirement by introducing slack variables.

These ideas will be discussed in the following. For now, it is assumed that the training set $\mathcal{D} = \{\mathbf{m}_1, \ldots, \mathbf{m}_N\}$, $\mathbf{m} \in \mathbb{R}^d$ is linearly separable. This assumption will be relaxed in the fifth step.

7.7.1 Linear separation with maximum margin

As was already seen in Section 7.2, typically more than one linear discriminant can separate a given dataset. Yet, some of the discriminants are intuitively “better” than others, because these discriminants generalize better than others. One method of finding such discriminants is to impose additional conditions on the separating hyperplane. With SVMs, the goal is to find those hyperplane parameters $(\mathbf{w}, b)$ such that the *margin* γ, the distance between the hyperplane and the closest training samples, is maximized:

$$(\mathbf{w}, b) = \arg\max_{\mathbf{w}', b'} \left\{\gamma(\mathbf{w}', b', \mathcal{D})\right\}. \tag{7.49}$$

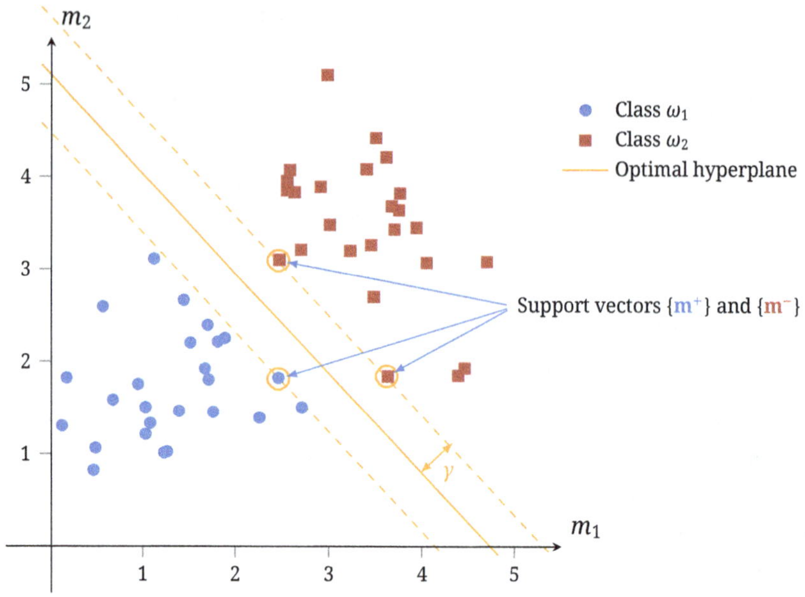

Fig. 7.20: Classification with maximum margin.

The intuition is that a larger margin results in better generalization. This intuition was already expressed in Figure 7.4, where the final hyperplane of the perceptron does separate the training data, but the margin is very small. The dashed line marks a hyperplane with a larger margin.

Interestingly, there is exactly one hyperplane that satisfies Equation (7.49), provided that $\mathcal{D}$ is linearly separable. This hyperplane is fully defined by the support vectors $\{\mathbf{m}^+\}$ and $\{\mathbf{m}^-\}$, i.e., the vectors that are closest to the hyperplane (see Figure 7.20). The SVM concentrates only on the boundaries between classes and therefore only on the most difficult samples.

But how does one estimate the parameters $\mathbf{w}$ and b from a training set $\mathcal{D}$ and how does the margin γ relate to $\mathbf{w}$ and b? To derive an answer, recall the linear decision function

$$k(\mathbf{m}) = \mathbf{w}^\mathsf{T}\mathbf{m} + b = \left(\sum_{i=1}^{d} w_i m_i\right) + b = \langle \mathbf{w}, \mathbf{m} \rangle + b, \tag{7.50}$$

where $\mathbf{m}$ is assigned to the class ω_1 if $k(\mathbf{m}) > 0$ and to ω_2 otherwise; that is, the classification depends on the sign of the decision function $k(\mathbf{m})$. Observe further that the sign of the decision function does not change when the parameters are scaled by some factor $\beta > 0$, that is, $\operatorname{sign}(\mathbf{w}^\mathsf{T}\mathbf{m} + b) = \operatorname{sign}(\beta\mathbf{w}^\mathsf{T}\mathbf{m} + \beta b)$. By definition, the support vectors $\{\mathbf{m}^+\}$ of ω_1 and $\{\mathbf{m}^-\}$ of ω_2 lie exactly on the margin and the optimal separating hyperplane has the same distance from all support vectors (see Figure 7.20). It follows that, without loss of generality, one can write

$$\left.\begin{aligned} \mathbf{w}^\mathsf{T}\mathbf{m}^+ + b &= 1 \\ \mathbf{w}^\mathsf{T}\mathbf{m}^- + b &= -1 \end{aligned}\right\} \Rightarrow \mathbf{w}^\mathsf{T}\mathbf{m}^+ - \mathbf{w}^\mathsf{T}\mathbf{m}^- = 2. \tag{7.51}$$

Normalizing the vector $\mathbf{w}$ (i.e., $\beta = \|\mathbf{w}\|^{-1}$) yields

$$\left.\begin{aligned} \frac{\mathbf{w}^\mathsf{T}}{\|\mathbf{w}\|}\mathbf{m}^+ + \frac{b}{\|\mathbf{w}\|} &= \gamma \\ \frac{\mathbf{w}^\mathsf{T}}{\|\mathbf{w}\|}\mathbf{m}^- + \frac{b}{\|\mathbf{w}\|} &= -\gamma \end{aligned}\right\} \Rightarrow \frac{1}{\|\mathbf{w}\|}\left(\mathbf{w}^\mathsf{T}\mathbf{m}^+ - \mathbf{w}^\mathsf{T}\mathbf{m}^-\right) = 2\gamma. \tag{7.52}$$

In other words, the margin $\gamma = \|\mathbf{w}\|^{-1}$ assumes its maximum iff $\|\mathbf{w}\|$ becomes minimal. This translates into the following optimization problem:

$$\begin{aligned} &\text{Minimize:} && \langle \mathbf{w}, \mathbf{w} \rangle = \mathbf{w}^\mathsf{T}\mathbf{w} = \|\mathbf{w}\|^2 \\ &\text{subject to:} && z_i\left(\langle \mathbf{w}, \mathbf{m}_i \rangle + b\right) \geq 1 \quad i = 1, \ldots, N, \end{aligned} \tag{7.53}$$

where z_i denotes membership in the class: $z_i = 1$, if $\mathbf{m}_i$ belongs to class ω_1 and $z_i = -1$, if $\mathbf{m}_i$ belongs to class ω_2. This formulation of the SVM optimization problem is called the *primal form*.

The main goal ensures that the hyperplane $\mathbf{w}^\mathsf{T}\mathbf{m}+b = 0$ maximizes the margin, while the constraints ensure that every training sample is classified correctly. The solution to this problem can be computed using Lagrange multipliers $\alpha_i \geq 0$, $i = 1, \ldots, N$,

$$L(\mathbf{w}, b, \boldsymbol{\alpha}) = \frac{1}{2}\langle \mathbf{w}, \mathbf{w} \rangle - \sum_{i=1}^{N} \alpha_i\left[z_i\left(\langle \mathbf{w}, \mathbf{m}_i \rangle + b\right) - 1\right] \overset{!}{\to} \min. \tag{7.54}$$

Taking the derivative of L with respect to $\mathbf{w}$ and b and setting the partial derivatives to 0 yields

$$\frac{\partial L}{\partial \mathbf{w}} = \mathbf{w} - \sum_{i=1}^{N} z_i\alpha_i\mathbf{m}_i \overset{!}{=} 0 \Leftrightarrow \mathbf{w} = \sum_{i=1}^{N} z_i\alpha_i\mathbf{m}_i \quad \text{and} \tag{7.55}$$

$$\frac{\partial L}{\partial b} = \sum_{i=1}^{N} z_i\alpha_i \overset{!}{=} 0. \tag{7.56}$$

7.7.2 Dual formulation

The primal formulation of the decision function in Equation (7.50) suggests that there are $(d + 1)$ parameters to estimate: d parameters for $\mathbf{w}$ and one parameter for b. If $\mathbf{w}$ is normalized, the number of free parameters reduces to d. Yet, above it was hinted that the hyperplane is fully determined by the support vectors. If the number of support vectors is smaller than d, this means that there are actually fewer parameters that need to be estimated.

Indeed, Equation (7.55) shows that the weight vector $\mathbf{w}$ can be written as a linear combination of the training samples (the same result is used in the perceptron algorithm, see line 5 in Algorithm 7.1). Substituting $\mathbf{w} = \sum_{i=1}^{N} \alpha_i z_i \mathbf{m}_i$ in Equation (7.50) yields

$$k(\mathbf{m}) = \langle \mathbf{w}, \mathbf{m} \rangle + b = \left\langle \sum_{i=1}^{N} \alpha_i z_i \mathbf{m}_i, \mathbf{m} \right\rangle + b$$
$$= \sum_{i=1}^{N} \alpha_i z_i \langle \mathbf{m}_i, \mathbf{m} \rangle + b. \tag{7.57}$$

Equation (7.57) is called the *dual form* of Equation (7.50). In this formulation, the number of free parameters is N: $(N-1)$ parameters for the α_i (recall Equation (7.56)) and one parameter for b. This number does not depend on the dimensionality d of the feature space! Below it will be seen that the α_i are nonzero only for the support vectors. Also note that the feature vectors $\mathbf{m}_i$ in Equation (7.57) only appear inside of inner products, but not on their own. This will become important in Section 7.7.4.

Substituting Equations (7.55) and (7.56) in Equation (7.54) yields the dual formulation of the Lagrange function:

$$L(\boldsymbol{\alpha}) = \sum_{i=1}^{N} \alpha_i - \frac{1}{2} \sum_{(i,j)=(1,1)}^{(N,N)} z_i z_j \alpha_i \alpha_j \langle \mathbf{m}_i, \mathbf{m}_j \rangle \overset{!}{\longrightarrow} \max. \tag{7.58}$$

The dual formulation depends only on the dual variables α_i and must be maximized instead of minimized. That is, if $\boldsymbol{\alpha}^*$ solves the following dual constrained quadratic optimization problem,

$$\begin{aligned} \text{maximize:} \quad & \sum_{i=1}^{N} \alpha_i - \frac{1}{2} \sum_{(i,j)=(1,1)}^{(N,N)} z_i z_j \alpha_i \alpha_j \langle \mathbf{m}_i, \mathbf{m}_j \rangle \\ \text{subject to:} \quad & \sum_{i=1}^{N} z_i \alpha_i = 0 \quad \text{and} \\ & \alpha_i \geq 0, \quad i = 1, \ldots, N, \end{aligned} \tag{7.59}$$

then the vector $\mathbf{w}^* = \sum_{i=1}^{N} z_i \alpha_i^* \mathbf{m}_i$ realizes the linear classifier with maximum margin $\gamma = \|\mathbf{w}^*\|^{-1}$. For b^* it follows that

$$b^* = -\frac{1}{2} \left(\max_{z_i=-1} \{\langle \mathbf{w}^*, \mathbf{m}_i \rangle\} + \min_{z_i=1} \{\langle \mathbf{w}^*, \mathbf{m}_i \rangle\} \right). \tag{7.60}$$

The above constrained quadratic optimization problem can be solved very efficiently, e.g., with quadratic programming techniques. Note that the solution $(\mathbf{w}^*, b^*)$ is unique. Non-support vector samples $\mathbf{m}_i$, i.e., the $\mathbf{m}_i$ that do not fall on the boundary of the margin, do not influence the solution at all. During the optimization, the corresponding weights α_i^* vanish: only the α_j^* that correspond to support vectors are $\neq 0$. This is, in fact, the reason for calling these samples *support vectors*.

Algorithm 7.2: Dual form of the perceptron algorithm.

Data: Training set $\mathcal{D}$, learning rate $\eta > 0$
Result: Sample weight $\boldsymbol{\alpha}$, offset b
$\boldsymbol{\alpha} \leftarrow \mathbf{0} \in \mathbb{R}^N$
$b \leftarrow 0$
$R \leftarrow \max_{1 \leq i \leq N} \|\mathbf{m}_i\|$
repeat
  forall $\mathbf{m}_i \in \mathcal{D}$ **do**
    if $z_i \left(\sum_{j=1}^{N} \alpha_j z_j \langle \mathbf{m}_j, \mathbf{m}_i \rangle + b \right) \leq 0$ **then**
      $\alpha_i \leftarrow \alpha_i + 1$
      $b \leftarrow b + \eta z_i R^2$
until *no training error in inner loop*
return $\boldsymbol{\alpha}, b$

Following this reasoning, the number of parameters that need to be estimated for the SVM classifier is neither d (as suggested by Equation (7.50)) nor N (as suggested by Equation (7.57)). Rather, the number of non-vanishing parameters is equal to the number of support vectors, $|\mathcal{SV}|$, where $\mathcal{SV}$ denotes the set of support vectors. This explains why an SVM is able to find a separating hyperplane in very high- and even infinite-dimensional spaces.

Complementary: Dual form of the perceptron algorithm

As mentioned above, the final hyperplane of the perceptron algorithm may be written as a linear combination of the training samples and the initial parameters. The dual algorithm is given in Algorithm 7.2.

The more difficult it is to classify a sample $\mathbf{m}_i$, the bigger the sample weight α_i (i.e., number of misclassifications) will become. In this sense, α_i is a measure of the information content of $\mathbf{m}_i$ with regard to the classification task. In the dual form of the algorithm, the training samples appear only in scalar products $\langle \mathbf{m}_i, \mathbf{m}_j \rangle$, which makes it possible to apply the kernel trick (see Section 7.7.4) to the perceptron algorithm.

7.7.3 Nonlinear mapping

As already discussed in Section 7.1.2, linear separability can be enforced by lifting the features into a higher-dimensional feature space. More specifically, the original features are mapped from $\mathbb{M} \subseteq \mathbb{R}^d$ to a high-dimensional vector space $\Phi \subseteq \mathbb{R}^{d^*}$:

$$\boldsymbol{\varphi} : \begin{cases} \mathbb{M} \to \Phi \\ \mathbf{m} \mapsto \boldsymbol{\varphi}(\mathbf{m}) = (\varphi_1(\mathbf{m}), \ldots, \varphi_{d^*}(\mathbf{m}))^\mathsf{T} \end{cases} , \tag{7.61}$$

where in general the $\varphi_i(\cdot)$ are nonlinear functions.

Let us further assume that Φ is equipped with an inner product $\langle \boldsymbol{\varphi}_1, \boldsymbol{\varphi}_2 \rangle$, where $\boldsymbol{\varphi}_1, \boldsymbol{\varphi}_2 \in \Phi$. The nonlinear (in $\mathbb{M}$) decision function in the primal form can be written as

$$k(\mathbf{m}) = \sum_{i=1}^{d^*} w_i \varphi_i(\mathbf{m}_i) + b = \mathbf{w}^\mathsf{T} \boldsymbol{\varphi}(\mathbf{m}) + b = \langle \mathbf{w}, \boldsymbol{\varphi}(\mathbf{m}) \rangle + b, \tag{7.62}$$

where $\mathbf{w} \in \mathbb{R}^{d^*}$. In this formulation, the number of free parameters is $(d^* + 1)$, which means that additional parameters have to be estimated and all the discussed drawbacks apply (see Sections 6.1 and 7.1.2). Note that although the separation is nonlinear in $\mathbb{M}$, it is linear in Φ. In essence, the linear separation with maximum margin to the nearest samples $\{\boldsymbol{\varphi}(\mathbf{m})\}$ is conducted in the space Φ instead of the space $\mathbb{M}$. If $d^* > d$ (which is generally, but not necessarily, the case), the mapping $\boldsymbol{\varphi}(\mathbf{m})$ determines a d-dimensional sub-manifold in Φ.

Note that in a two-class problem ($c = 2$), a dataset $\mathcal{D} = \{\mathbf{m}_1, \ldots, \mathbf{m}_N\}$ consisting of d-dimensional feature vectors $\mathbf{m}_i$ can *always* be linearly separated if $d \geq N - 1$ and the $\{\mathbf{m}_i\}$ do not reside in a $(d-1)$-dimensional subspace of $\mathbb{M}$. As a consequence, linear separation may always be achieved by a suitable mapping $\boldsymbol{\varphi}(\cdot)$.

Recall the dual form of the decision function in Equation (7.57). When applying the mapping $\boldsymbol{\varphi}(\cdot)$, the decision function becomes

$$k(\mathbf{m}) = \sum_{i=1}^{N} \alpha_i z_i \langle \boldsymbol{\varphi}(\mathbf{m}_i), \boldsymbol{\varphi}(\mathbf{m}) \rangle + b. \tag{7.63}$$

In this formulation, the number of free parameters α_i is N. In particular, it is independent of the dimension d^* of the high-dimensional vector space Φ! If it was possible to calculate the inner products $\langle \boldsymbol{\varphi}(\mathbf{m}_i), \boldsymbol{\varphi}(\mathbf{m}) \rangle$ directly as functions of the feature vectors $\mathbf{m}_i$ and $\mathbf{m}$, both steps—the transformation $\boldsymbol{\varphi}(\cdot)$ and the inner product in Φ—could be computed simultaneously without explicitly lifting the features into Φ.

As it turns out, this is indeed possible by using the so-called *kernel trick*.

7.7.4 The kernel trick

Definition 7.3 (Kernel function). A function K is said to be a *kernel function* (or *kernel* for short) if for all $\mathbf{m}, \mathbf{m}' \in \mathbb{M}$

$$K(\mathbf{m}, \mathbf{m}') = \left\langle \boldsymbol{\varphi}(\mathbf{m}), \boldsymbol{\varphi}(\mathbf{m}') \right\rangle, \tag{7.64}$$

where $\boldsymbol{\varphi}(\cdot)$ is a mapping from $\mathbb{M}$ to Φ.

Replacing the inner product in Equation (7.63) by a kernel yields

$$k(\mathbf{m}) = \sum_{i=1}^{N} \alpha_i z_i K(\mathbf{m}_i, \mathbf{m}) + b. \tag{7.65}$$

This is the basic insight of the kernel trick: an inner product of mapped feature vectors may be replaced by a kernel function that computes both in one step. The necessary and sufficient conditions for an arbitrary bivariate function $K(\cdot,\cdot)$ to be a kernel function are given by Mercer's theorem (Mercer [1909]).

Theorem 7.4 (Mercer's theorem). A symmetric function K in L_2 has an expansion

$$K(\mathbf{m},\mathbf{m}') = \sum_{j=1}^{\infty} \lambda_j \varphi_j(\mathbf{m})\varphi_j(\mathbf{m}') \tag{7.66}$$

with positive coefficients $\lambda_j > 0$ (i.e., K denotes an inner product in the feature space Φ associated with K) iff

$$\int\limits_{\mathbb{M}}\int\limits_{\mathbb{M}} K(\mathbf{m},\mathbf{m}')\, f(\mathbf{m})\, f(\mathbf{m}')\, \mathrm{d}\mathbf{m}\, \mathrm{d}\mathbf{m}' > 0 \tag{7.67}$$

for all $f \not\equiv 0$ with $\int f^2(\mathbf{m})\, \mathrm{d}\mathbf{m} < \infty$. The λ_j and $\varphi_j(\cdot)$ are the solutions to the eigenvalue problem

$$\int\limits_{\mathbb{M}} K(\mathbf{m},\mathbf{m}')\, \varphi(\mathbf{m}')\, \mathrm{d}\mathbf{m}' = \lambda\, \varphi(\mathbf{m}). \tag{7.68}$$

Given a function $K(\cdot,\cdot)$ that satisfies the hypotheses of Mercer's theorem (and hence is a kernel function), one can use this function in the dual formulation of the classifier in Equation (7.65) to *implicitly* use the possibly infinite-dimensional transformed feature space Φ without needing to explicitly compute the corresponding feature vectors $\{\varphi_j\}$. In other words: the kernel function K *induces* the feature space Φ.

Note, again, that even though the feature vector $\varphi(\mathbf{m})$ may have a very high, possibly even infinite dimensionality d^*, the classifier in Equation (7.65) is still fully determined by only N free parameters.

Examples of common kernel functions

As the first example, consider the trivial kernel

$$K(\mathbf{m},\mathbf{u}) := \langle \mathbf{m}, \mathbf{u} \rangle = \mathbf{m}^{\mathsf{T}}\mathbf{u}. \tag{7.69}$$

The corresponding mapping is the identity function $\varphi(\mathbf{m}) = \mathbf{m}$. Therefore, this kernel is also called the linear kernel.

A more interesting kernel arises by squaring the scalar product:

$$\begin{aligned} K(\mathbf{m},\mathbf{u}) := \langle \mathbf{m}, \mathbf{u} \rangle^2 &= \left(\sum_{i=1}^{d} m_i u_i\right)^2 = \left(\sum_{i=1}^{d} m_i u_i\right)\left(\sum_{j=1}^{d} m_j u_j\right) \\ &= \sum_{i=1}^{d}\sum_{j=1}^{d} m_i m_j\, u_i u_j = \sum_{(i,j)=(1,1)}^{(d,d)} (m_i m_j)(u_i u_j). \end{aligned} \tag{7.70}$$

The corresponding mapping produces feature vectors of the form

$$\varphi(\mathbf{m}) = (m_i m_j)_{(i,j)=(1,1)}^{(d,d)} = \left(m_1^2, m_1 m_2, m_2^2, m_2 m_3, \ldots, m_d^2\right)^{\mathrm{T}}, \tag{7.71}$$

i.e., the vector of all monomials of degree 2 of the entries $\{m_i\}$. The dimensionality of Φ is $d^* = \frac{1}{2}(d+1)d$, because the terms $m_i m_j$ appear twice for every $i \neq j$, but must only be counted once.

The above kernel function can be modified by adding a constant $c \in \mathbb{R}$ before squaring:

$$\begin{aligned} K(\mathbf{m},\mathbf{u}) := (\langle \mathbf{m}, \mathbf{u}\rangle + c)^2 &= \left(\sum_{i=1}^{d} m_i u_i + c\right)\left(\sum_{j=1}^{d} m_j u_j + c\right) \\ &= \sum_{i=1}^{d}\sum_{j=1}^{d} m_i m_j\, u_i u_j + 2c\sum_{i=1}^{d} m_i u_i + c^2 \\ &= \sum_{(i,j)=(1,1)}^{(d,d)} (m_i m_j)(u_i u_j) + \sum_{i=1}^{d}\left(\sqrt{2c}\, m_i\right)\left(\sqrt{2c}\, u_i\right) + c^2. \end{aligned} \tag{7.72}$$

The implicit mapping produces vectors of all monomials of the $\{m_i\}$ of degree ≤ 2. The dimensionality of Φ is

$$d^* = \binom{d+2}{2} = \frac{1}{2}(d+2)(d+1). \tag{7.73}$$

This kernel can be generalized by raising to the power of $q \in \mathbb{N}$,

$$K(\mathbf{m},\mathbf{u}) := (\langle \mathbf{m}, \mathbf{u}\rangle + c)^q. \tag{7.74}$$

Akin to the reasoning above, the transformed feature vectors contain all monomials of degree $\leq q$. For this reason, this kernel function is also known as the *polynomial kernel*. Together with the linear kernel, the polynomial kernel is one of the standard kernels often used with SVMs.

Another popular kernel is the Gaussian kernel or radial basis function (RBF) kernel:

$$K(\mathbf{m},\mathbf{u}) := \exp\left\{-\frac{\|\mathbf{m}-\mathbf{u}\|^2}{\sigma^2}\right\}. \tag{7.75}$$

Unlike the linear and polynomial kernels, this kernel produces a mapping into an infinite-dimensional space Φ, that is, the eigenfunction decomposition in Theorem 7.4 has an infinite number of solutions (Rasmussen and Williams [2006]).

A decision function can also be found when a kernel function is used. The optimization problem simply swaps the scalar product in Equation (7.59) for a kernel:

$$\begin{aligned} \text{Maximize:} \quad & \sum_{i=1}^{N} \alpha_i - \frac{1}{2}\sum_{(i,j)=(1,1)}^{(N,N)} z_i z_j\, \alpha_i \alpha_j K(\mathbf{m}_i, \mathbf{m}_j) \\ \text{subject to:} \quad & \sum_{i=1}^{N} z_i \alpha_i = 0 \quad \text{and} \\ & \alpha_i \geq 0, \quad i = 1, \ldots, N. \end{aligned} \tag{7.76}$$

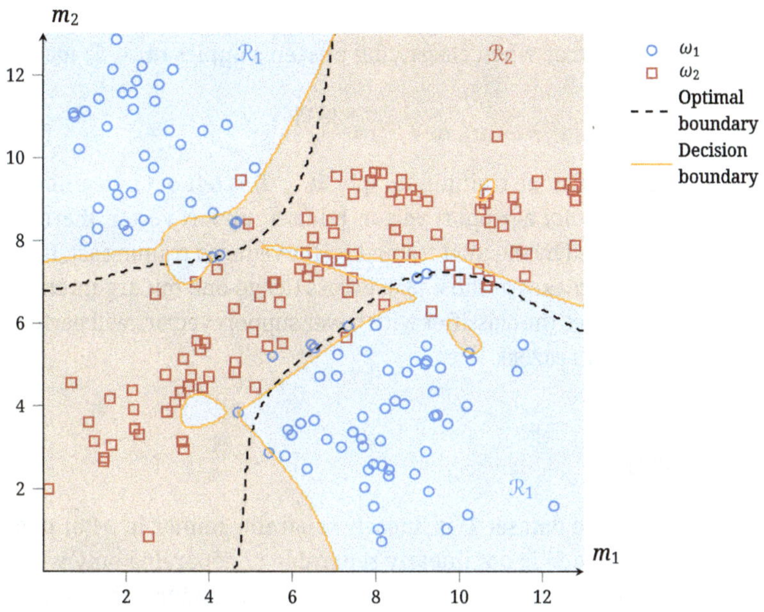

Fig. 7.21: Application to the reference example of Section 3.3.2. Decision regions of a hard margin SVM classifier with Gaussian kernel ($\sigma = 4$). The training error is $e_{\text{train}} = 3.5\,\%$ and the testing error is $e_{\text{test}} = 11\,\%$. The latter asymptotically approaches $e_{\text{test}} \approx 11.2\,\%$. The training set is the same as in Figure 3.8. Test samples are shown with hollow marks.

The decision function

$$k(\mathbf{m}) = \sum_{i \in SV} z_i a_i^* K(\mathbf{m}_i, \mathbf{m}) + b^* \tag{7.77}$$

is equivalent to the hyperplane in the induced space Φ with maximum margin (Cristianini and Shawe-Taylor [2000])

$$\gamma = \frac{1}{\sqrt{\sum_{\mathbf{m}_i \in \mathcal{SV}} a_i^*}}. \tag{7.78}$$

Note that it follows from Mercer's theorem that the (kernel) matrix $(K(\mathbf{m}_i, \mathbf{m}_j))_{i,j=1}^N$ is positive definite. This means that the optimization problem in Equation (7.76) is convex and has a unique global optimum that can be found using, e.g., quadratic programming.

Figure 7.21 shows the decision regions of a hard margin SVM with a Gaussian kernel ($\sigma = 4$) learned from the ongoing reference dataset. The decision regions are very complicated and rugged. Clearly, this classifier overfits the data and does not generalize well. This can also be seen in the low training error of $e_{\text{train}} = 3.5\,\%$, but rather large testing error of $e_{\text{test}} = 11\,\%$. Note that in theory a hard margin SVM should have no training error ($e_{\text{train}} = 0$). In practice this is rarely achieved, as the optimization of Equation (7.76) is usually terminated before the (true) optimum is reached.

Probability of error

The probability of making an error when classifying unseen samples $\mathbf{m}_i \notin \mathcal{D}$ may be eye-balled as

$$P(\hat{\omega}(\mathbf{m}) \neq \omega(\mathbf{m}), \mathbf{m} \notin \mathcal{D}) \approx \frac{|\mathcal{SV}|}{N}. \tag{7.79}$$

The reasoning goes as follows. If some training sample $\mathbf{m}_i \in \mathcal{D}$ is left out in training, it will be correctly classified if it is *not* a support vector. If it is a support vector, there is a chance that it will be misclassified by the SVM trained on the reduced training set. If this is repeated for all $\mathbf{m}_i \in \mathcal{D}$, one makes at most $|\mathcal{SV}|$ errors (leave-one-out argument). It follows that for a fixed training set, the classifier with fewer support vectors will perform better (consistently with Occam's razor).

7.7.5 No linear separability

So far we have assumed that the dataset $\mathcal{D}$ is linearly separable (either in $\mathbb{M}$ or in Φ). However, there are cases where $\mathcal{D}$ is not linearly separable or separable only with a small margin. To allow the SVM to work well in these cases, one introduces the so called *slack variables* $\xi_i \geq 0, i = 1, \ldots, N$ that measure how much a training sample $\mathbf{m}_i$ violates the margin or even how far $\mathbf{m}_i$ lies on the wrong side of the separating hyperplane, see Figure 7.22. For the linear classifier with maximum margin, the optimization goal becomes:

$$\begin{aligned} \text{Minimize:} \quad & \langle \mathbf{w}, \mathbf{w} \rangle + C \sum_{i=1}^{N} \xi_i^2 \\ \text{subject to:} \quad & z_i \left(\langle \mathbf{w}, \mathbf{m}_i \rangle + b \right) \geq 1 - \xi_i \quad \text{and} \\ & \xi_i \geq 0, \quad i = 1, \ldots, N. \end{aligned} \tag{7.80}$$

The design parameter $C > 0$ defines how much emphasis should be put on correct classification (i.e., C is large) versus a large margin (i.e., C is small). A dual formulation of the above and the use of a kernel K instead of the scalar product $\langle \mathbf{w}, \mathbf{m} \rangle$ leads to the *soft margin SVM*:

$$\begin{aligned} \text{Maximize:} \quad & \sum_{i=1}^{N} \alpha_i - \frac{1}{2} \sum_{(i,j)=(1,1)}^{(N,N)} z_i z_j \, \alpha_i \alpha_j \left(K(\mathbf{m}_i, \mathbf{m}_j) + \frac{1}{C} \delta_{ij} \right) \\ \text{subject to:} \quad & \sum_{i=1}^{N} z_i \alpha_i = 0 \quad \text{and} \\ & \alpha_i \geq 0, \quad i = 1, \ldots, N. \end{aligned} \tag{7.81}$$

With $\boldsymbol{\alpha}^*$ denoting the solution vector, the decision function becomes

$$k(\mathbf{m}) = \sum_{i=1}^{N} z_i \alpha_i^* K(\mathbf{m}_i, \mathbf{m}) + b^*, \tag{7.82}$$

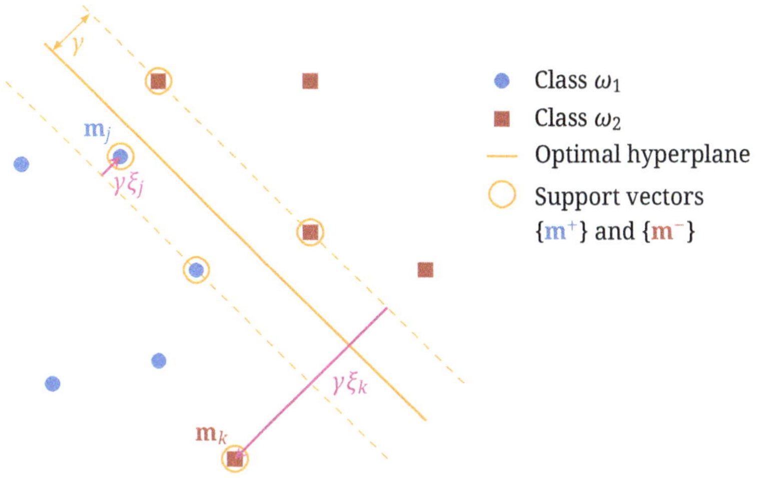

Fig. 7.22: Geometric interpretation of the slack variables ξ_i, $i = 1, \ldots, N$: The slack variables measure (w.r.t. to the margin γ) whether and how far the training samples penetrate the margin.

where the offset b^* is chosen so that $z_i k(\mathbf{m}) = 1 - \frac{\alpha_i^*}{C}$ for all i with $\alpha_i^* \neq 0$. A proof of the above can be found, for example, in Cristianini and Shawe-Taylor [2000].

The decision regions of a soft margin SVM can be seen in Figure 7.23. Again, the classifier was trained using the ongoing reference dataset from Section 3.3.2. As with Figure 7.21, the SVM uses a Gaussian kernel, albeit here the kernel parameter was chosen to be $\sigma = 1$. The design parameter was chosen to be $C = 1$. Unlike with a hard margin, the soft margin SVM does not overfit the training data, but generalizes well. The decision boundaries are reasonably close to the decision regions of the Bayesian optimal classifier and the asymptotic testing error is only 0.7 percentage points above the optimal Bayes error rate. Different choices for σ and C vary the shape of the decision boundary: higher values of σ generally lead to a smooth decision boundary, whereas higher values of C lead to a more complicated boundary.

In practice, one can estimate the hyperparameter C and the kernel parameters (σ in the example) using the validation set $\mathcal{V}$. To this end, an SVM classifier with fixed hyperparameters is trained using the training set $\mathcal{D}$ and the classification performance on $\mathcal{V}$ is recorded. The process is repeated for various combinations of the parameters, where the parameters are often determined using a rule (e.g., grid search, the parameters are drawn from a regular grid) or are randomly sampled (randomized search). Finally, the parameters that yield the highest classification performance are kept.

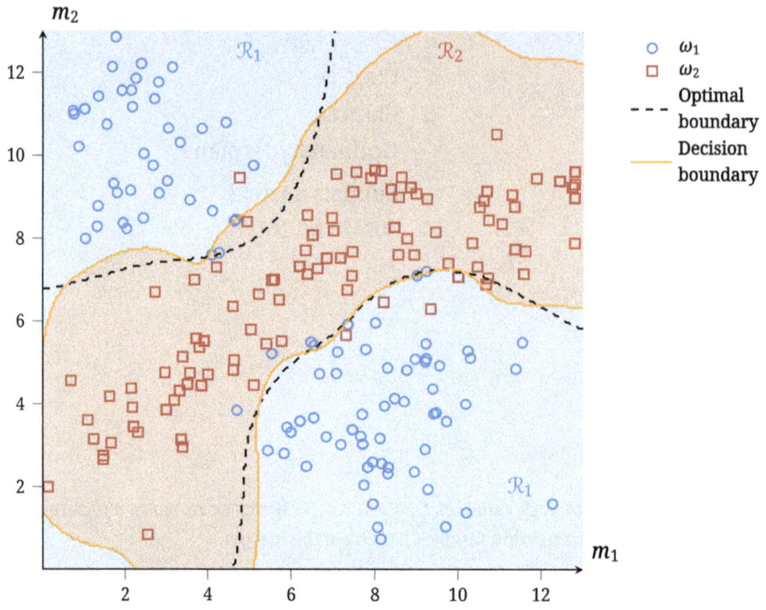

Fig. 7.23: Application to the reference example of Section 3.3.2. Decision regions of a soft margin SVM classifier with Gaussian kernel ($\sigma = 1$) and $C = 1$. The training and testing errors are $e_{\text{train}} = e_{\text{test}} = 6.5\,\%$. The testing error asymptotically approaches $e_{\text{test}} \approx 6.8\,\%$. The training set is the same as in Figure 3.8. Test samples are shown with hollow marks.

7.7.6 Discussion

SVMs are very powerful classifiers, which are applicable to a wide range of problems. As discussed in Section 7.7.4, the complexity of an SVM is determined not by the dimensionality of the feature space, but only by the number of support vectors. Because of this, an SVM is generally less prone to overfitting than other classifiers. Unlike artificial neural networks, which often yield suboptimal classifiers, an SVM classifier is uniquely determined by the training data and the learning algorithm will always produce a globally optimal classifier. Furthermore, the SVM algorithm allows for a geometric interpretation and is easy to apply without the need of prior knowledge about the problem. Using appropriate kernel functions, an SVM classifier can even be used to classify complicated objects like genome sequences or the words of a (natural) language. Most of all, it is based on a very well developed theoretical foundation (see Boser et al. [1992], Cortes and Vapnik [1995], Schölkopf and Burges [1999]). One major drawback, however, is that the presented SVM is only a binary classifier. Extension to multiple classes typically requires training at least one SVM for each class (see Section 7.1.1), but extensions to true multi-class SVMs exist as well (e.g., Crammer and Singer [2001]).

Figure 7.24 shows an example of the different decision boundaries that a hard margin SVM (left) and a soft margin SVM (right) can derive. The boundaries are shown in

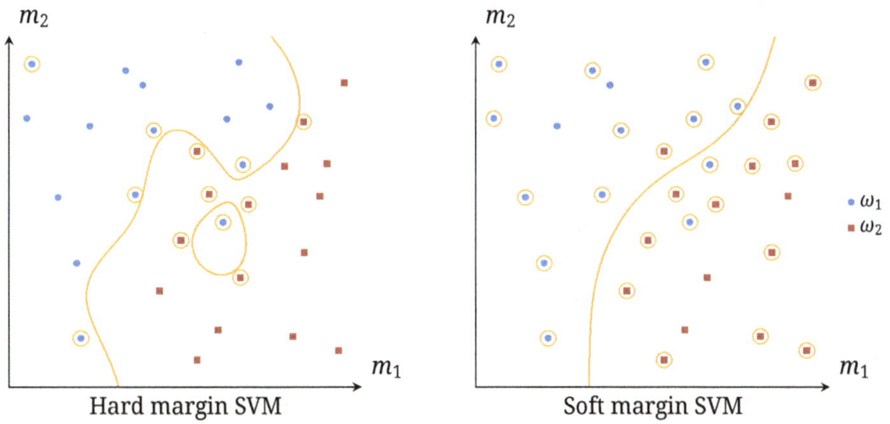

Fig. 7.24: Decision boundaries (shown in the original feature space) of a hard margin and soft margin SVM with Gaussian kernel on the same dataset. The shown feature vectors are the training data $\mathcal{D}$. The support vectors are marked with circles around them. Example according to Cristianini and Shawe-Taylor [2000].

the original, two-dimensional feature space. Both SVMs use a Gaussian kernel (see Equation (7.75)) with the same kernel parameter $\sigma^2 = 0.5$ and were trained on the same data. The shown data are the training data $\mathcal{D}$. Support vectors are marked with orange circles. The hard margin SVM shows perfect classification on the training data, but has a relatively complicated decision boundary. The decision region of the soft margin SVM, on the other hand, is smooth and relatively simple, but results in three errors on the training set. Furthermore, the number of support vectors is significantly higher with the soft margin SVM than with the hard margin SVM.

7.8 Matched filters

Often the goal is not only classifying an object, but also locating that object within an image. Consider the toy example in Figure 7.25. Here, the goal is to find the location of three characters *A*, *B*, and *C* against a noisy background. In other words: the image not only contains the object to be classified, but also unwanted noise. Matched filters, also known as template matching, are a popular tool to achieve just that.

The idea behind matched filters is that the objects to be found are known in advance—which is always the case in a classification setting—and that a prototypical template can be derived for each of the objects. Matched filters provide a mathematical mechanism that assumes extremal values in places where the image matches the template. In the above example, a matched filter for the character “A” produces an image that is (nearly) black everywhere except at the center of the “A” in the original image. In other words, objects are found within an image (or any other type of signal) by moving

(a) Noisy image with three objects (the letters A, B, and C)

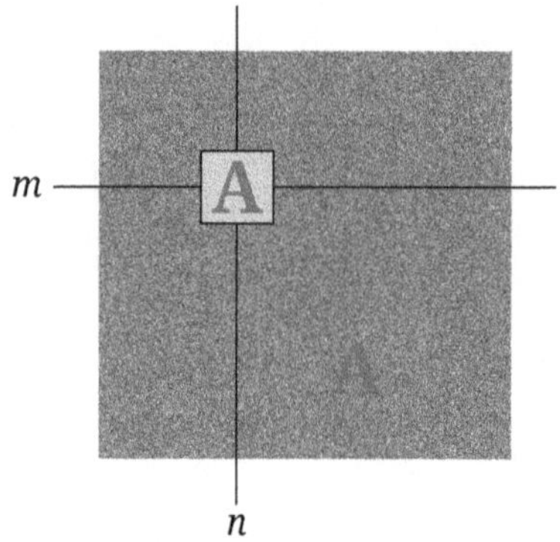

(b) A matched filter for the letter A is moved across the image

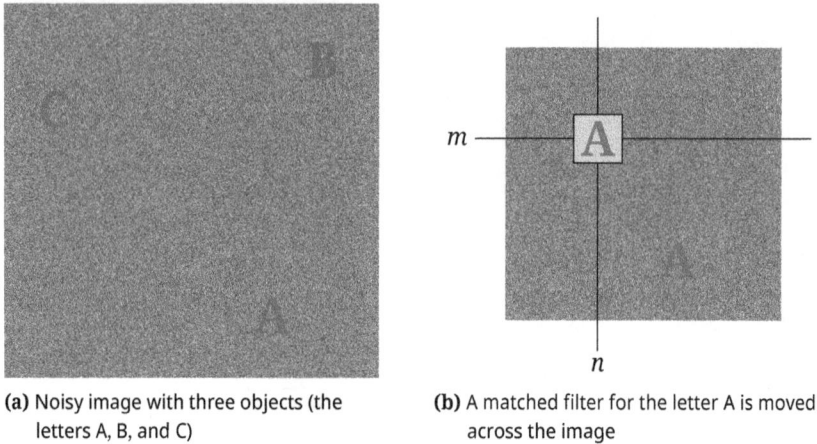

Fig. 7.25: Toy example of a matched filter.

the template over the image and recording the positions where the template matches and the resulting image is maximal.

In the following, this intuitive description is formalized in a discrete notation, that is, the images are considered to be two-dimensional discrete signals, as opposed to continuous signals. In particular, let g_{ij} denote the value of the image at the pixel position (i,j) and let

$$\mathbf{g}_{mn} := (\ldots, g_{m-i,n-j}, \ldots)^\mathsf{T}, \quad \forall (i,j) \in \mathcal{U} \tag{7.83}$$

denote the image patch around the position (m,n). In other words, $\mathbf{g}_{mn}$ is the *vector* of all image pixels in the region $\mathcal{U}$ around the patch origin (m,n). The image patch is modeled as being composed of a true, underlying object image $\mathbf{o}_{mn}$ and additive stationary noise $\mathbf{r}_{mn}$ with $\mathrm{E}\{\mathbf{r}_{mn}\} = \mathbf{0}$,

$$\mathbf{g}_{mn} = \mathbf{o}_{mn} + \mathbf{r}_{mn}. \tag{7.84}$$

The object and noise terms are defined in the same way as the image patch, which means that $\mathbf{g}_{mn}, \mathbf{o}_{mn}, \mathbf{r}_{mn} \in \mathbb{R}^{|\mathcal{U}|}$ are all of the same size. Given a filter $\mathbf{v} \in \mathbb{R}^{|\mathcal{U}|}$, the response of the filter to the image patch $\mathbf{g}_{mn}$ is obtained by taking the inner product of the two vectors:

$$k_{mn} = \mathbf{v}^\mathsf{T}\mathbf{g}_{mn} = \mathbf{v}^\mathsf{T}\mathbf{o}_{mn} + \mathbf{v}^\mathsf{T}\mathbf{r}_{mn}. \tag{7.85}$$

In the following, the indices mn are dropped for the sake of notational brevity. For example, the above equation will be written as simply $k = \mathbf{v}^\mathsf{T}\mathbf{o} + \mathbf{v}^\mathsf{T}\mathbf{r}$.

The question now is how to find a suitable filter. Many different approaches are possible. For example, one could train a linear SVM classifier and take the weight vector as a filter. With matched filters, however, the filter $\mathbf{v} := (\ldots, v_{ij}, \ldots)^\mathsf{T}$ is chosen so that it maximizes the signal to noise ratio

$$\mathrm{SNR} := \frac{P_1}{P_2} = \frac{\text{power of wanted signal}}{\text{power of noise signal}} \tag{7.86}$$

of the resulting output image k_{mn}. As is common in signal processing, P_1 is defined as the square of the local signal power, while P_2 is defined as the mean square noise power:

$$P_1 := (\mathbf{v}^\mathsf{T}\mathbf{o})^2 \quad \text{and} \tag{7.87}$$

$$P_2 := \mathrm{E}\left\{(\mathbf{v}^\mathsf{T}\mathbf{r})^2\right\} = \mathrm{E}\left\{(\mathbf{v}^\mathsf{T}\mathbf{r})(\mathbf{r}^\mathsf{T}\mathbf{v})\right\} = \mathbf{v}^\mathsf{T}\,\mathrm{E}\left\{\mathbf{r}\mathbf{r}^\mathsf{T}\right\}\mathbf{v} = \mathbf{v}^\mathsf{T}\mathbf{K}_{rr}\mathbf{v}. \tag{7.88}$$

Putting these into Equation (7.86) yields

$$\mathrm{SNR} = \frac{P_1}{P_2} = \frac{(\mathbf{v}^\mathsf{T}\mathbf{o})^2}{\mathbf{v}^\mathsf{T}\mathbf{K}_{rr}\mathbf{v}} \overset{!}{\longrightarrow} \max, \tag{7.89}$$

which we wish to maximize by choosing the free parameter $\mathbf{v}$ accordingly. To arrive at a closed form solution, observe that the covariance of the noise $\mathbf{K}_{rr}$ is symmetric and positive semidefinite. Therefore, there exists a decomposition $\mathbf{K}_{rr} = \mathbf{Q}^\mathsf{T}\mathbf{Q}$ and the inverse covariance factors as $\mathbf{K}_{rr}^{-1} = \mathbf{Q}^{-1}(\mathbf{Q}^\mathsf{T})^{-1}$. With the introduction of $\mathbf{w} := \mathbf{Q}\mathbf{v}$ and consequently $\mathbf{v} = \mathbf{Q}^{-1}\mathbf{w}$ as well as $\mathbf{v}^\mathsf{T} = \mathbf{w}^\mathsf{T}(\mathbf{Q}^{-1})^\mathsf{T}$, the signal to noise ratio can be written as

$$\mathrm{SNR} = \frac{(\mathbf{v}^\mathsf{T}\mathbf{o})^2}{\mathbf{v}^\mathsf{T}\mathbf{Q}^\mathsf{T}\mathbf{Q}\mathbf{v}} = \frac{\left(\mathbf{w}^\mathsf{T}(\mathbf{Q}^{-1})^\mathsf{T}\mathbf{o}\right)^2}{\mathbf{w}^\mathsf{T}\mathbf{w}}. \tag{7.90}$$

Without loss of generality, it can be assumed that $\|\mathbf{w}\| = 1$, because $\mathbf{w}^\mathsf{T}$ appears squared both in the numerator and denominator, and the length $\|\mathbf{w}\|$ cancels out. Therefore, with $\mathbf{w}^\mathsf{T}\mathbf{w} = 1$, Equation (7.90) is maximized by

$$\begin{aligned} &\mathrm{SNR} = \left(\mathbf{w}^\mathsf{T}(\mathbf{Q}^{-1})\mathbf{o}\right)^2 \overset{!}{\longrightarrow} \max \\ \Leftrightarrow\ &\mathbf{w} \| (\mathbf{Q}^{-1})^\mathsf{T}\mathbf{o} && (\mathbf{w} \text{ is parallel to } (\mathbf{Q}^{-1})^\mathsf{T}\mathbf{o}) \\ \Leftrightarrow\ &\mathbf{w} = c(\mathbf{Q}^{-1})^\mathsf{T}\mathbf{o} \end{aligned} \tag{7.91}$$

with some constant $c \in \mathbb{R}$. Using the relation between $\mathbf{v}$ and $\mathbf{w}$ defined above finally yields the discrete matched filter

$$\mathbf{v} = c\mathbf{Q}^{-1}\left(\mathbf{Q}^{-1}\right)^\mathsf{T}\mathbf{o} = c\mathbf{K}_{rr}^{-1}\mathbf{o}. \tag{7.92}$$

As c is just a linear factor in the maximization of the SNR, it is usually defined as $c := 1$, that is, the matched filter is fully defined by the object image $\mathbf{o}$ and the noise covariance matrix $\mathbf{K}_{rr}$. In the special case that $\mathbf{r}$ is white noise, i.e., in the case that $\mathbf{K}_{rr} \propto \mathbf{I}$, it follows that $\mathbf{v} \propto \mathbf{o}$. In other words: the matched filter *is* the object image.

To get an intuition of how a matched filter works, we substitute Equation (7.92) in Equation (7.85) to derive the filter response

$$k_{mn} = \mathbf{v}^\mathsf{T}\mathbf{g}_{mn} = \mathbf{o}^\mathsf{T}\mathbf{Q}^{-1}\left(\mathbf{Q}^{-1}\right)^\mathsf{T}\mathbf{g}_{mn}. \tag{7.93}$$

Here, $\left(\mathbf{Q}^{-1}\right)^{\mathrm{T}}$ acts as a whitening filter that de-correlates the noise in the image patch $\mathbf{g}_{mn}$. However, the transformed image patch resides now in a different space than the object image $\mathbf{o}$. To correct for this, $\mathbf{Q}^{-1}$ modifies $\mathbf{o}$ to match after the whitening of the image.

Note that while matched filters correct for (pixel) noise, they are still very sensitive to rotation, scale, and other distortions of the input image. Since these perturbations are very common in detection tasks, the image has to be normalized before applying the filter.

In order to use matched filters for classification, one matched filter $\mathbf{v}_i$ is created for each class ω_i to be recognized. The filters are moved over the image and for each position the best match is recorded. More formally, the feature vector at position $\mathbf{x} = (x,y)^{\mathrm{T}}$ is given by

$$\mathbf{m}(\mathbf{x}) := \operatorname{col}\left\{\left\{g(\mathbf{x}') \mid \mathbf{x}' = \mathbf{x} - \alpha, \forall \alpha \in \mathcal{U}\right\}\right\}, \tag{7.94}$$

i.e., the column vector of image pixels within the (shifted) region $\mathcal{U}$ around $\mathbf{x}$. The decision function for each filter $\mathbf{v}_i$, $i = 1, \ldots, c$, is given by $k_i\left(\mathbf{m}(\mathbf{x})\right) = \mathbf{v}_i^{\mathrm{T}}\mathbf{m}(\mathbf{x})$ and the decision vector becomes

$$\mathbf{k}\left(\mathbf{m}(\mathbf{x})\right) = \begin{pmatrix} \mathbf{v}_1^{\mathrm{T}} \\ \vdots \\ \mathbf{v}_c^{\mathrm{T}} \end{pmatrix} \mathbf{m}(\mathbf{x}) = \mathbf{V}\,\mathbf{m}(\mathbf{x}). \tag{7.95}$$

This decision vector is evaluated at *every pixel* of the input image. As there is always a maximal entry in the decision vector, there will be a match *at every pixel*, even though this certainly cannot be the case. In practice, match candidates with low responses will be discarded as “none of the classes”. This maximum criterion is an example of classification with rejection, which will be explored in more detail in Section 9.5.

More details about matched filters can be found, for example, in Beyerer et al. [2016].

7.9 Classification of sequences

An underlying assumption in the discussion so far was that the data are independent and identically distributed, i.e., the feature vector $\mathbf{m}_i$ does not depend on the feature vectors seen before. So far this assumption has been valid, but it does not hold, e.g., for videos, where the content of the next frame depends on the content of the current frame, for speech, where grammar restricts which words can follow another word in a valid sentence, or for games, where the next move depends on the moves that have been played before. In general, any sequence where the probability of drawing an object depends on which objects have been drawn before, i.e., sequences that depend on some state, violate the i.i.d. assumption.

As an example, consider the recognition of spoken words, or more specifically, the classification of utterances into characters that make up a word. In this scenario, each character constitutes a class, i.e., $\omega_1 \mathrel{\hat{=}} \mathrm{A}$, $\omega_2 \mathrel{\hat{=}} \mathrm{B}$, $\omega_3 \mathrel{\hat{=}} \mathrm{C}$, etc. Words are generated by some source that sequentially attains one of the classes as its internal state and produces the corresponding character. Clearly, the characters of a word are not independent of the surrounding characters. For example, if the letters observed so far are "T" and "E", the characters "A", "N" and "D" are more probable to be observed next than the character "X". A classifier will be more powerful if these dependences are modeled into it.

In our example, however, it is not possible to observe the characters directly. In other words, the classes are *hidden* from our view. It is possible, after some suitable signal processing, to observe associated phonemes—the smallest indivisible parts of speech—v_i, e.g., (in IPA phoneme notation) $v_1 \mathrel{\hat{=}}$ /iː/, $v_2 \mathrel{\hat{=}}$ /e/, $v_3 \mathrel{\hat{=}}$ /æ/, etc. Given a sequence of such phonemes, the goal is to recognize the corresponding word, i.e., the sequence of states (characters) that will produce the observed sequence of phonemes.

More concretely, consider the word "sequence". This word is produced by reaching the states S, E, Q, U, E, N, C, and E one after another. When spoken, however, one can only observe the phoneme sequence /ˈsikwəns/. The goal is to work this process backwards, that is, to map the observed phonemes to the word "sequence" by virtue of a model of the generation process. Note that here the number of phonemes is the same as the number of characters. This does not always have to be the case: "model" has five characters, but the (US English) pronunciation /ˈmɑdl̩/ contains only four phonemes.

7.9.1 Markov models

Sequences of any type can be modeled using discrete Markov models. A Markov model describes the probability distribution to switch to the state $\omega(t)$ at time t, given the system's states $\omega(t-1), \omega(t-2), \ldots$ in the previous time steps $(t-1), (t-2), \ldots$ If all of these time steps were to be taken into account, such a model would not be very useful: inference and learning would take place in a very high-dimensional space, with all the associated problems (see Section 6.1). A discrete Markov model of the l-th order therefore assumes that the probability of going into a state depends on the l preceding states, but not more:

$$P(\omega(t) \mid \omega(t-1), \omega(t-2), \ldots) = P(\omega(t) \mid \omega(t-1), \omega(t-2), \ldots, \omega(t-l)) \\ \forall t \in \mathbb{Z} \quad \text{and} \quad \omega(t) \in \Omega/\!\sim\, = \{\omega_1, \ldots, \omega_c\}. \tag{7.96}$$

In practical applications, the term "Markov model" is often used as synonymous with a first order Markov model, where the probability of a state only depends on the previous state:

$$P(\omega(t) \mid \omega(t-1), \omega(t-2), \omega(t-3), \ldots) = P(\omega(t) \mid \omega(t-1)). \tag{7.97}$$

Here, the above is referred to as the state transition probability and the probability of switching from state ω_i to state ω_j is abbreviated as

$$a_{ij} := P(\underline{\omega}(t+1) = \omega_j \mid \underline{\omega}(t) = \omega_i), \quad \text{where} \quad \sum_j a_{ij} = 1. \tag{7.98}$$

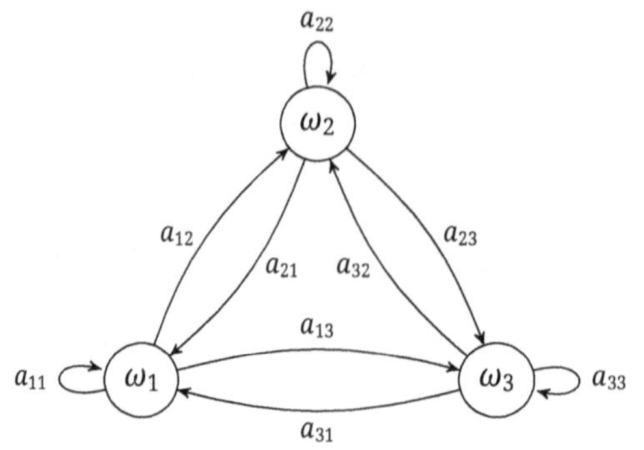

Fig. 7.26: Discrete first order Markov model with three states ω_i. The probability of a transition from state ω_i to state ω_j is denoted a_{ij}. Image according to Duda et al. [2001].

Finally, a Markov model comprises the a priori state probabilities $\pi_i = P(\omega_i(0)) = P(\omega_i)$, which encode the probability of starting in the state ω_i.

First order Markov models may be represented by a stochastic automaton, where the states correspond to the classes ω_i and the state transition probabilities are given by the a_{ik}. An example is shown in Figure 7.26.

Markov models can be, and have been, used to generate sequences of characters. Table 7.1 presents some examples of sequences that were generated by Markov models of increasing order. The state transition probabilities and the a priori probabilities were estimated from large corpora of German, English, and Russian texts.

The sequences generated by the 0th order models were fully determined by the prior probabilities and do not resemble words from the corresponding languages very much. Higher order models, on the other hand, include the transition probabilities and even generate valid words like "IN" and "WHEY"—even though the model does not explicitly encode the concept of a word!

7.9.2 Hidden Markov models

So far we have assumed that the model state at a time t is known with absolute certainty, but this is not always the case. In the introductory speech recognition example at the beginning of Section 7.9, the underlying states (characters) were only indirectly observable via the associated phonemes. A hidden Markov model (HMM) is an extension to a Markov model that can deal with such situations.

In addition to states and state transition probabilities, an HMM consists of observations v_k and emission probabilities that denote the probability of seeing an observable given a chain of states $\underline{\omega}(t), \dots, \underline{\omega}(t-l)$, where l is the order of the model. In a first order HMM, the emission probability of observing $\underline{v}(t) = v_k$ depends only on the current state $\underline{\omega}(t)$ and is denoted by

$$b_{jk} := P(\underline{v}(t) = v_k \mid \underline{\omega}(t) = \omega_j), \quad \text{where} \quad \sum_k b_{jk} = 1. \tag{7.99}$$

Tab. 7.1: Character sequences generated by Markov models of different order. Table reproduced from Hoffmann [1998].

Order	Generated examples
	German (Küpfmüller [1954])
0	EME GKNEET ERS TITBL BTZENFNDBGD EAI E LASZ BETEATR IASMIRCH EGEOM
1	AUSZ KEINU WONDINGLIN DUFRN ISAR STEISBERER ITEHM ANORER
2	PLANZEUDGES PHIN INE UNDEN VEBEICHT GES AUF ES SO UNG GAN DICH WANDERSO
3	ICH FOLGEMAESZIG BIS STEHEN DISPONIN SEELE NAMEN
	English (Shannon [1948])
0	OCRO HLI RGWR NMIELWIS EULL NBNESEBYA TH EEI ALHENHTTPA OOBTTVA NAH BRL
1	ON IE ANTSOUTINYS ARE T INCTORE ST BE S DEAMY ACHIN D ILONASIVE TUCOOWE AT TEASONARE FUSO TIZIN ANOY TOBE SEACE CTISBE
2	IN NO IST LAT WHEY CRACTICT FROURE BIRS GROCID PONDENOME OF DEMONSTRURES OF THE REPTAGIN IS REGOACTIONA OF CRE
	Russian (Jaglom and Jaglom [1960])
0	EYNT CIJA'A OERV ODNG 'UEMLOLJK Z-JA ENVTŠA
1	UMARONO KAČ VSVANNYJ ROSJA NYCH KOVKROV NEDARE
2	PKAK POT DURNOSKAKA NAKONEPNO ZNE STVOLOVIL SE TVOJ O-NIL'
3	VESEL VRAT'SJA NE SUCHOM I NEP I KORKO

Note that with a hidden Markov model, the states are not directly observable. Instead, the state sequence can only be inferred from the sequence of observations v_k. Figure 7.27 shows a first order hidden Markov model with three hidden states and four observations.

There are three important tasks when working with hidden Markov models:

1. Evaluation (forward problem): Given an HMM with transition probabilities a_{ij} and emission probabilities b_{jk}, how probable is it to observe the sequence $v(1), \ldots, v(T)$?
2. Decoding (backward problem): Given an HMM with probabilities a_{ij} and b_{jk}, and a sequence of observations $v(1), \ldots, v(T)$, what is the most probable sequence of states $\omega(1), \ldots, \omega(T)$ that generated this sequence?
3. Learning (parameter estimation): Given the number of states, the set of possible observations, and a set of training sequences $v(1), \ldots, v(T)$, how can the parameters a_{ij} and b_{jk} be estimated?

The first task is straightforward, as it only involves known quantities and is mostly of theoretical importance. The second and third tasks, on the other hand, are very important in the context of pattern recognition. The second task is analogous to classification in the conventional setting: observations correspond to feature vectors, whereas the states correspond to classes. At first, it seems that this task is not much more complicated than the forward problem, but unfortunately it is much more complicated. The usual approach to the backward problem is the Viterbi algorithm (Viterbi [1967], Forney [1973]),

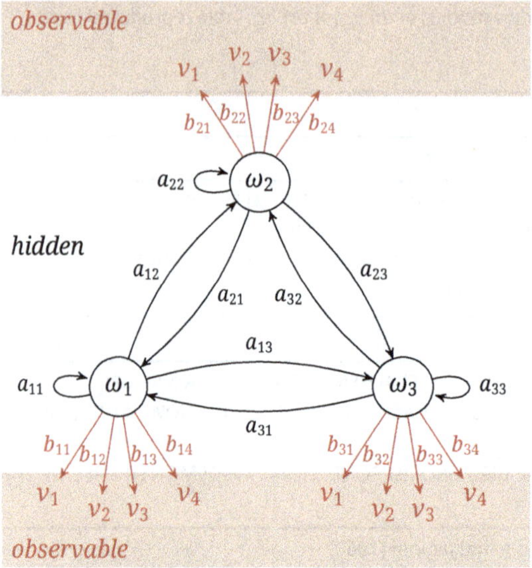

Fig. 7.27: Discrete first order hidden Markov model with three hidden states ω_i and four possible observations v_k. The a_{ij} denote the state transition probabilities, while the b_{jk} denote the probability that v_k is observed when entering state ω_j. Image according to Duda et al. [2001].

a variant of dynamic programming. The last task corresponds to learning a conventional classifier. Here, the goal is to construct an HMM from the data. Most common approaches estimate the parameters a_{ij} and b_{jk} using an expectation maximization (EM) algorithm (see Section 4.5).

Before going into details of the three aforementioned tasks, a formal definition of a discrete first order HMM is given. Furthermore, differences between *discrete, continuous*, and *semi-continuous* HMMs are described.

Definition 7.5 (Discrete first order hidden Markov model). A discrete first order HMM λ is defined as 5-tuple $\lambda = (\Omega, \mathrm{N}, \mathbf{A}, \mathbf{B}, \boldsymbol{\pi})$ with a

1. set of states $\Omega = \{\omega_1, \ldots, \omega_c\}$,
2. set of observations $\mathrm{N} = \{v_1, \ldots, v_d\}$,
3. matrix of state transition probabilities $\mathbf{A} = (a_{ij}),\ i,j = 1 \ldots c$,
4. matrix of emission probabilities $\mathbf{B} = (b_{jk}),\ j = 1 \ldots c,\ k = 1 \ldots d$, and
5. vector of state start probabilities $\boldsymbol{\pi} = (\pi_i),\ i = 1 \ldots c$.

Different from Definition 7.5, the observations in continuous HMM are in general vectors $\mathbf{x}$ whose probability distributions are given by density functions:

$$b_{j\mathbf{x}} = b_j(\mathbf{x}) = p(\mathbf{x}|\omega_j). \tag{7.100}$$

For those observation densities of continuous HMMs, Gaussian mixture models (GMM) are often used due to their universal approximation capabilities (cf. Section 3.4):

$$b_j(\mathbf{x}) = \sum_{l=1}^{L_j} c_{jl} g_{jl}(\mathbf{x}) = \sum_{l=1}^{L_j} c_{jl} \mathcal{N}(\mathbf{x}|\boldsymbol{\mu}_{jl}, \boldsymbol{\Sigma}_{jl}). \tag{7.101}$$

Recall that μ_{jl} and Σ_{jl} are the expectation vector and covariance matrix of the lth GMM component $g_{jl}(\mathbf{x})$. The GMM $b_j(\mathbf{x})$ represents the emission probability distribution of $\mathbf{x}$ from state j. Each GMM comprises its own set of L_j components. Obviously, the number of parameters of the formalism is much higher for continuous HMMs w.r.t. their discrete counterpart. Therefore, methods to reduce the number of parameters referred to as *tying* have been developed. They are out of the scope of this book but can be studied in Fink [2003], for example.

Another variant that has less parameters than the continuous HMM is called semi-continuous HMM. The observations are also modelled with GMMs but there is only one set of basic components $\{\mathcal{N}(\mathbf{x}|\mu_l, \Sigma_l)\}_{l=1\ldots L}$ for all emission densities:

$$b_j(\mathbf{x}) = \sum_{l=1}^{L} c_{jl}\mathcal{N}(\mathbf{x}|\mu_l, \Sigma_l). \tag{7.102}$$

Note that in contrast to the continuous case, there is no state dependency in the parameters of the Gaussian distributions μ_l and Σ_l. Despite the different modelling of observations, the algorithms for the evaluation, decoding, and learning problem follow the same principles independent from the HMM variant. The three different problems are now examined in detail.

Evaluation (forward problem)

Given a first order HMM λ according to Definition 7.5 and a sequence of observations $V = [v(t)]_{t=1}^{T}$ with length T, the forward problem is to find the *production probability* $P(V|\lambda)$ that the sequence V is generated by the HMM λ. In a first order HMM, the observations $v(t)$ depend only on the states $\omega(t)$. Hence, there are $N = c^T$ corresponding state sequences $W = [\omega(t)]_{t=1}^{T}$ that could produce a specific observation sequence V (ignoring the fact that some transition or emission probabilities might be zero). The conditioned probability $P(V|W,\lambda)$ depends only on the emission probabilities b:

$$P(V|W,\lambda) = \prod_{t=1}^{T} b_{\omega(t)}(v(t)). \tag{7.103}$$

Here and in the following, an alternative formulation for the emission probabilities $b_{\omega(t)}(v(t)) := b_{jk}$ from Equation (7.99) is used to take the time variable t explicitly into account. For the same reason, the transition probabilities are now denoted by $a_{\omega(t-1),\omega(t)} := a_{ij}$ (see Equation (7.98)). In addition, one can define the start probabilities for states by $a_{\omega(0),\omega(1)} := \pi_{\omega(1)}$. With these definitions, the state sequence probability $P(W|\lambda)$ is given by

$$P(W|\lambda) = \pi_{\omega(1)} \prod_{t=2}^{T} a_{\omega(t-1),\omega(t)} = \prod_{t=1}^{T} a_{\omega(t-1),\omega(t)}. \tag{7.104}$$

Combining Equations (7.103) and (7.104), the joint probability of observation and state sequence $P(V,W|\lambda)$ can be computed:

$$P(V,W|\lambda) = P(V|W,\lambda)P(W|\lambda) = \prod_{t=1}^{T} a_{\omega(t-1),\omega(t)} b_{\omega(t)}(v(t)). \tag{7.105}$$

Finally, summing over all state sequences $W^n = [\omega^n(t)]_{t=1}^T$ yields the searched production probability

$$P(V|\lambda) = \sum_{n=1}^{N} P(V,W^n|\lambda) = \sum_{n=1}^{N}\prod_{t=1}^{T} a_{\omega^n(t-1),\omega^n(t)} b_{\omega^n(t)}(v(t)). \tag{7.106}$$

This *brute force* approach has a high computational complexity of $\mathcal{O}(Tc^T)$ and thus is not practicable, especially for large sequence lengths T. Fortunately, there exists an efficient solution to calculate the production probability $P(V|\lambda)$, which is called *forward algorithm* and described in the following.

The forward algorithm for the evaluation problem of HMMs makes use of the Markov property that the current state depends only on the previous state in a first order model (see Equation (7.97)). The central entity in the algorithm is the *forward variable* $\alpha_t(i)$ that represents the probability of emitting the first t observations $v(1), \ldots, v(t)$ and reaching state ω_i of the HMM λ at time t:

$$\alpha_t(i) := P(v(1), \ldots, v(t), \omega(t) = \omega_i|\lambda). \tag{7.107}$$

At first, the initial forward variable $\alpha_1(i)$ can be calculated for each state ω_i with emission probabilities $b_i(v(1))$ and start probabilities π_i:

$$\alpha_1(i) = \pi_i b_i(v(1)). \tag{7.108}$$

Then, the forward variables $\alpha_{t+1}(j)$ can recursively be computed as

$$\alpha_{t+1}(j) = \sum_{i=1}^{c} \alpha_t(i) a_{ij} b_j(v(t+1)) \quad \text{for} \quad t = 1, \ldots, T-1. \tag{7.109}$$

Note that the forward variables for the different states can be calculated in parallel at each time step. Finally, summation over the forward variables of all states for $t = T$ yields the searched production probability

$$P(V|\lambda) = \sum_{i=1}^{c} \alpha_T(i). \tag{7.110}$$

The forward algorithm is summarized in Algorithm 7.3. Due to its recursive nature, a lot of computations of the brute force approach need not to be performed which makes it much more efficient with a computational complexity of $\mathcal{O}(Tc^2)$.

Algorithm 7.3: The forward algorithm.

Data: HMM λ with states $\omega_1, \dots, \omega_c$ and transition, emission, start probabilities $a_{ij}, b_j(v), \pi_i$, observation sequence $V = [v(1), \dots, v(T)]$
Result: Production probability $P(V|\lambda)$
forall $i \in \{1, \dots, c\}$ **do**
  $\alpha_1(i) \leftarrow \pi_i b_i(v(1))$
for $t = 1, \dots, T-1$ **do**
  forall $j \in \{1, \dots, c\}$ **do**
    $\alpha_{t+1}(j) \leftarrow \sum_{i=1}^{c} \alpha_t(i) a_{ij} b_j(v(t+1))$
$P(V|\lambda) \leftarrow \sum_{i=1}^{c} \alpha_T(i)$
return $P(V|\lambda)$

Another interesting evaluation problem for HMMs is to determine the so-called *optimal production probability* $P^*(V|\lambda)$ which quantifies the modelling quality of an observation sequence V by a HMM λ:

$$P^*(V|\lambda) = P(V, W^*|\lambda) = \max_W P(V, W|\lambda). \tag{7.111}$$

In contrast to the overall production probability $P(V|\lambda)$, where all state sequences W that can generate V are considered, only the *optimal* path W^* with maximum single production probability P^* is evaluated. The procedure of calculating $P^*(V|\lambda)$ is similar to the forward algorithm but even more efficient. Again, an auxiliary variable is introduced—the *partial path probability* $\delta_t(i)$—that gives the maximum probability of emitting the first t observations and reaching state ω_i at time t:

$$\delta_t(i) := \max_{\omega(1),\dots,\omega(t-1)} \{P(v(1), \dots, v(t), \omega(1), \dots, \omega(t-1), \omega(t) = \omega_i|\lambda)\}. \tag{7.112}$$

The algorithm to compute the optimal production probability $P^*(V|\lambda)$ with help of the partial path probabilities $\delta_t(i)$ is given in Algorithm 7.4. Besides its greater efficiency, the optimal production probability can be a good approximation of the overall production probability if the probability of the optimal path W^* dominates the summation over all path probabilities. Note that the described algorithm determines only the probability of the optimal path W^* but the path itself remains unknown. However, the algorithm can be extended to the famous *Viterbi algorithm* that is designed to decode an observed sequence to the most probable state sequence which has generated the observations. This backward problem is treated in the following.

Decoding (backward problem)

Given a HMM λ and an observation sequence $V = [v(1), \dots, v(T)]$, the backward problem is to determine the path of states $W^* = [\omega^*(1), \dots, \omega^*(T)]$ with maximum single production probability $P^*(V|\lambda) = P(V, W^*|\lambda)$. Solving this problem is also referred to

Algorithm 7.4: Algorithm for the optimal production probability.

Data: HMM λ with states $\omega_1, \dots, \omega_c$ and transition, emission, start probabilities $a_{ij}, b_j(v), \pi_i$, observation sequence $V = [v(1), \dots, v(T)]$

Result: Optimal production probability $P^*(v|\lambda)$

```
forall i ∈ {1, ..., c} do
 | δ_1(i) ← π_i b_i(v(1))
for t = 1, ..., T − 1 do
 | forall j ∈ {1, ..., c} do
 | | δ_{t+1}(j) ← max_{i=1...c}{δ_t(i) a_ij} b_j(v(t + 1))
P*(V|λ) ← max_{i=1...c}{δ_T(i)}
return P*(V|λ)
```

decoding and can be achieved with the Viterbi algorithm. According to Algorithm 7.4, the state $\omega^*(T)$ which maximizes $\delta_T(i)$ is given by

$$\omega^*(T) = \arg\max_{i=1,\dots,c} \{\delta_T(i)\}. \tag{7.113}$$

Another type of auxiliaries termed *backward pointers* $\psi_t(j)$ is introduced. For each partial path probability $\delta_t(i)$, a corresponding backward pointer $\psi_t(j)$ saves the optimal previous state such that

$$\psi_t(j) = \arg\max_{i=1,\dots,c} \{\delta_{t-1}(i)a_{ij}\} \tag{7.114}$$

holds. One can then recursively calculate the states $\omega^*(t)$ of the optimal path W^* with

$$\omega^*(t) = \psi_{t+1}(\omega^*(t+1)) \quad \text{for} \quad t = T-1, \dots, 1, \tag{7.115}$$

whereby the recursion starts at the end of the sequence ($t = T - 1$) and ends at its beginning ($t = 1$). The complete Viterbi algorithm including the calculation of partial path probabilities $\delta_t(i), t = 1, \dots, T$, backward pointers $\psi_t(j), t = 2, \dots, T$, and optimal path states $\omega^*(t), t = 1, \dots, T$ is depicted in Algorithm 7.5.

Learning (parameter estimation)

In contrast to the two previously discussed problems of evaluation and decoding, the parameters a_{ij} and b_{jk} of a HMM are not given for the learning task, also known as parameter estimation. The goal is to determine the transition and emission probabilities with the help of a training set of observed sequences $\mathcal{D} = \{V_1, \dots, V_N\}$ with $V_i = [v_i(1), \dots, v_i(T)], i = 1, \dots, N$. The set of states $\Omega = \{\omega_1, \dots, \omega_c\}$ and set of observations $\mathcal{V} = \{v_1, \dots, v_d\}$ are also given.

The main idea of the parameter estimation algorithms of HMMs is to monitor the HMM at the generation of observations and estimate the transition and emission proba-

Algorithm 7.5: Viterbi algortihm.

Data: HMM λ with states $\omega_1, \dots, \omega_c$ and transition, emission, start probabilities $a_{ij}, b_j(v), \pi_i$, observation sequence $V = [v(1), \dots, v(T)]$
Result: Optimal path $W^* = [\omega^*(1), \dots, \omega^*(T)]$
forall $i \in \{1, \dots, c\}$ **do**
  $\delta_1(i) \leftarrow \pi_i b_i(v(1))$
for $t = 1, \dots, T-1$ **do**
  forall $j \in \{1, \dots, c\}$ **do**
    $\delta_{t+1}(j) \leftarrow \max_{i=1\dots c}\{\delta_t(i)a_{ij}\}b_j(v(t+1))$
    $\psi_{t+1}(j) \leftarrow \arg\max_{i=1\dots c}\{\delta_t(i)a_{ij}\}$
$P^*(V|\lambda) \leftarrow \max_{i=1\dots c}\{\delta_T(i)\}$
$\omega^*(T) \leftarrow \arg\max_{i=1\dots c}\{\delta_T(i)\}$
for $t \in T-1, \dots, 1$ **do**
  $\omega^*(t) \leftarrow \psi_{t+1}(\omega^*(t+1))$;
$W^* \leftarrow [\omega^*(1), \dots, \omega^*(T)]$
return W^*

bilities as relative frequencies:

$$\hat{a}_{ij} = \frac{\text{expected number of transitions from state } \omega_i \text{ to state } \omega_j}{\text{expected total number of transitions starting from state } \omega_i}, \tag{7.116}$$

$$\hat{b}_j(v_k) = \frac{\text{expected number of observations of } v_k \text{ in state } \omega_j}{\text{expected total number of observations in state } \omega_j}. \tag{7.117}$$

Note that in the above equations, only discrete HMMs without separate estimation equalities for start state probabilities $\hat{\pi}_i$ (which can be modelled as additional transition probabilities) are considered for the sake of simplicity.

Different algorithms with various evaluation measures exist to estimate the parameters of HMMs. In the following, the so-called *Baum–Welch algorithm*, a special case of the EM algorithm presented in Section 4.5, that utilizes the overall production probability $P(V|\lambda)$ as evaluation measure is described in detail. Other algorithms for learning in HMMs are the *Viterbi training* and the *segmental-k-means algorithm* which both use the optimal production probability $P^*(V|\lambda)$ as evaluation criterion. Those algorithms can be studied in Fink [2003].

Before delving into the Baum–Welch algorithm, another basic algorithm for HMMs is introduced—the *backward algorithm*. It is the counterpart of the forward algorithm that has been described earlier. Similar to the forward variable, the *backward variable* $\beta_t(i)$ is defined as the probability of generating the observations $v(t+1), \dots, v(T)$ starting from a state ω_i of a HMM λ at time t:

$$\beta_t(i) = P(v(t+1), \dots, v(T), \omega(t) = \omega_i|\lambda). \tag{7.118}$$

Algorithm 7.6: Backward algorithm.

Data: HMM λ with states $\omega_1, \dots, \omega_c$ and transition, emission, start probabilities $a_{ij}, b_j(v), \pi_i$, observation sequence $V = [v(1), \dots, v(T)]$
Result: Production probability $P(V|\lambda)$
forall $i \in \{1, \dots, c\}$ **do**
  $\beta_1(i) \leftarrow 1$
for $t = T - 1, \dots, 1$ **do**
  forall $i \in \{1, \dots, c\}$ **do**
    $\beta_t(i) \leftarrow \sum_{j=1\dots c} a_{ij} b_j(v(t+1))\beta_{t+1}(j)$
$P(V|\lambda) \leftarrow \sum_{i=1\dots c} \pi_i b_i(v(1))\beta_1(i)$
return $P(V|\lambda)$

Naturally, the probability $\beta_T(i)$ of being in state ω_i at time $t = T$ and generating the empty set of observations $\{v_t | t > T\}$ is 1:

$$\beta_t(i) = 1. \tag{7.119}$$

The backward variables for $t < T$ can again recursively be computed according to

$$\beta_t(i) = \sum_{j=1}^{c} a_{ij} b_j(v(t+1))\beta_{t+1}(j) \quad \text{for} \quad t = T - 1, \dots, 1. \tag{7.120}$$

Finally, one can get the production probability $P(V|\lambda)$ of a sequence V with summation over all backward variables $\beta_1(i)$ weighted by the start probabilities π_i and emission probabilities $b_i(v(1))$:

$$P(V|\lambda) = \sum_{i-1}^{c} \pi_i b_i(v(1))\beta_1(i). \tag{7.121}$$

The backward algorithm is summarized in Algorithm 7.6. The combination of forward and backward algorithm is often referred to as *forward–backward algorithm* and is the basis of the Baum–Welch algorithm.

The Baum–Welch algorithm is an iterative optimization technique for parameter estimation in HMMs that uses the production probability $P(V|\lambda)$ as optimization criterion. Given a sequence from the training set $V \in \mathcal{D}$ and an initial parameter configuration of a HMM λ, the algorithm changes those parameters such that the production probability of the sequence V is increased:

$$P(V|\hat{\lambda}) \geq P(V|\lambda). \tag{7.122}$$

Again, some auxiliary variables are defined for the following calculations. First, the a posteriori state probability $\gamma_t(i)$ of being in state ω_i at time t after observing V can be derived with Bayes' formula (Equation (3.7)) to

$$\gamma_t(i) := P(\omega(t) = \omega_i | V, \lambda) = \frac{P(\omega(t) = \omega_i, V|\lambda)}{P(V|\lambda)} = \frac{\alpha_t(i)\beta_t(i)}{P(V|\lambda)}, \tag{7.123}$$

whereby the forward and backward variables $\alpha_t(i)$ and $\beta_t(i)$ are used. Next, the a posteriori transition probabilities $\gamma_t(i,j)$ are defined as

$$\gamma_t(i,j) := \frac{P(\omega(t) = \omega_i, \omega(t+1) = \omega_j, V|\lambda)}{P(V|\lambda)} = \frac{\alpha_t(i) a_{ij} b_j(\mathbf{x}(t)) \beta_{t+1}(j)}{P(V|\lambda)}. \tag{7.124}$$

Note that the general formulation, i.e., for continuous HMMs with probability density functions modelling the emission probabilities b_j is given. In this general case, the emission densities are typically Gaussian mixture models for which the underlying parameters have to be estimated (c_{jl}, $\boldsymbol{\mu}_{jl}$, and $\boldsymbol{\Sigma}_{jl}$ as per Equation (7.101)). For these estimations, another auxiliary variable $\xi_t(j,l)$ is defined which denotes the a posteriori probability that the observation $\mathbf{x}(t)$ has been generated at time t in state ω_j by the lth component L of the Gaussian mixture model:

$$\xi_t(j,l) = P(\omega(t) = \omega_j, L(t) = l|V, \lambda) = \frac{\sum_{i=1}^{c} \alpha_{t-1}(i) a_{ij} c_{jl} g_{jl}(\mathbf{x}(t)) \beta_t(j)}{P(V|\lambda)}. \tag{7.125}$$

The three a posteriori probabilities $\gamma_t(i)$, $\gamma_t(i,j)$, and $\xi_t(j,l)$ are used in the optimization step of the Baum–Welch algorithm to improve the estimated model parameters $\hat{\lambda} = (\hat{\boldsymbol{\pi}}, \hat{\mathbf{A}}, \hat{\mathbf{B}})$. First, an initial set of parameters is chosen either randomly or with the integration of a priori information about the parameters if available. The set of parameters comprises the start probabilities $\hat{\pi}_i$, the transition probabilities $\hat{a}_{ij}$, as well as the parameters $\hat{c}_{jl}$, $\hat{\boldsymbol{\mu}}_{jl}$ and $\hat{\mathbf{K}}_{jl}$ of the Gaussian mixture models that represent the emission densities $\hat{b}_j(\mathbf{x}(t))$. As mentioned before, the number of parameters as well as the complexity of the computations decreases for the semi-continuous and discrete case. In the following, the formulas for the general continuous case are given.

During the optimization, the Baum–Welch algorithm iteratively improves the estimated parameters of the HMM as follows:

$$\hat{\pi}_i = \gamma_1(i) \tag{7.126}$$

$$\hat{a}_{ij} = \frac{\sum_{t=1}^{T-1} \gamma_t(i,j)}{\sum_{t=1}^{T-1} \gamma_t(i)} \tag{7.127}$$

$$\hat{c}_{jl} = \frac{\sum_{t=1}^{T} \xi_t(j,l)}{\sum_{t=1}^{T} \gamma_t(j)} \tag{7.128}$$

$$\hat{\boldsymbol{\mu}}_{jl} = \frac{\sum_{t=1}^{T} \xi_t(j,l) \mathbf{x}(t)}{\sum_{t=1}^{T} \xi_t(j,l)} \tag{7.129}$$

$$\hat{\mathbf{K}}_{jl} = \frac{\sum_{t=1}^{T} \xi_t(j,l) \mathbf{x}(t) \mathbf{x}^{\mathsf{T}}(t)}{\sum_{t=1}^{T} \xi_t(j,l)} - \hat{\boldsymbol{\mu}}_{jl} \hat{\boldsymbol{\mu}}_{jl}^{\mathsf{T}}. \tag{7.130}$$

Note that the calculation of the parameters of the Gaussian functions $\hat{\boldsymbol{\mu}}_{jl}$ and $\hat{\mathbf{K}}_{jl}$ is a variant of the EM algorithm (cf. Section 4.5). The optimization is slightly simplified for the semi-continuous case as the a posteriori probabilities $\xi_t(l)$ become independent from the state j. For the discrete case, the optimization is greatly simplified because no density

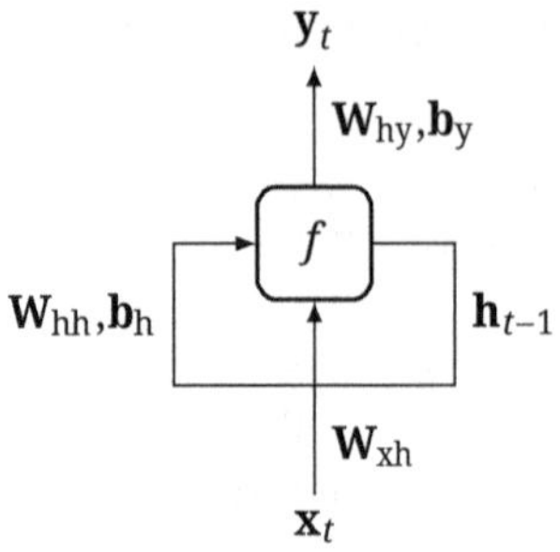

Fig. 7.28: Illustration of a recurrent neural network.

functions but only discrete probabilities $\hat{b}_{jk}$ have to be estimated:

$$\hat{b}_{jk} = \frac{\sum_{t:v_t=v_k} \gamma_t(j)}{\gamma_t(j)}. \tag{7.131}$$

The notation $\sum_{t:v_t=v_k}$ means that only the observations with $v_t = v_k$ are included in the summation over the time index t. For further details about parameter estimation in HMMs, interested readers are referred to Moon and Stirling [2000] and Fink [2003].

7.9.3 Recurrent neural networks

Since artificial neural networks (ANNs) are universal approximators, they can also be used for sequential input data. A special variant of an ANN for sequence modelling is called *recurrent neural network* (RNN) Rumelhart et al. [1986]. The main characteristic of RNNs is the presence of feedback loops within the network architecture. Thus, the corresponding graph is not acyclic as it is the case for feed-forward networks. The basic structure of an RNN is depicted in Figure 7.28.

Due to the feedback loops, a RNN has a time dependent internal state $\mathbf{h}_t$ that saves temporal information from the past. Given an input vector $\mathbf{x}_t \in \mathbb{R}^d$ at time t and a RNN with one hidden layer and n hidden neurons, the state $\mathbf{h}_t \in \mathbb{R}^n$ is updated leveraging the previous state $\mathbf{h}_{t-1}$, the activation function f, as well as the learnable network weights $\mathbf{W}_{\mathrm{xh}}$, $\mathbf{W}_{\mathrm{hh}}$ and bias parameters $\mathbf{b}_{\mathrm{h}}$ such that

$$\mathbf{h}_t = f(\mathbf{W}_{\mathrm{xh}}\mathbf{x}_t + \mathbf{W}_{\mathrm{hh}}\mathbf{h}_{t-1} + \mathbf{b}_{\mathrm{h}}) \tag{7.132}$$

holds. Note that in this formulation, the network weights $\mathbf{w}_1, \ldots, \mathbf{w}_n \in \mathbb{R}^d$ from the d input neurons to the n hidden neurons are summarized in the weight matrix $\mathbf{W}_{\mathrm{xh}} \in \mathbb{R}^{(d\times n)}$. Similarly, the weights stemming from the bias neurons are summarized by a vector $\mathbf{b}_{\mathrm{h}} \in \mathbb{R}^n$. The weight matrix that is multiplied with the internal state vector is denoted with $\mathbf{W}_{\mathrm{hh}} \in \mathbb{R}^{(n\times n)}$.

After updating the state $\mathbf{h}_t \in \mathbb{R}^n$, it is used together with the weight matrix $\mathbf{W}_{\mathrm{hy}} \in \mathbb{R}^{(n\times c)}$ and bias vector $\mathbf{b}_{\mathrm{y}} \in \mathbb{R}^c$ to calculate the output $\mathbf{y}_t \in \mathbb{R}^c$:

$$\mathbf{y}_t = f(\mathbf{W}_{\mathrm{hy}}\mathbf{h}_t + \mathbf{b}_{\mathrm{y}}). \tag{7.133}$$

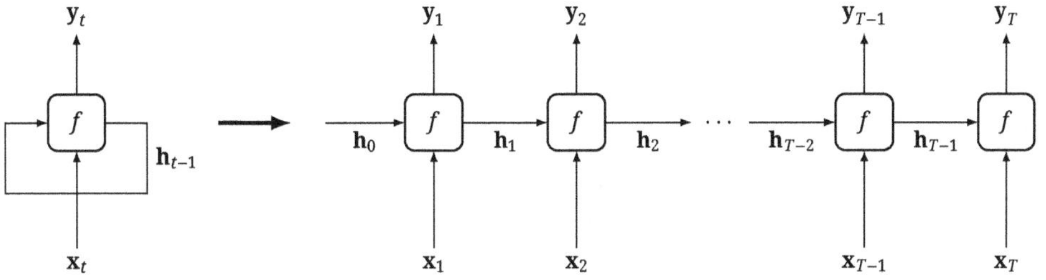

Fig. 7.29: Unrolling of a recurrent neural network.

A typical choice for the activation function f besides the Fermi function (Section 7.4) is the tangens hyperbolicus which ensures the outputs to be bounded in the interval $(-1,1)$. The number of learnable parameters N follows from the dimensions of $\mathbf{W}_{\mathrm{xh}}, \mathbf{W}_{\mathrm{hh}}, \mathbf{W}_{\mathrm{hy}}, \mathbf{b}_{\mathrm{h}}, \mathbf{b}_{\mathrm{y}}$ to $N = dn + n^2 + nc + n + c = n^2 + (d + c + 1)n + c$.

RNNs are dynamical systems that process a sequence of inputs $\mathbf{x}_1, \ldots, \mathbf{x}_T$ in consecutive iterations, whereby the output $\mathbf{y}_t$ depends on the past state $\mathbf{h}_{t-1}$. The initial state $\mathbf{h}_0$ is set to zero if no a priori information is available. Because of the sequential nature, the standard backpropagation algorithm (Section 7.4) cannot be used to train RNNs. Instead, a variant termed *backpropagation through time* (BPTT, Werbos [1990]) is leveraged in which the RNN is *unrolled* such that an acyclic network graph emerges. This process is illustrated in Figure 7.29. The error term ε in the BPTT algorithm can consider only the last time step of the sequence ($\varepsilon = \varepsilon_T$) or all time steps, for example by summing all time dependent error terms ($\varepsilon = \sum_{t=1}^{T} \varepsilon_t$).

As RNNs allow input sequences of arbitrary length T, they can suffer from the *vanishing gradient problem* when the sequence length is large. As a consequence, classical RNNs are not able to store long-term information. Fortunately, there exist advanced architectures that mitigate this problem. The most famous one is the *long short-term memory* which is treated in the following.

Long short-term memory

Standard RNNs only have a *short-term* memory. Since the state $\mathbf{h}_t$ is updated in each time step, the past state $h_{t-\Delta t}$ has a vanishing influence on the current state $\mathbf{h}_t$ if the time difference Δt is large. The idea of the long short-term memory (LSTM), proposed by Hochreiter and Schmidhuber [1997], is to introduce another state $\mathbf{c}_t$ for a *long-term* memory. One can think of the basic unit of LSTMs as a kind of cell that is capable of storing relevant information. Accordingly, $\mathbf{c}_t$ is referred to as *cell state*. The flow of information into and out of the LSTM cell is controlled via *gates*—the forget gate, the input gate, and the output gate. The interplay of those gates with the input $\mathbf{x}_t$ and the short-term and long-term states $\mathbf{h}_t$ and $\mathbf{c}_t$, respectively, is visualized in Figure 7.30 and described in the following.

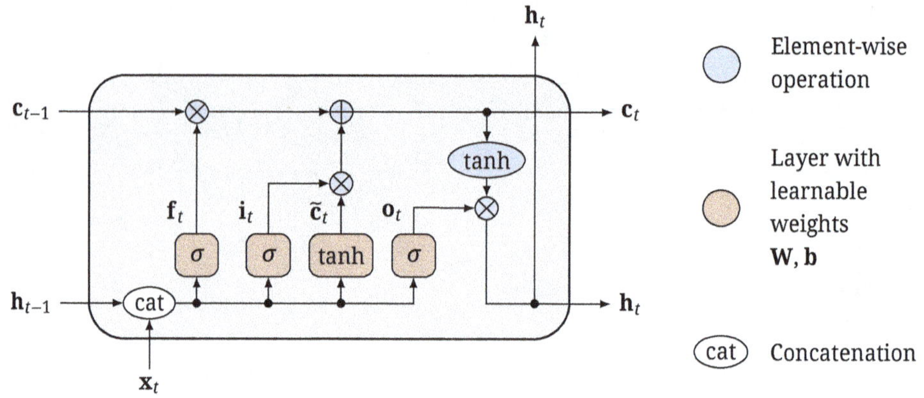

Fig. 7.30: Long short-term memory (LSTM) cell.

The forget gate (f) with parameters $\mathbf{W}_\mathrm{f}, \mathbf{b}_\mathrm{f}$ and Fermi function $\sigma(\psi) = \frac{1}{1+\exp(-\psi)}$ controls which information is kept within the cell:

$$\mathbf{f}_t = \sigma\left(\mathbf{W}_\mathrm{f}\left[\mathbf{h}_{t-1}^\mathrm{T}, \mathbf{x}_t^\mathrm{T}\right]^\mathrm{T} + \mathbf{b}_\mathrm{f}\right). \tag{7.134}$$

Note that the Fermi function is applied element-wise on its argument. The input gate (i) with parameters $\mathbf{W}_\mathrm{i}, \mathbf{W}_\mathrm{c}$ and $\mathbf{b}_\mathrm{i}, \mathbf{b}_\mathrm{c}$ controls the information flow into the cell which can be described with the two equations

$$\mathbf{i}_t = \sigma\left(\mathbf{W}_\mathrm{i}\left[\mathbf{h}_{t-1}^\mathrm{T}, \mathbf{x}_t^\mathrm{T}\right]^\mathrm{T} + \mathbf{b}_\mathrm{i}\right), \tag{7.135}$$

$$\tilde{\mathbf{c}}_t = \tanh\left(\mathbf{W}_\mathrm{c}\left[\mathbf{h}_{t-1}^\mathrm{T}, \mathbf{x}_t^\mathrm{T}\right]^\mathrm{T} + \mathbf{b}_\mathrm{c}\right), \tag{7.136}$$

where tanh denotes the tangens hyperbolicus that is also applied element-wise. Combining Equations (7.134) to (7.136), the cell state $\mathbf{c}_t$ can be updated:

$$\mathbf{c}_t = \mathbf{f}_t \otimes \mathbf{c}_{t-1} + \mathbf{i}_t \otimes \tilde{\mathbf{c}}_t. \tag{7.137}$$

Here, $\otimes$ represents the element-wise vector multiplication. The output gate (o) with parameters $\mathbf{W}_\mathrm{o}, \mathbf{b}_\mathrm{o}$ controls the information flow out of the LSTM cell according to

$$\mathbf{o}_t = \sigma\left(\mathbf{W}_\mathrm{o}\left[\mathbf{h}_{t-1}^\mathrm{T}, \mathbf{x}_t^\mathrm{T}\right]^\mathrm{T} + \mathbf{b}_\mathrm{o}\right). \tag{7.138}$$

Finally, the short-term memory state $\mathbf{h}_t$—which is also the output of the LSTM cell—can be computed by element-wise multiplication of the resulting vector from the output gate $\mathbf{o}_t$ and the tangens hyperbolicus of the updated cell state $\mathbf{c}_t$:

$$\mathbf{h}_t = \mathbf{o}_t \otimes \tanh\left(\mathbf{c}_t\right). \tag{7.139}$$

Hence, the output contains both long-term and short-term information as the name LSTM suggests. This output $\mathbf{h}_t$ is on the one hand used as input to the LSTM cell in the next iteration $t+1$ and on the other hand for computing the overall network output y_t (cf. Figure 7.29), which is indicated with the two arrows for h_t in Figure 7.30. Note that the dimension of the cell state $\mathbf{c}$ equals the dimension of the state $\mathbf{h}$. With $\mathbf{x} \in \mathbb{R}^d$ and $\mathbf{h}, \mathbf{c} \in \mathbb{R}^c$, it follows $\mathbf{W} \in \mathbb{R}^{(c,c+d)}$ and $\mathbf{b} \in \mathbb{R}^c$ for the four weight matrices and bias vectors, respectively. Thus, the number of parameters N of an LSTM layer can be computed as $N = 4(c(c+d)+c) = 4(c^2+dc+c)$.

One can show that with the help of the cell state $\mathbf{c}$, the gradient does hardly vanish even with a large sequence length T making the LSTM a great architecture for processing data with long-term dependencies, such as those needed for speech recognition or machine translation. Next to the presented original LSTM architecture, there exist variants with various numbers of gates or different types of activation functions. The perhaps most famous variation is the *gated recurrent unit* (GRU), introduced by Cho et al. [2014], that only comprises two instead of three gates and thus contains less parameters which can be beneficial for the training process if data is limited.

7.9.4 Transformers

In 2017, a new network architecture for processing sequential data named *transformer* has been proposed that is conceptually very different from previous approaches. While (hidden) Markov models and recurrent neural networks process each part of a sequence one after another, transformers are able to process a complete sequence at once. This allows to model arbitrary long temporal dependencies. The core concept is based on *attention*: for each sequence *token* (= single entry, e.g., a word in a sentence), dependencies to all other tokens are modelled with attention weights such that much weight is given to relevant information and that irrelevant information is suppressed. More details on this attention mechanism will be given later.

The original transformer proposed by Vaswani et al. [2017] is a *seq2seq* model, i.e., both network input and output is a sequence, and is thus well suited for natural language processing (NLP) tasks such as machine translation. The architecture consists of an encoder-decoder structure without recurrent paths, which is beneficial for the computation because many operations can run in parallel. The transformer architecture is depicted in Figure 7.31. The single components as well as basic ideas are now discussed.

Since one cannot directly perform calculations on the sequence tokens (words of a sentence), each token is transformed into an *embedding* with the help of a *vocabulary*. An embedding maps tokens into a vector space $\mathbb{R}^d$. Note that different methods exist to create such embeddings, e.g., a separate neural network could be applied to learn the mapping of a token to an embedding vector. This mapping is designed such that words with semantically similar meanings result in embeddings that are close to each other in the embedding space.

The attention mechanism which is described in a moment processes all embeddings of the sequence tokens simultaneously. In order to not lose the positional relations of the

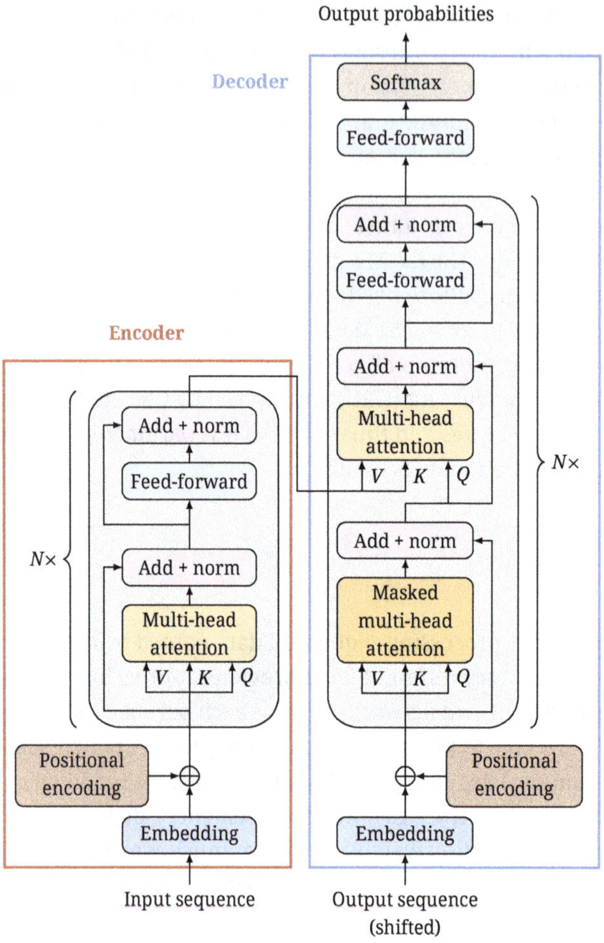

Fig. 7.31: Transformer architecture. Image adapted from Vaswani et al. [2017].

input sequence—that for sure contain important information—a positional encoding is applied to the embeddings. Without going into detail, each embedding is added a special pattern. By learning these patterns, the model is able to reason about the positional distances of tokens from the input sequence.

The main component of the transformer architecture is the *scaled dot-product attention*, a part of the *multi-head attention* that is both present in the encoder and decoder part of the network (see Figure 7.31). To understand the scaled dot-product attention, the concepts of *queries*, *keys*, and *values* are introduced. Given an embedding with positional encoding $\mathbf{x}_i \in \mathbb{R}^{d_\mathrm{e}}$ with dimension d_e of an input sequence $(\mathbf{x}_1, \ldots, \mathbf{x}_T)$, three different representations of this embedding are learned by the network: query $\mathbf{q}_i \in \mathbb{R}^{d_\mathrm{q}}$, key $\mathbf{k}_i \in \mathbb{R}^{d_\mathrm{k}}$ with $d_\mathrm{k} = d_\mathrm{q}$, and value $\mathbf{v}_i \in \mathbb{R}^{d_\mathrm{v}}$ according to

$$\mathbf{q}_i = \mathbf{W}_\mathrm{Q}\mathbf{x}_i, \qquad \mathbf{k}_i = \mathbf{W}_\mathrm{K}\mathbf{x}_i, \qquad \mathbf{v}_i = \mathbf{W}_\mathrm{V}\mathbf{x}_i, \tag{7.140}$$

where $\mathbf{W}_\mathrm{Q} \in \mathbb{R}^{d_\mathrm{k} \times d_\mathrm{e}}$, $\mathbf{W}_\mathrm{K} \in \mathbb{R}^{d_\mathrm{k} \times d_\mathrm{e}}$, and $\mathbf{W}_\mathrm{V} \in \mathbb{R}^{d_\mathrm{v} \times d_\mathrm{e}}$ are the learnable weight matrices of the neural network layer of the scaled dot-product attention. For an efficient computation of the following attention mechanism, queries, keys, and values are summarized in matrices **Q**, **K**, and **V**, respectively:

$$\mathbf{Q} = \begin{pmatrix} \mathbf{q}_1^\mathrm{T} \\ \vdots \\ \mathbf{q}_T^\mathrm{T} \end{pmatrix}, \qquad \mathbf{K} = \begin{pmatrix} \mathbf{k}_1^\mathrm{T} \\ \vdots \\ \mathbf{k}_T^\mathrm{T} \end{pmatrix}, \qquad \mathbf{V} = \begin{pmatrix} \mathbf{v}_1^\mathrm{T} \\ \vdots \\ \mathbf{v}_T^\mathrm{T} \end{pmatrix}. \tag{7.141}$$

Finally, the scaled dot-product attention can be calculated as

$$\mathrm{attention}(\mathbf{Q},\mathbf{K},\mathbf{V}) = \mathrm{softmax}\left(\frac{\mathbf{Q}\mathbf{K}^\mathrm{T}}{\sqrt{d_\mathrm{k}}}\right)\mathbf{V} = \mathbf{W}\mathbf{V} \tag{7.142}$$

with d_k used as normalization factor. One can interpret $\mathbf{W} := \mathrm{softmax}(\frac{\mathbf{Q}\mathbf{K}^\mathrm{T}}{\sqrt{d_\mathrm{k}}})$ as weights that model to which extent a token of the sequence (**Q**) is influenced by all other tokens of the sequence (**K**). These attention weights lie in [0,1] and sum up to 1 due to the softmax function. They are then multiplied with the tokens of the input sequence (**V**). As long as queries, keys, and values stem from the same input, the process is termed *self-attention*. Self-attention is present both in the encoder and decoder part. Note that there exists another variant of attention in the transformer architecture called *encoder-decoder attention* which is treated later. The module that computes the scaled dot-product attention can also be referred to as single *attention head*. It is the basic building block of the *multi-head attention*.

In order to enhance the model capacity, h attention heads are combined to form a multi-head attention module. For each attention head, different transformation matrices **Q**, **K**, and **V** are learned. This has the advantage that the model learns various representations of the input embeddings. While the dimension of values and keys has not to be the same, the original paper sets these hyperparameters identically: $d_\mathrm{v} = d_\mathrm{k} = 64$. The input embedding has dimension $d_\mathrm{e} = 512$ and h is set to 8 such that $d_\mathrm{q} = d_\mathrm{k} = d_\mathrm{v} = \frac{d_\mathrm{e}}{h}$ holds. Thus, the computational cost of a multi-head attention is similar to that of a single attention head if the original dimension of embeddings would be kept. After parallel computation of the scaled dot-product attentions, those are concatenated followed by a feed-forward layer to get the final results of the multi-head attention. The feed-forward layer projects the outputs onto the original dimension d_e and uses ReLU (Section 7.6.4) as activation function. In addition, dropout is leveraged to prevent overfitting.

Besides multi-head attention, a building block of the transformer encoder comprises residual connections (Section 7.6.6), normalization layers, as well as further feed-forward layers for an enhanced model capacity. Furthermore, the potential performance of the transformer can be tuned with the hyperparameter N that denotes how many building blocks are stacked on top of each other (see Figure 7.31). Comparable to convolutional neural networks, the deeper the network, the more powerful the learned encodings. Note that residual connections can also be referred to as *skip* connections because they allow the interaction of representations among layers of different depth.

The structure of the transformer decoder is very similar to the one of the encoder. The differences are discussed in the following. Similar to the input sequence, an output embedding is generated for the output sequence, for example, the translated input sequence in the case of a machine translation task. However, the output sequence is shifted by one position. Additionally, the multi-head self-attention layers are masked in a way that the influence of future positions is zero by setting attention values to $-\infty$ before the softmax function. This masking together with the shifted output sequence secures an autoregressive behavior, i.e., that predictions for a position i of a sequence depend only on the known outputs at positions smaller than i.

As mentioned previously, besides self-attention, there are encoder-decoder attentions within the transformer architecture. The structure is identical to the self-attention modules, however, not all inputs come from the same source. Keys $\mathbf{K}$ and values $\mathbf{V}$ come from the encoder, whereas queries $\mathbf{Q}$ come from the decoder. This establishes a connection between encoder and decoder such that the focus of the decoder is put on the relevant positions of the encoder outputs.

The N building blocks that are stacked on top of the positional encoded embeddings consist of a masked multi-head attention layer, an encoder-decoder multi-head attention layer, a feed-forward layer, as well as residual connections and normalization layers (see Figure 7.31). The last attention block is followed by a final feed-forward layer that projects the encoding vector of length d_e to a *logits* vector with length equal to the size of the vocabulary, which can contain tens or hundreds of thousands of words. A logits vector represents the non-normalized predictions of the network. Finally, the softmax function is applied on these logits resulting in output probabilities for the predicted token, that lie in [0,1] and sum to 1.

During inference, the encoder is passed through only once resulting in an encoding of the complete input sequence. After that, the decoder is passed through multiple times with the output sequence containing only a so-called *start-of-sentence-token* in the first run. In the consecutive iterations, the output sequence is extended with the predicted token from the previous iteration. The inference stops whenever an *end-of-sentence-token* is predicted by the network. Note that start-of-sentence-token and end-of-sentence-token belong to the vocabulary.

Nowadays, transformer networks are successfully trained on huge datasets with millions of sequences and have become the state-of-the-art architecture for all kind of natural language processing tasks. Furthermore, transformers have been adapted also to computer vision tasks such as image classification, for example.

7.9.5 Generative adversarial networks

The so far treated architectures for sequence modelling are not only able to analyze sequential data but also to generate new data. One architecture that has been designed especially for the task of data generation is the *generative adversarial network* (GAN),

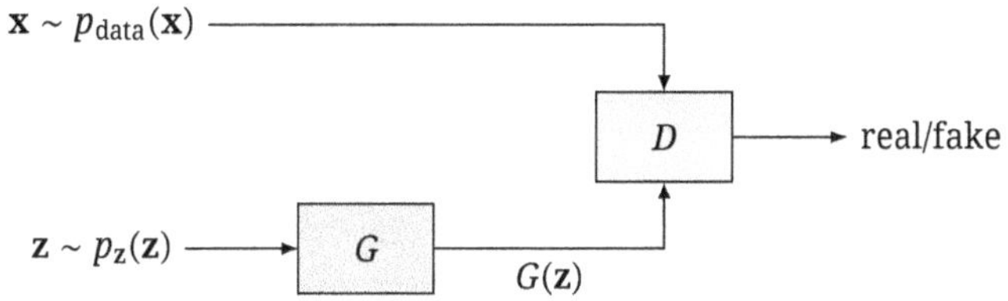

Fig. 7.32: Concept of a generative adversarial network.

a neural network trained in an unsupervised fashion. While the first GAN proposed by Goodfellow et al. [2014] was designed to generate synthetic images, meanwhile, there exist GANs for generating sequential data like videos, audio, or other time series. Before dealing with GANs for sequential data, the basic concept of a GAN is explained, which is independent from the specific data type.

Figure 7.32 illustrates the basic components of a GAN. It consists of a generator G and a discriminator D. Given a random vector $\mathbf{z}$ drawn from a probability distribution $p_{\mathbf{z}}(\mathbf{z})$, the goal of the generator—an artificial neural network (ANN)—is to generate a synthetic datum $G(\mathbf{z})$ that is similar to real data $\mathbf{x} \sim p_{\text{data}}(\mathbf{x})$. The distribution of the random vector $\mathbf{z}$ is often chosen to be a standard normal distribution: $p_{\mathbf{z}}(\mathbf{z}) = \mathcal{N}(\mathbf{0},\mathbf{1})$.

The goal of the discriminator D—another ANN—is to classify its input into real/fake, i.e., it has to tell whether a real datum $\mathbf{x}$ or a synthetic generated datum $\mathbf{z}$ from the generator is present.

While the generator tries to fool the discriminator and wants to *maximize* the classification error, the discriminator aims at *minimizing* the classification error. This results in a *minimax game* in that both modules are continuously improved during the training process. More precisely, given the objective function

$$V(D,G) := \mathrm{E}_{\mathbf{x}\sim p_{\text{data}}(\mathbf{x})}\{\log(D(\mathbf{x}))\} + \mathrm{E}_{\mathbf{z}\sim p_{\mathbf{z}}(\mathbf{z})}\{\log(1 - D(G(\mathbf{z})))\} \tag{7.143}$$

with $D(\mathbf{x})$ denoting the probability that $\mathbf{x}$ is real ($\mathbf{x} \sim p_{\text{data}}$), the minimax game can mathematically be expressed as

$$\min_G \max_D V(D,G). \tag{7.144}$$

When the training has been successfully converged, $G(\mathbf{z}) \sim p_{\text{data}}$ holds such that the discriminator is not able to distinguish real and synthetic data anymore. The generator can then be used to generate an arbitrary amount of synthetic data by passing random vectors $\mathbf{z}$ through the generator network G.

Two known problems in training GANs are vanishing gradients and so-called *mode collapse* which means that the generator learns only a few modes of the real data distribution. Further developments of the original architecture have been proposed to mitigate these problems but are out of the scope of this book.

Depending on the application area and type of data to be generated, different types of ANNs are leveraged within GANs including feed-forward networks, CNNs, and RNNs. Next to the aforementioned problems in the training process of GANs for static data,

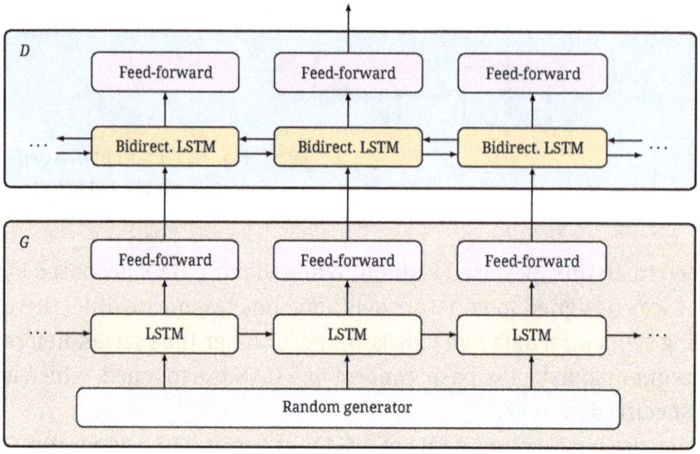

Fig. 7.33: Scheme of C-RNN-GAN. Illustration adapted from Mogren [2016].

GANs for generating sequential data have to deal with additional challenges. On the one hand, modelling of temporal dependencies is necessary and on the other hand, the input data can be of different length. For these reasons, RNNs and LSTMs are a good choice for the ANN in the generator and discriminator.

GANs for generating time series can be further classified based on the type of time series. GANs for *discrete* time series generate sequences in which the entries are symbols like characters for example. In contrast, sequences produced by GANs for *continuous* time series, i.e., continuous-valued series at discrete time steps, include real values, e.g., temperature measured in °C.

Some applications for GANs specialized on sequential data are the enlargement of time series datasets, the imputation of missing data points within a query sequence, and the prediction of time series (weather, share prices, etc.). In the following, an example of a GAN for continuous time series trained to synthesize classical music is given.

Example: GAN for continuous time series

Mogren [2016] proposed a continuous recurrent neural network with adversarial training (C-RNN-GAN). Its architecture is schematically shown in Figure 7.33. The generator comprises LSTM and feed-forward layers. The LSTM cells get a concatenation of random vector **z** and cell output from the previous time step as input. The discriminator uses *bidirectional LSTM* layers, a further development of the classical LSTM to store not only past but also future information.

The loss functions for the generator and discriminator L_G and L_D, respectively, sum over losses from all time steps of the sequence with length T:

$$L_G = \frac{1}{T} \sum_{i=1}^{T} \log\left(1 - D(G(\mathbf{z}_i))\right) \tag{7.145}$$

$$L_D = \frac{1}{T}\sum_{i=1}^{T}\left[-\log(D(\mathbf{x}_i)) - \log(1 - D(G(\mathbf{z}_i)))\right]. \qquad (7.146)$$

Here, $\mathbf{z}_i$ is a random vector of dimension k and $\mathbf{x}_i$ is a real datum of dimension l from a sequence $(\mathbf{x}_1, \ldots, \mathbf{x}_T)$, whereby k and l can be set identical but don't have to.

Backpropagation through time (Section 7.9.3) is used to train C-RNN-GAN. Furthermore, two tricks are applied in the training process. First, the generator is pre-trained with a kind of *curriculum learning* increasing the input length with progressive training of the LSTM cells. Second, the weights of either the generator or the discriminator are *frozen* during the adversarial training whenever one of the two counterparts becomes too strong, i.e., when the loss terms L_G and L_D differ too much.

In C-RNN-GAN, the datum $\mathbf{x}$ of a sequence $(\mathbf{x}_1, \ldots, \mathbf{x}_T)$ is a 4-dimensional vector ($l = 4$) which represents a sound of an audio sequence characterized by length, frequency, intensity, and point in time. After training the GAN with classical audio sequences, it is able to synthesize new classical music. Although the generated audio can (still) easily be classified as fake by a human discriminator, the work shows the great potential of GANs for generating sequential data of all kinds.

7.10 Exercises

(7.1) Is $K(\mathbf{m},\mathbf{m}') = \|\mathbf{m}\|_2$ a kernel function?

(7.2) Is $K(\mathbf{m},\mathbf{m}') = 4\mathbf{m}^\mathsf{T}\mathbf{m}' - \mathbf{m}^\mathsf{T}\mathbf{m} - (\mathbf{m}')^\mathsf{T}\mathbf{m}'$ a kernel function?

(7.3) Construct the kernel function $K(\mathbf{p},\mathbf{q})$ with $\mathbf{p} = (p_1,p_2,p_3)^\mathsf{T}$ and $\mathbf{q} = (q_1,q_2,q_3)^\mathsf{T}$ that implicitly performs the following feature mapping $\varphi : \mathbb{R}^3 \to \mathbb{R}^4$:

$$\varphi(\mathbf{m}) = \varphi\left(\begin{pmatrix} m_1 \\ m_2 \\ m_3 \end{pmatrix}\right) = \begin{pmatrix} m_1 m_2 \\ m_2 m_3 \\ m_3 m_1 \\ (m_1 + m_2 + m_3)^3 \end{pmatrix}.$$

(7.4) Suppose given the following sample of 8 two-dimensional feature vectors:

$$\mathbf{m}_1 = \begin{pmatrix} 0.5 \\ 3.5 \end{pmatrix}, \mathbf{m}_2 = \begin{pmatrix} 2.5 \\ 4.0 \end{pmatrix}, \mathbf{m}_3 = \begin{pmatrix} 2.0 \\ 6.0 \end{pmatrix}, \mathbf{m}_4 = \begin{pmatrix} 6.0 \\ 6.0 \end{pmatrix},$$

$$\mathbf{m}_5 = \begin{pmatrix} 1.5 \\ 1.5 \end{pmatrix}, \mathbf{m}_6 = \begin{pmatrix} 4.0 \\ 1.0 \end{pmatrix}, \mathbf{m}_7 = \begin{pmatrix} 3.5 \\ 2.0 \end{pmatrix}, \mathbf{m}_8 = \begin{pmatrix} 5.0 \\ 2.0 \end{pmatrix},$$

where $\omega(\mathbf{m}_i) = \omega_1$ for $i = 1, \ldots, 4$ and $\omega(\mathbf{m}_i) = \omega_2$ for $i = 5, \ldots, 8$.

1. Sketch a diagram of the features and the decision boundary of a hard margin SVM classifier. Mark the margin and the support vectors.

2. Determine the parameters of the decision boundary $\mathbf{w}^\top \mathbf{m} + b = 0$.
3. Give an estimate $\hat{P}_e$ of the error probability for this classifier.
4. How does the estimate $\hat{P}_e$ change, if the class ω_2 is enlarged by 12 additional samples, where none of the sample is a support vector and all the samples are classified correctly by the existing classifier?

8 Classification with nominal features

The techniques we have discussed so far implicitly assume that the feature space is continuous. This means that the classifiers operate on features with an interval scale, ratio scale, or absolute scale (see Section 2.1). However, this is not always the case.

Consider, for example, the classification of plant leaves according to a high level botanical description of (a) the shape of the bud of the plant and (b) the morphology of the leaf. Both are nominal features, i.e., features without a quantitative meaning and without any ordering. The only meaningful relation is equivalence, that is, it is only possible to tell whether two occurrences of the feature are the same or not. Nominal features are discrete and can only assume a finite number of values. Therefore, this two-dimensional feature space to describe plants is also discrete, which means that a linear classifier like the SVM cannot be used.

Note, however, that the Bayesian classifier framework

$$P(\omega|m) = \frac{P(m|\omega)P(\omega)}{P(m)} = \frac{P(m|\omega)P(\omega)}{\sum_{i=1}^{c} P(m|\omega_i)P(\omega_i)}, \tag{8.1}$$

where $\omega \in \{\omega_1, \ldots, \omega_c\} = \Omega/\sim$ and $m \in \{m_1, \ldots, m_z\} = \mathbb{M}$ is a nominal feature, is still applicable if one can find suitable estimates for the class-specific probabilities $P(m \mid \omega)$. As such, the methods from Chapters 3 and 4 apply without limitations to the case of nominal features and of mixed feature vectors of nominal and other feature types.

In this chapter, we will address three additional approaches that allow a classification based on nominal features: decision trees and random forests, string matching, and grammars, i.e., syntactic pattern recognition.

8.1 Decision trees

Consider again the classification of an unknown plant. A botanist will ask a sequence of questions regarding different nominal properties like the shape of the leaf, the formation of the buds of the plant, the shape of the blossom, the color of the blossom, etc. Every question rules out certain plant candidates, until there is only one option left—the final decision.

In such situations it is typical that the next question depends on the answer to the previous question. Formally, this technique of asking questions can be represented by a decision tree. In this sense, a floral field guide is a "manual" decision tree classifier. Other examples of such decision trees are medical diagnoses, error detection procedures for technical equipment, and codified business procedures.

A decision tree is a tree structure, where the inner nodes correspond to questions, the links between the nodes represent the answers, and the leaves represent the decisions or classes. An example of a decision tree to classify fruit is shown in Figure 8.1. In the

https://doi.org/10.1515/9783111339207-008

example, which was taken from Duda et al. [2001], the classes are

$$\begin{aligned}\Omega/\sim &= \{\omega_1 \ldots, \omega_7\} \\ &= \{\text{Apple, Watermelon, Grape, Grapefruit, Lemon, Cherry, Banana}\}\end{aligned}$$

and the discrete, four dimensional space of nominal features is given by

$$\begin{aligned}\mathbf{m} \in \mathbb{M} &\subseteq \mathbb{M}_1 \times \mathbb{M}_2 \times \mathbb{M}_3 \times \mathbb{M}_4 \\ &= \{\text{green, yellow, red}\} \times \{\text{big, medium, small}\} \times \{\text{round, thin}\} \times \{\text{sweet, sour}\}.\end{aligned}$$

Decision trees are easy to understand and can, unlike most of the classifiers discussed so far, be intuitively interpreted by humans. Of course, the interpretability breaks down with deeper, more complex trees, but in principle every classification decision can be reproduced and understood by a human.

Figure 8.1 also shows some key properties of decision tree classifiers: The branches of a node must be mutually exclusive (no answer may appear on two branches) and exhaustive (all answers that are possible at a particular node must be covered). Furthermore, a question in a node must have a deterministic and unambiguous answer. The classification of an unknown object is achieved by sequential decisions along a path through the tree, until a leaf node is reached. The path itself is determined by the answers to the questions in the inner nodes. Note that the same question may appear at multiple points in the decision tree, even on a single path through the tree. For example, the question "Size?" appears three times in the example in Figure 8.1. Depending on where the question is asked, it may have a different number of possible answers (outgoing edges). In Figure 8.1, "Size?" has different possible answers depending on the node that asks this question. Leaf nodes represent classes. An object that reaches a given leaf node is assigned to the class that is represented by that node. Multiple leaf nodes may (and usually do) represent the same class. The tree in Figure 8.1, for example, contains two leaf nodes each for the classes "Apple" and "Grape".

Decision trees are generally very fast classifiers, provided the tree is well structured and not too deep. Decision trees allow easily incorporating prior knowledge about the pattern recognition task into the classifier. For example, one can augment a decision tree learned from a sample (see Section 8.1.1) by rules that represent expert knowledge. Decision trees are also applicable to features of a higher scale, e.g., the interval scale. Then, the nodes typically quantize the features into two or more possible subsets of values. If, for example, the size were measured in mm, the question "Size?" in the third node from the left in the second level of Figure 8.1 may be quantized so that "small" means a size of $\leq$ 20 mm and "medium" means a size of $>$ 20 mm. Then, the question would become "Size $\leq$ 20 mm?".

Using grouping and quantization, every decision tree can be transformed into a binary decision tree, i.e., a tree where each internal node has exactly two outgoing edges. In a binary decision tree, the nodes usually represent yes/no questions. A binarized tree of Figure 8.1 is shown in Figure 8.2. In the following, we will only consider binary trees.

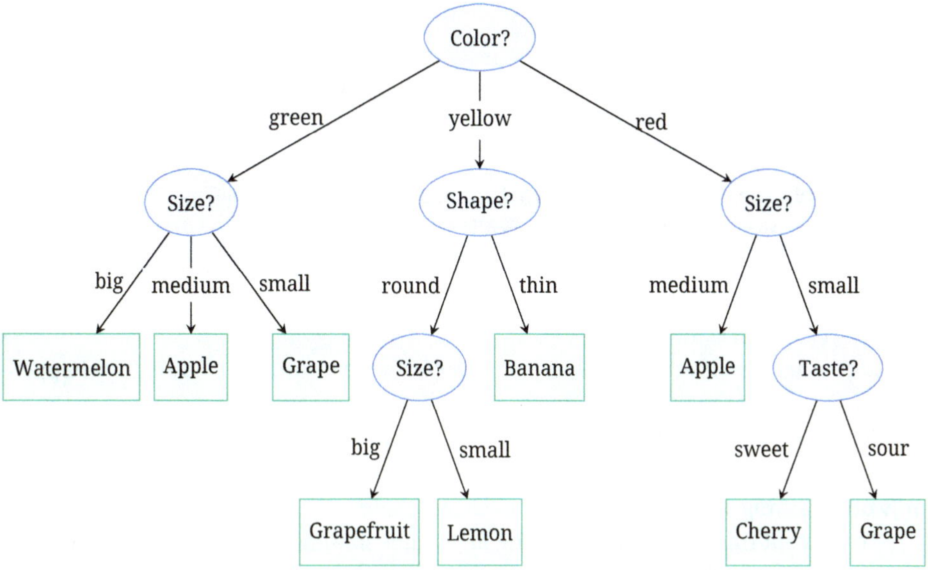

Fig. 8.1: Decision tree to classify fruit. Inner nodes represent questions about the features, edges represent possible answers, and leaf nodes represent classes. Recreated according to Duda et al. [2001].

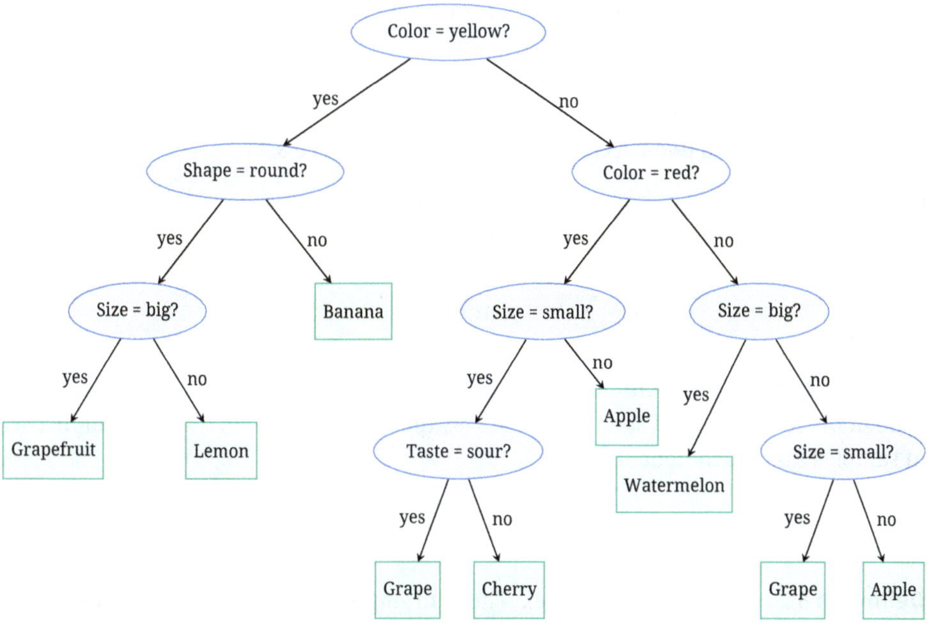

Fig. 8.2: Binarized version of the decision tree in Figure 8.1.

8.1.1 Decision tree learning

Learning a decision tree corresponds to determining the structure, that is, the nodes and branches, of a tree using the training set $\mathcal{D}$. The training procedure is recursive. At each step, a node is constructed according to some splitting criterion (see below). The resulting branches split the training set into two disjoint subsets, $\mathcal{D} = \mathcal{D}_{\text{Yes}} \uplus \mathcal{D}_{\text{No}}$. The subset $\mathcal{D}_{\text{Yes}}$ is associated with the left branch of the split and $\mathcal{D}_{\text{No}}$ is associated with the right branch. On each branch, a node is again constructed according to the splitting criterion, but only using the samples that reach that node. This procedure is repeated recursively until a stopping criterion is met.

Like other classifiers, decision trees may overfit the training set. Overfitting happens if the structure of the tree is too fine grained and the corresponding decision paths are too detailed. This can be spotted in the learning phase, when too few samples of the training set reach the deeper nodes and leaves of the decision tree. In the extreme case, only one training sample is allotted to each leaf node.

To prevent this scenario, a learning algorithm should create a compact tree, i.e., a tree with as few nodes as possible. A common greedy approach chooses a question for the current node so that the training sets in the resulting split $\mathcal{D}_{\text{Yes}}$ and $\mathcal{D}_{\text{No}}$ are as pure as possible. The pureness of a dataset is measured using a heterogeneity or impurity measure, denoted by $i(\cdot)$. The impurity measure $i(n)$ should assume a minimum if the dataset $\mathcal{D}_n$ at node n consists of samples of only one class and should attain a maximum if the classes are uniformly distributed, i.e., if each class is represented by the same number of samples in $\mathcal{D}_n$.

There are three standard measures that fulfill these properties: the *entropy* measure, the *Gini impurity* measure, and the *misclassification* measure. A qualitative comparison of the measures for a two-class scenario is shown in Figure 8.3.

Definition 8.1 (Entropy measure). The entropy measure corresponds to the entropy of the empirical class distribution in the training set,

$$i(n) = -\sum_{k=1}^{c} \hat{P}(\omega_k \mid \mathbf{m}, n) \log_2 \left(\hat{P}(\omega_k \mid \mathbf{m}, n) \right). \tag{8.2}$$

Here, n denotes the current node and the probability distribution is estimated as the ratio of the number N_n^k of samples of class ω_k that reach node n to the total number N_n of samples that reach node n:

$$\hat{P}(\omega_k \mid \mathbf{m}, n) := \frac{N_n^k}{N_n}. \tag{8.3}$$

Definition 8.2 (Gini impurity measure). The Gini impurity measure (or simply the Gini measure) estimates the expected error probability if the class were to be randomly as-

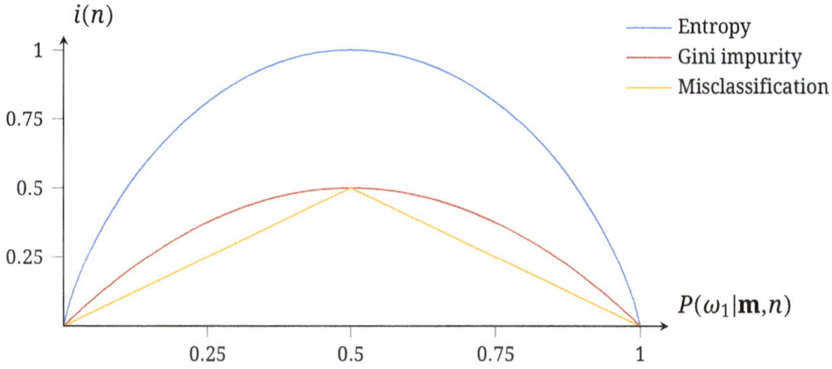

Fig. 8.3: Qualitative comparison of impurity measures in dependence on the class probability $P(\omega(\mathbf{m}) = \omega_1)$ in a two-class scenario.

signed according to the class distribution at the node n:

$$i(n) = \sum_{k=1}^{c} \sum_{\substack{l=1 \\ l \neq k}}^{c} \hat{P}(\omega_l \,|\, \mathbf{m},n)\, \hat{P}(\omega_k \,|\, \mathbf{m},n) = 1 - \sum_{k=1}^{c} \left(\hat{P}(\omega_k \,|\, \mathbf{m},n)\right)^2. \tag{8.4}$$

In the binary case ($c = 2$), the Gini measure simplifies to

$$i(n) = 2 \cdot \hat{P}(\omega_1 \,|\, \mathbf{m},n) \cdot \hat{P}(\omega_2 \,|\, \mathbf{m},n). \tag{8.5}$$

Definition 8.3 (Misclassification measure). Lastly, the error probability of the dominant class at node n is estimated by the misclassification measure:

$$i(n) = 1 - \max_k \left\{\hat{P}(\omega_k \,|\, \mathbf{m},n)\right\}. \tag{8.6}$$

Impurity minimization

Given an impurity measure $i(n)$, the question to ask at node n is chosen so that the impurity of the split is minimized, i.e., chosen to maximize the decrease in impurity

$$\Delta i(n) := i(n) - \hat{P}_{\text{Yes}}\, i(n_{\text{Yes}}) - \hat{P}_{\text{No}}\, i(n_{\text{No}}). \tag{8.7}$$

This process is iterated, each time using the corresponding partitions of the training data, until the decrease in impurity falls below a threshold ($\Delta i(n) < \tau_i$) or until the number of training samples that reach the node n becomes too low. In addition to stopping the training procedure on these conditions, a fully grown decision tree may be post-processed by merging or pruning nodes after training. These topics are, however, outside the scope of this textbook.

In Figure 8.4, the reference dataset from Section 3.3.2 was used to train a decision tree with the greedy strategy outlined above. Gini impurity was used as split criterion

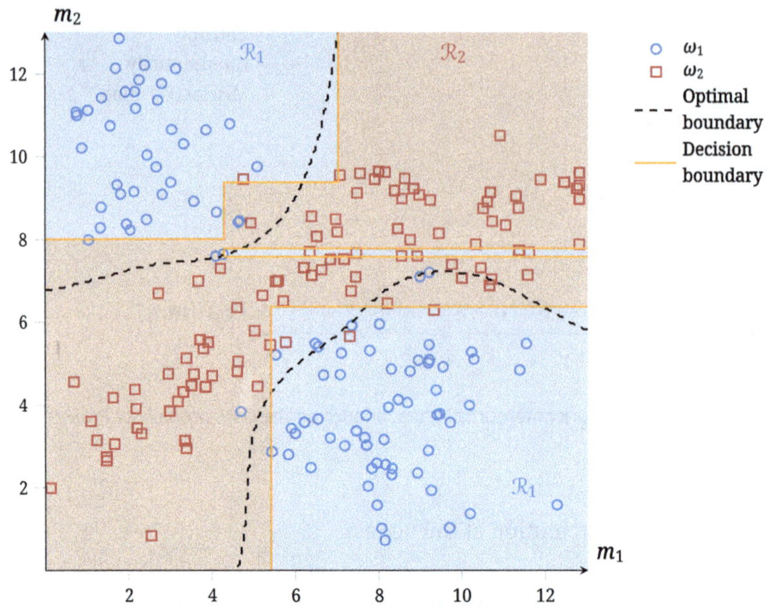

Fig. 8.4: Application to the reference example of Section 3.3.2. Decision regions of a decision tree classifier. Training was stopped when there were fewer than 15 samples available for a split. The training and testing errors are $e_{\text{train}} = 7\,\%$ and $e_{\text{test}} = 12\,\%$, respectively. The testing error asymptotically approaches $e_{\text{test}} \approx 14.6\,\%$. The training set is the same as in Figure 3.8. Test samples are shown with hollow marks.

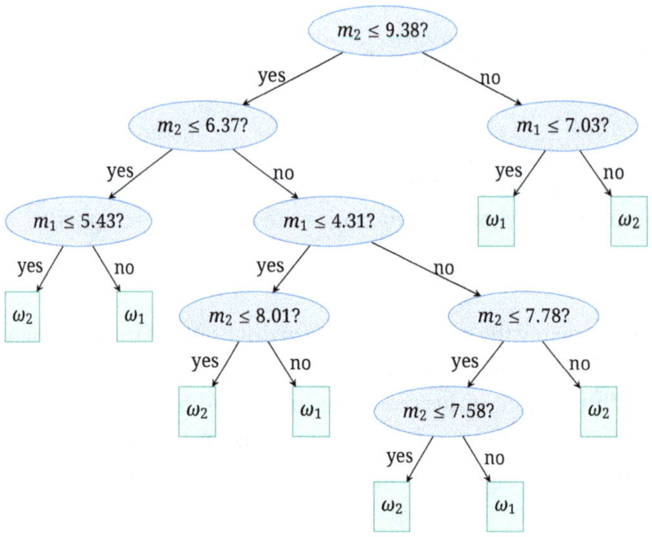

Fig. 8.5: Structure of the decision tree of Figure 8.4.

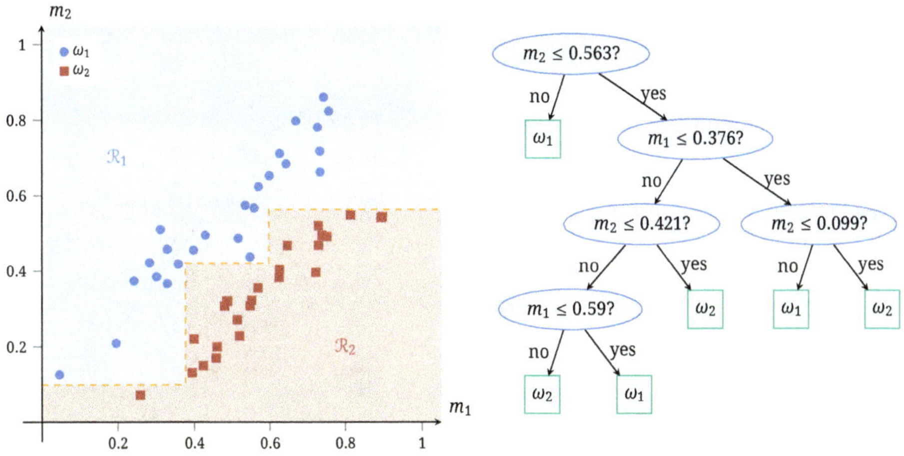

(a) Inadequate features lead to unnecessarily complicated decision trees.

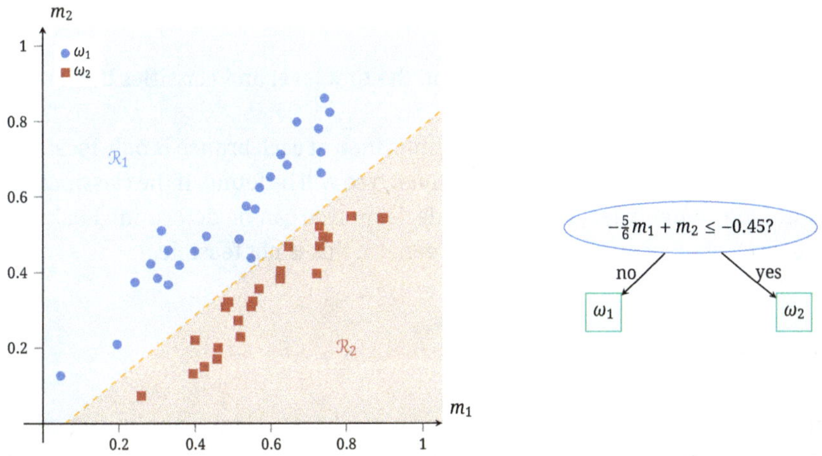

(b) Feature transformation results in a minimal decision tree.

Fig. 8.6: Impact of the features used in decision tree learning. This example descriptively underlines what is meant by qualifying the boundary between feature extraction and classification as "blurry" in Figure 1.2.

and recursion was stopped when less than 15 training samples were available for a split. The decision rules of the tree are shown in Figure 8.5. The testing error is relatively large and from a visual inspection of the decision regions it is evident that the tree is more complicated than it should be. The reasons for this will be explored in the next section.

Decision trees can also be interpreted as meta-algorithms for combining different classifiers, which solve subproblems of the classification task in the inner nodes. Indeed, an arbitrary classifier can be deployed at each inner node. Furthermore, hierarchical classifiers can be easily realized with decision trees. For example, imagine a classifier that

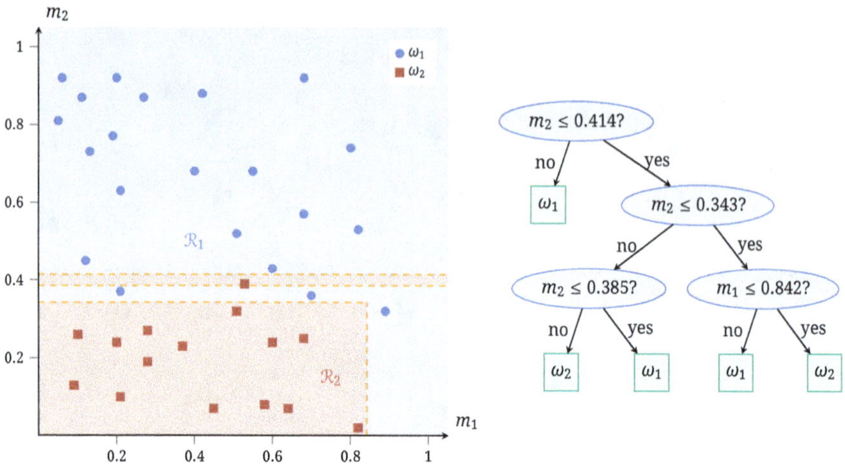

Fig. 8.7: A decision tree that does not generalize well. The filled marks show the training sample $\mathcal{D}$.

distinguishes between letters and numerals on the first level and classifies the distinct letters and numerals on the second level.

Note that with this greedy strategy, the optimization at each branch is only local. As a result, there is no guarantee that a globally optimal tree will be found. If the classification problem is not too complicated, the optimal decision tree can be determined using an exhaustive search of all possible trees, but in general this is not feasible.

8.1.2 Influence of the features used

As mentioned in Chapter 2, the choice of features can have a significant impact on the performance of the classifier. Since decision trees are transparent classifiers, they are well suited to exemplify this point. Figure 8.6 shows the impact of inadequate features on the classifier. Here, the features are on the quantitative scale and each node of the decision tree splits the feature space parallel to the axes. As can be seen in Figure 8.6a, the features m_1 and m_2 produce an unnecessarily complicated and overfitted decision tree. Yet, a simple feature transformation produces the minimal decision tree shown in Figure 8.6b. Similarly, the decision tree in Figure 8.7 does not generalize well. Allowing the misclassification of one training sample would eliminate the thin decision region around $m_2 = 0.4$ and produce a much simpler decision tree without overfitting.

8.2 Random forests

In practice, decision trees are often observed to overfit the data, especially when the trees are relatively deep. As mentioned above, one method to address this issue is to prune the tree after construction. An alternative solution is to train an ensemble of several classifiers and average their predictions. The idea is that while every classifier in the ensemble may give inaccurate predictions, the random part of the pertaining error will average out over the collective predictions. In other words: averaging over the ensemble will reduce the variance of the classification system. We will state this intuition more precisely below.

Methods that take this approach, i.e., methods that derive a decision from a set of classifiers, are called ensemble methods. Ensemble methods subsume a broad range of techniques, much more than can be covered in this book. Rokach [2010] and Zhou [2012] give a much more thorough discussion of this subject. Here, we will restrict ourselves to the following general understanding: ensemble methods predict the class of a sample according to a weighted average

$$\mathbf{k}(\mathbf{m}) = \sum_{j=1}^{M} \alpha_j \mathbf{k}^j(\mathbf{m}), \tag{8.8}$$

where the $\mathbf{k}^j(\mathbf{m})$ denote the base classifiers in the ensemble, the $\alpha_j \in \mathbb{R}$ denote weights associated with each classifier, and $\mathbf{k}(\mathbf{m})$ is the overall decision function of the ensemble.

One particularly successful instance of ensemble methods is the method of random forests, sometimes also called random decision forests or randomized forests. As the name suggests, a random forest is composed of decision trees $\mathbf{k}^j(\mathbf{m}), j = 1, \dots, M$, where each tree is weighted equally, $\alpha_j = \frac{1}{M}$. We will discuss another ensemble method, AdaBoost, in Section 9.4, where the weights α_j of the classifiers are adapted during the training phase. Interestingly, under certain conditions, AdaBoost can be interpreted as a special case of a random forest (Breiman [2001]).

Similar to the SVM, random forests have shown remarkable classification performance out of the box in many practical classification tasks. This success can largely be attributed to a key idea that lends random forests the first half of their name: randomization.

Random forests use randomization during training in two ways: First, each tree in the ensemble is trained on a random subsample of the training set. Second, at each node in each tree, only a random subspace of the feature space is considered for a split.

More formally, let $\mathcal{D} = \{\mathbf{m}_1, \dots, \mathbf{m}_N\}$ be the training set of N d-dimensional training samples $\mathbf{m}_i = (m_{i1}, \dots, m_{id})^\mathsf{T}$ with known class memberships $\omega(\mathbf{m}_i), i = 1, \dots, N$. To train the decision tree $\mathbf{k}^j(\mathbf{m})$, first a new training set $\widetilde{\mathcal{D}}^j = \{\mathbf{m}_{r(1)}, \dots, \mathbf{m}_{r(B)}\}$ is constructed by randomly sampling from $\mathcal{D}$ with replacement. Here, $r(l)$ is a function that maps l to a random integer $1 \leq k \leq N$ and $B \leq N$ is the size of the sub-sampled training set $\widetilde{\mathcal{D}}^j$ (typically $B \approx 0.7N$). Note that $r(\cdot)$ may map to the same integer more than once, meaning that one and the same training sample $\mathbf{m}_i$ may occur more than once in $\widetilde{\mathfrak{D}}^j$. The motivation behind the resampling is that each tree will be sensitive to a slightly different version of the classification problem. As a result, each tree will make different errors in classification, but these errors will presumably be corrected by the other trees in the ensemble. In a broader context, $\widetilde{\mathfrak{D}}^j$ is called a *bootstrap sample* (*bootstrapping* is a statistical technique to deal with small datasets) and the aggregation of decision trees trained on different bootstrap samples is referred to as bootstrap aggregating or bagging for short.

Still, the trees are likely to choose the same features in the first few splits, since these splits favor features that are highly correlated with the class memberships $\omega(\mathbf{m}_i)$. To circumvent this issue, only a random subset of $d' \leq d$ (typically with $d' \approx \sqrt{d}$) features are considered when finding each split. In other words, during the decision tree learning, the question to ask at node n is chosen to minimize the impurity of the split according to Equation (8.7), but only d' randomly selected features are considered as candidates. Note that each split considers a different set of features, that is, the d' feature candidates are chosen anew in each individual iteration of the decision tree learning.

The main effect of bagging and feature sub-sampling is that the trees will become decorrelated, because they specialize on different training samples and emphasize different features. As a result, this minimizes the variance of the whole ensemble. This result can be seen from the following calculation, where the weights $\alpha_j = \frac{1}{M}$ in Equation (8.8) are pulled in front of the sum (Hastie et al. [2001]). Note that the discussion uses real-valued decision functions $k^j(\cdot)$ instead of the vectorial $\mathbf{k}^j(\cdot)$ in Equation (8.8) to simplify the notation, but the argument still holds for vector-valued decision functions.

With the assumptions $\mathrm{E}\{k^j(\underline{\mathbf{m}})\} = 0$, $\mathrm{Var}\{k^j(\underline{\mathbf{m}})\} = \sigma^2$ and $\mathrm{Cov}\{k^j(\underline{\mathbf{m}}), k^l(\underline{\mathbf{m}})\} = \rho\sigma^2$ for all $j, l = 1, \ldots, M$:

$$
\begin{aligned}
\mathrm{Var}\left\{\frac{1}{M}\sum_{j=1}^{M} k^j(\underline{\mathbf{m}})\right\} &= \frac{1}{M^2}\sum_{j=1}^{M}\sum_{l=1}^{M}\mathrm{Cov}\{k^j(\underline{\mathbf{m}}), k^l(\underline{\mathbf{m}})\} \\
&= \frac{1}{M^2}\sum_{j=1}^{M}\left(\sum_{\substack{l=1\\ l\neq j}}^{M}\mathrm{Cov}\{k^j(\underline{\mathbf{m}}), k^l(\underline{\mathbf{m}})\} + \mathrm{Var}\{k^j(\underline{\mathbf{m}})\}\right) \\
&= \frac{1}{M^2}\sum_{j=1}^{M}\left((M-1)\rho\sigma^2 + \sigma^2\right) \\
&= \frac{M(M-1)\rho\sigma^2 + M\sigma^2}{M^2} \\
&= \frac{(M-1)\rho\sigma^2}{M} + \frac{\sigma^2}{M} \\
&= \rho\sigma^2 + \sigma^2\frac{1-\rho}{M}.
\end{aligned}
\tag{8.9}
$$

It can be seen that the variance of $k(\underline{\mathbf{m}})$ is reduced when

- the correlation ρ between the trees is reduced or
- M is increased, i.e., more trees are added to the ensemble.

Of course, adding more trees to the ensemble will probably increase the correlation between the trees and thereby increase the first term of the above equation. At the same time, removing too many trees will increase the second term. The "correct" number of trees in the ensemble depends on the classification performance, but in many applications a number M on the order of tens will provide a good baseline.

What remains to be discussed is classification with random forests, i.e., how to implement Equation (8.8). Breiman [2001] suggests deriving a class probability estimate $\hat{P}(\omega_k \mid \mathbf{m})$ from a majority vote. Each tree $\mathbf{k}^j(\cdot)$ in the ensemble will classify the sample $\mathbf{m}$ and the probability estimate for class ω_k is the fraction of trees that voted for ω_k,

$$\hat{P}(\omega_k \mid \mathbf{m}) = \frac{1}{M} \sum_{j=1}^{M} \delta_{\left[\arg\max_{\omega} \hat{P}^j(\omega \mid \mathbf{m}) = \omega_k\right]}, \tag{8.10}$$

where $\hat{P}^j(\omega \mid \mathbf{m})$ is the a posteriori probability derived from the decision tree $\mathbf{k}^j(\mathbf{m})$ and $\delta_{[\cdot]}$ denotes the generalized Kronecker symbol.

In practice, Equation (8.10) does not require probability estimates $\hat{P}^j(\omega \mid \mathbf{m})$, but only a class assignment $\hat{\omega}^j(\mathbf{m})$. In other words, one can simply use the class assignment stored in the leaf nodes of the tree $\mathbf{k}^j(\cdot)$, as described in Section 8.1.

An alternative approach is due to Ho [1995], where the class probability estimate is the average of the probability estimates $\hat{P}^j(\omega \mid \mathbf{m})$ of the trees,

$$\hat{P}(\omega_k \mid \mathbf{m}) = \frac{1}{M} \sum_{j=1}^{M} \hat{P}^j(\omega_k \mid \mathbf{m}). \tag{8.11}$$

Here, simple class assignments are not enough: instead, this method requires full probability estimates. A straightforward approach to obtain these estimates is to store the class membership probabilities $\hat{P}^j(\omega_k \mid \mathbf{m}, n)$ in each leaf node n of each tree $\mathbf{k}^j(\cdot)$. The overall estimate of the tree $\hat{P}^j(\omega_k \mid \mathbf{m})$ is then given by the leaf node n that is reached by $\mathbf{m}$. The $\hat{P}^j(\omega_k \mid \mathbf{m}, n)$ can be estimated as in Equation (8.3), i.e., with N_{jn}^k denoting the number of training samples of class ω_k that reach node n of tree $\mathbf{k}^j(\cdot)$ and N_{jn} denoting the total number of training samples that reach that node,

$$\hat{P}^j(\omega_k \mid \mathbf{m}, n) := \frac{N_{jn}^k}{N_{jn}}. \tag{8.12}$$

In either case, the final class estimate is given by the maximum a posteriori classifier as in Equation (3.23),

$$\hat{\omega}(\mathbf{m}) = \underset{\omega_k \in \Omega/\sim}{\arg\max}\, \hat{P}(\omega_k \mid \mathbf{m}), \tag{8.13}$$

or by minimizing the a posteriori risk as described in Section 3.3.

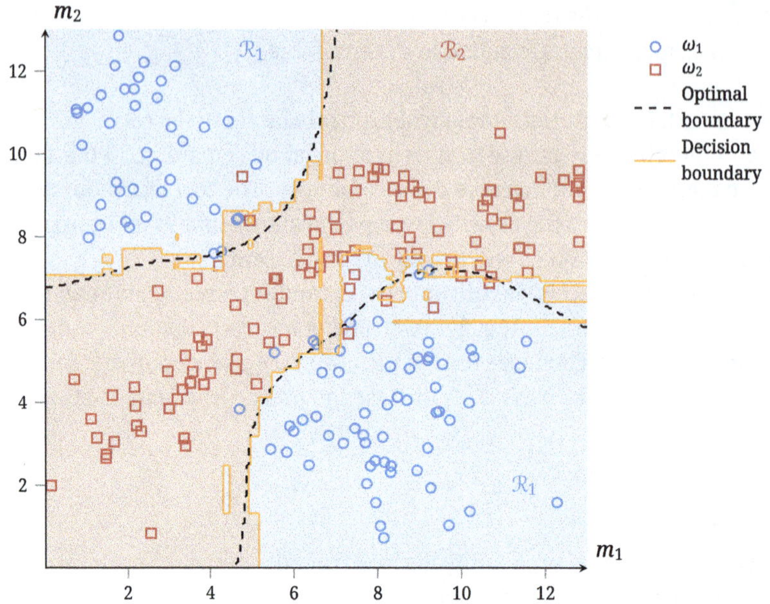

Fig. 8.8: Application to the reference example of Section 3.3.2. Decision regions of a random forest classifier. The forest was composed of 10 decision trees. Training was stopped when there were fewer than two samples available for a split. The training and testing errors are $e_{\text{train}} = 1\,\%$ and $e_{\text{test}} = 8.9\,\%$, respectively. The testing error asymptotically approaches $e_{\text{test}} \approx 8.31\,\%$, which is very close to the 6.16 % asymptotic testing error of the optimal classifier. The training set is the same as in Figure 3.8. Test samples are shown with hollow marks.

In Figure 8.8, the reference dataset from Section 3.3.2 was used to train a random forest of ten decision trees. Feature sub-sampling was omitted (i.e., $d' = d = 2$), since the example uses only a two-dimensional feature vector. As with the decision tree example, the Gini impurity measure was used as the split criterion and the recursion was stopped when fewer than two training samples were available for a split. Compared to Figure 8.4, the decision regions are much more complicated, but also much closer to the decision regions of the Bayes optimal classifier.

To conclude this section, consider the following remarks:

- Learning the decision trees of a random forest classifier is generally fast and simple, since only $d' \leq d$ features need to be considered at each split.
- Learning and classification can also be accelerated by exploiting the inherent parallelism of the approach: each decision tree can be trained and evaluated in a separate thread.
- Bagging and random feature sub-sampling lead to better generalization properties compared to a single decision tree. This effect is commonly observed in ensemble methods.

- Random forests can also be used for regression, cluster analysis, and density estimation (see, e.g., Criminisi et al. [2012]).
- Since they build on decision trees, random forests can be used with features on all measurement scales and mixtures thereof.
- Random forests can be used to reduce the dimensionality of the feature space by assessing each feature's importance using the left-out samples from bootstrapping and removing features with little importance (see Breiman [2001]: feature selection is embedded in a random forest, cf. Section 2.7.7).
- Random forests may consume a large amount of memory, since the classifier needs to store several decision tree classifiers.

8.3 String matching

Another specialized technique may be used when the pattern is a sequence of symbols. As an example, consider the classification of DNA sequences in biomedical applications. DNA can be thought of a series of pairs of the bases Adenine, Thymine, Cytosine, and Guanine. However, as Thymine always pairs with Adenine and Cytosine always pairs with Guanine, DNA can be sufficiently described as a sequence of *symbols* in the *alphabet* $\mathcal{A} = \{A,C,G,T\}$. Here, A stands for Adenine, C for Cytosine, G for Guanine, and T for the base Thymine. A short DNA sequence may be described by the *symbol sequence* $AGCTTCGAATC$. Long sequences of symbols (where the meaning of "long" depends on the context) are also called a *text* and a substring of a sequence is denoted a *factor*. Symbols may represent either nominal or ordinal features. With the DNA example, the symbols are nominal.

The task in string matching is to find a sequence $\mathbf{m}$ in a given text $\mathbf{t}$, where $\mathbf{t}$ is usually much longer than $\mathbf{m}$ (see Figure 8.9). In other words, the task is to answer the question: is the sequence $\mathbf{m}$ a factor of the text $\mathbf{t}$ and where is it located? With the DNA example, string matching can be used to find markers for genetic disorders in the DNA of a patient.

Often, it is not necessary or even possible to find exact matches. With genetic markers, random mutations may cause individual genes to change, without affecting the overall behavior of the factor. One method to deal with such situations is to use the nearest neighbor classifier (see Section 5.3), where the distance between two sequences $\mathbf{m}_1$ and $\mathbf{m}_2$ is the edit distance of the sequences. The edit distance between $\mathbf{m}_1$ and $\mathbf{m}_2$ is

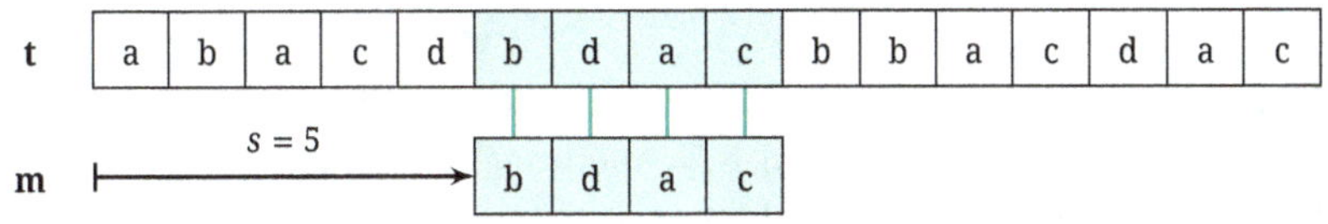

Fig. 8.9: Strict string matching. Figure according to Duda et al. [2001].

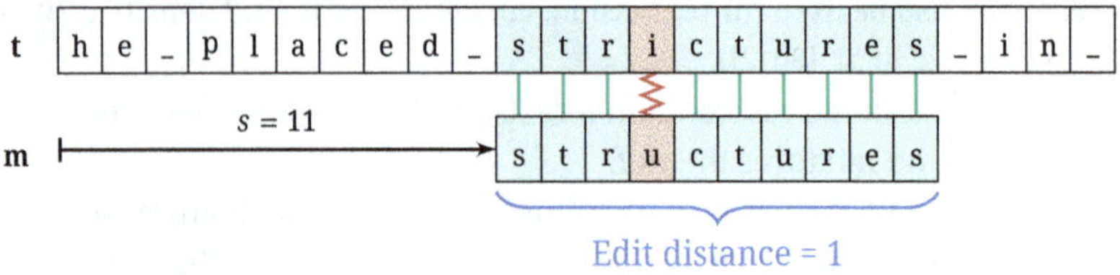

Fig. 8.10: Approximate string matching. Figure according to Duda et al. [2001].

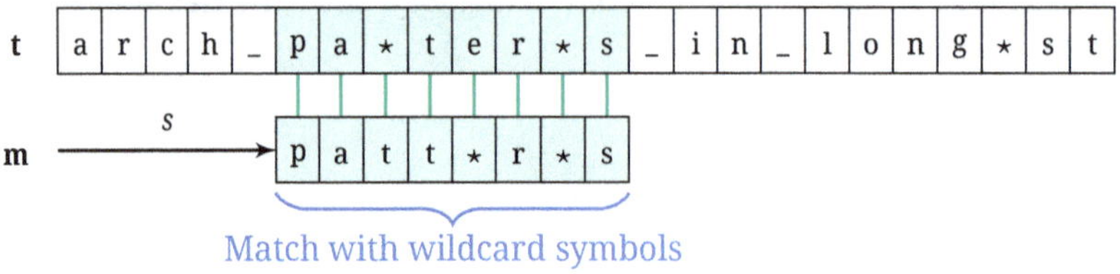

Fig. 8.11: String matching with wildcard symbol $\star$. Figure according to Duda et al. [2001].

the minimum number of string operations—insertions, deletions and substitutions—to transform the sequence $\mathbf{m}_1$ into the sequence $\mathbf{m}_2$ (see Figure 8.10).

During the training of the classifier, all given strings (factors) and their class memberships are stored. When classifying an unknown sequence **m**, all edit distances of **m** to the stored factors are computed and the class of the sequence with the minimum distance to **m** is assigned.

Another strategy is to include in the alphabet the special *wildcard* symbol $\star$, which matches with any character in $\mathcal{A}$. The wildcard symbol may appear both in the text and the target sequence **m**. An example of string matching with wildcard symbols is shown in Figure 8.11.

8.4 Grammars

Besides string matching, another approach to classifying sequences is the use of grammars, where every class is represented by a grammar G_i, $i = 1, \ldots, c$. All sequences that are generated by the grammar G_i are considered equivalent. In this sense, a grammar G_i corresponds to a model of the patterns of the class ω_i. Classification with grammars corresponds to parsing: a pattern is assigned to the class whose grammar generates the pattern.

Formally, a grammar G is a quadruple $G = (\mathcal{A}, \mathcal{V}, S, \mathcal{P})$ of an alphabet $\mathcal{A}$ of terminal symbols, variables $\mathcal{V}$, the starting variable $S \in \mathcal{V}$, and a set of production rules $\mathcal{P}$ that replace variables $v \in \mathcal{V}$ with strings of variables or terminal symbols. $L(G)$ denotes the language of the grammar G, where the language is the set of all sequences of terminal symbols that can be produced by G.

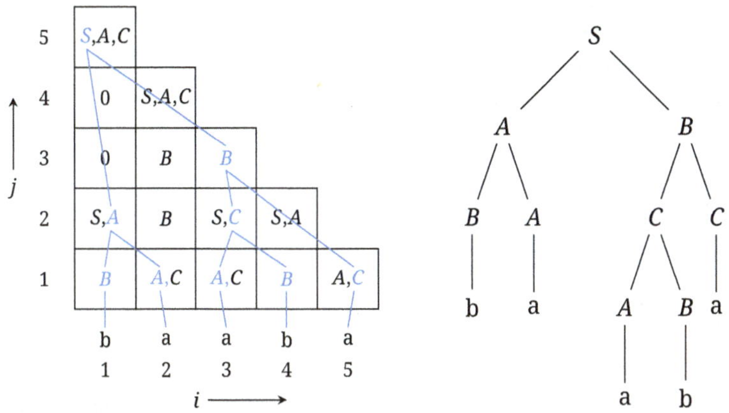

Fig. 8.12: Bottom up (left) and top down (right) parsing of the sequence "baaba" given the example grammar (see text). In both cases, the sequence is accepted.

An example of a grammar (according to Duda et al. [2001]) is given by the alphabet $\mathcal{A} := \{a,b,c\}$, the variables $\mathcal{V} = \{A,B,C,S\}$ and the production rules

$$\mathcal{P} = \left\{ \begin{array}{llll} p_1: & S \to AB & \text{or} & BC \\ p_2: & A \to BA & \text{or} & \mathrm{a} \\ p_3: & B \to CC & \text{or} & \mathrm{b} \\ p_4: & C \to AB & \text{or} & \mathrm{a} \end{array} \right\} .$$

Parsing a grammar can be approached bottom–up or top–down. Bottom–up parsing starts at the sequences and applies rules in reverse until the starting variable S is reached (see Figure 8.12, left). Top–down parsing starts with S and applies rules until the sequence is matched (Figure 8.12, right). The details of both approaches are outside the scope of this textbook.

The learning problem corresponds to the construction of a grammar from the data in $\mathcal{D}$ and is also known as *grammar induction*. However, there are generally infinitely many grammars that are consistent with a finite number of sample sequences. A solution is again offered by Ockham's razor: the simplest grammar that is consistent with the data is to be preferred over more complex grammars that also fit the data. Like parsing, methods to learn a grammar from data are outside the scope of this textbook.

8.5 Exercises

(8.1) Use the features animal, weight, and speed to construct a balanced decision tree that classifies the following sample without error:

Animal	Weight	Speed	Class
Dog	medium	medium	ω_1
Snail	light	slow	ω_1
Kangaroo	medium	medium	ω_1
Hawk	light	fast	ω_1
Tortoise	heavy	slow	ω_2
Cheetah	medium	fast	ω_2
Whale	heavy	medium	ω_3
Hare	light	medium	ω_3

(8.2) Given below is a training sample of two-dimensional features for two classes ω_1 and ω_2 and a decision tree learned from the data.
Sketch the decision boundary of this classifier. Which leaves are a symptom of overfitting? Can the tree be simplified to eliminate the overfitting?

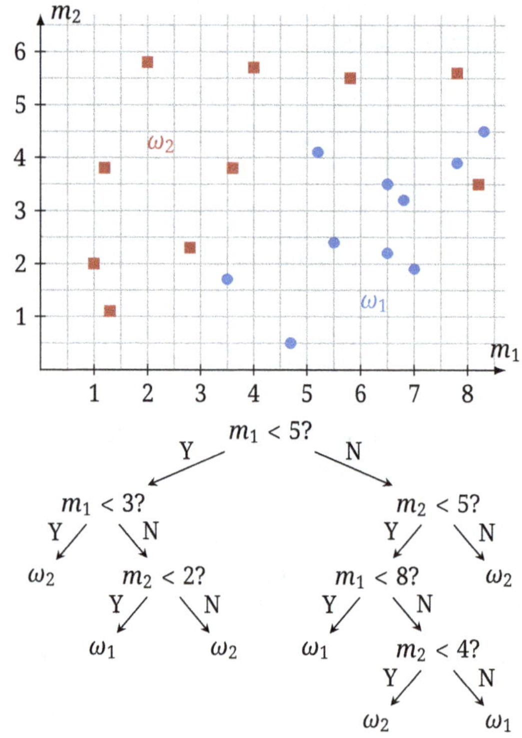

9 Classifier-independent concepts

In the final chapter of this book, we will explore topics that are, in a sense, orthogonal to the classifiers we have seen so far. The first section deals with fundamental concepts and the limits of all statistical learning. The second section presents the no-free-lunch theorem stating that without prior knowledge about the classification problem, no classifier can be deemed superior to any other one. The third section will give an overview of methods for empirically evaluating a classifier's performance. The last two sections will introduce boosting, a meta-technique for combining the predictions of several weak classifiers into one strong classifier, and will discuss techniques for classifying with the option of rejecting a sample.

9.1 Learning theory

So far we have gained an understanding of Bayesian classification that uses probability density estimates to derive a classification decision as well as classifiers that directly determine the decision boundary. Despite the differences between their approaches, all statistical learners have a common core. This is the subject of (statistical) learning theory.

Recall from Chapter 1 that the full dataset $\mathcal{S}$ is composed of a learning set $\mathcal{D}$ and a test set $\mathcal{T}$: $\mathcal{S} = \mathcal{D} \uplus \mathcal{T}$ (for convenience, the validation set $\mathcal{V}$ is set to $\mathcal{V} := \emptyset$). *Learning* means, in effect, to determine a function $\hat{\omega}(\mathbf{m} \mid \mathcal{D})$ from $\mathcal{D}$, such that $\hat{\omega}$ estimates the true class membership of a feature vector $\mathbf{m}$ with as little expected error as possible. In *supervised learning*, the class memberships of the samples in $\mathcal{D}$ are known, whereas in *unsupervised learning* they are not. The goal of the latter is not as well defined as with supervised learning, but generally one wishes to understand the underlying process that generated the dataset. In either case, without considering the required computational costs, there are four major sources of problems:

1. the training set is too small;
2. the training set is not representative of the true distribution;
3. the features are inappropriate or non-discriminative; and
4. a small training error (on $\mathcal{D}$) does not guarantee a small validation error (on $\mathcal{T}$), e.g., due to overfitting.

The first two issues are part of the topic of proper sampling and so of lesser interest in the context of learning theory. The third point addresses the quality of the features and the problem of extracting the relevant information from the patterns. The fourth point, which here is reformulated compared to point 4 on page 5 in Section 1.4, gives rise to the central problem of statistical learning. Some of the potential problems are illustrated in Figure 9.1, where the training and test sets $\mathcal{D}$ and $\mathcal{T}$ are depicted as subsets of some world

https://doi.org/10.1515/9783111339207-009

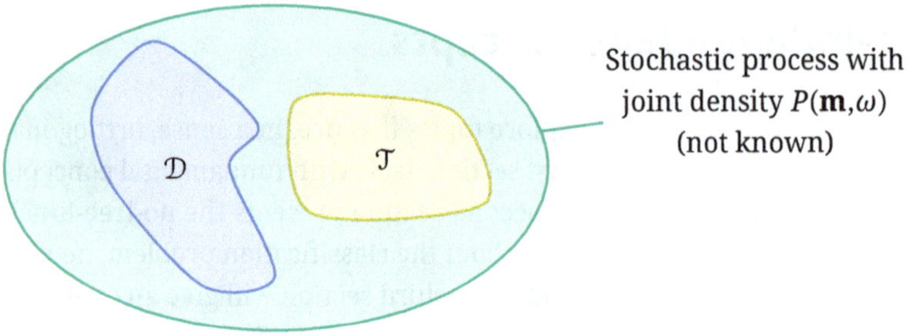

Fig. 9.1: Relation of the world model $P(\mathbf{m},\omega)$ and training and test sets $\mathcal{D}$ and $\mathcal{T}$. The training set $\mathcal{D}$ and the test set $\mathcal{T}$ are drawn from a stochastic process with unknown joint probability distribution $P(\mathbf{m},\omega)$. The training error $e_{\text{training}}(\mathbf{m},\theta)$ and the test error $e_{\text{test}}(\mathbf{m},\theta)$ of a model θ are estimated on the sets $\mathcal{D}$ and $\mathcal{T}$.

model $P(\mathbf{m},\omega)$. Since this world model is not known, the occurrence of aforementioned problems is strongly related to the shape, size, and location of the training and test sets. In the following, we will discuss the fourth problem in detail using a binary classification problem (i.e. $c = 2$). In this case, it is sufficient to consider a single decision function $k(\mathbf{m} \mid \theta)$ instead of $\hat{\omega}(\mathbf{m} \mid \theta)$.

9.1.1 The central problem of statistical learning

Under what conditions does a small training error lead to a small test error? More generally, statistical learning theory is concerned with the ability of a classifier to *generalize*, that is, to classify unseen samples with little error. Note that the two are inversely related: if the ability to generalize is great, the mean test error will be small and vice versa.

More formally, given a decision function $k(\mathbf{m},\theta)$ governed by a controllable parameter vector θ, what can be known about the expected test error $\varepsilon(\theta) = \mathrm{E}\{e_{\text{test}}(\underline{\mathbf{m}},\theta)\}$? Under what conditions will $\varepsilon(\theta)$ be minimal and what is the probably approximately correct (PAC, Valiant [1984]) lower bound on $\varepsilon(\theta)$?

9.1.2 Vapnik–Chervonenkis learning theory

Vapnik and Chervonenkis provided a theoretical framework to tackle these questions (Vapnik [1998]). They observed that the following bound on the test error holds with probability $(1 - \eta)$:

$$\varepsilon(\theta) \leq e_{\text{training}}(\theta) + \Phi(\nu,N,\eta). \tag{9.1}$$

Here, $N = |\mathcal{D}|$ is the number of samples in the training set, ν is the *VC dimension* (see below) of the set $\mathcal{K} = \{k(\mathbf{m},\theta)|\theta \in \Theta\}$ of decision functions and $\Phi(\nu,N,\eta)$ denotes the *VC*

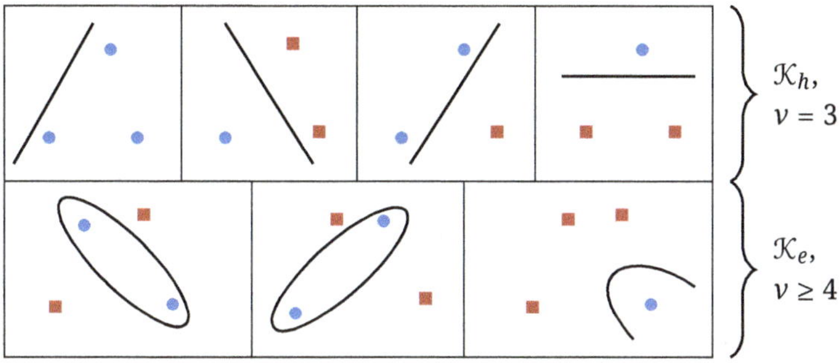

Fig. 9.2: Sketch of different class assignments to a sample using the model families $\mathcal{K}_h$ of two-dimensional hyperplanes and $\mathcal{K}_e$ of ellipses for separation. The VC dimension of $\mathcal{K}_h$ is $\nu = 3$, whereas that of $\mathcal{K}_e$ is $\nu \geq 4$.

confidence (Vapnik [1998]), defined by

$$\Phi(\nu,N,\eta) = \sqrt{\frac{\nu\left(\log\frac{2N}{\nu}+1\right)-\log\left(\frac{\eta}{4}\right)}{N}}. \tag{9.2}$$

This bound holds regardless of the underlying distribution of the data if this distribution is the same for all samples and all the samples were produced independently, that is, if the data is i.i.d. A key quantity in Equation (9.2) is the *Vapnik–Chervonenkis dimension (VC dimension)* ν of the set of decision functions $\mathcal{K}$. Briefly, ν acts as a measure of the complexity of the model family represented by $\mathcal{K}$. Since a rigorous definition requires concepts that are outside the scope of this text, we will give only an informal, intuitive description of ν:

Consider a given set of N samples to be assigned to two classes. Because each sample can belong to either class, but not to both, there are 2^N constellations of possible class assignments in total. The VC dimension ν of a set $\mathcal{K}$ of decision functions is defined as the maximum number of samples that can be separated by $\mathcal{K}$ for all possible class assignments, independently of the spatial distribution of the samples. An example with two-dimensional features $\mathbf{m} \in \mathbb{R}^2$ is shown in Figure 9.2. In the first row, the set $\mathcal{K}$ is the set of two-dimensional hyperplanes, $\mathcal{K}_h = \{\mathbf{w}^\mathsf{T}\mathbf{m} - b = 0\}$. In the second row, $\mathcal{K}$ is the set of ellipses, $\mathcal{K}_e = \{(\mathbf{m} - \boldsymbol{\mu})^\mathsf{T}\boldsymbol{\Lambda}(\mathbf{m} - \boldsymbol{\mu}) - b = 0\}$, where $\boldsymbol{\mu}$ denotes the center of the ellipse and $\boldsymbol{\Lambda} \in \mathbb{R}^{2\times 2}$ is a positive semidefinite matrix. Note that in general a set of four two-dimensional samples cannot be separated by a hyperplane (e.g., the configuration in the first panel of the second row of Figure 9.2—the XOR problem), but three samples can. Thus, the VC dimension is $\nu = 3$. In general, if $\mathcal{K}$ denotes the set of hyperplanes in $\mathbb{R}^d$, then $\nu = d + 1$. For polynomials, ν grows with the degree of the polynomial.

Since the VC dimension ν depends on the model parameters $\boldsymbol{\theta}$, all three quantities in Equation (9.1), the expected test error $\varepsilon(\boldsymbol{\theta})$, the empirical training error $e_{\text{training}}(\boldsymbol{\theta})$, and the VC confidence $\Phi(\nu,N,\eta)$, change with varying ν. Figure 9.3 outlines how the three quantities change with increasing VC dimension ν. It can be seen that with larger ν, the

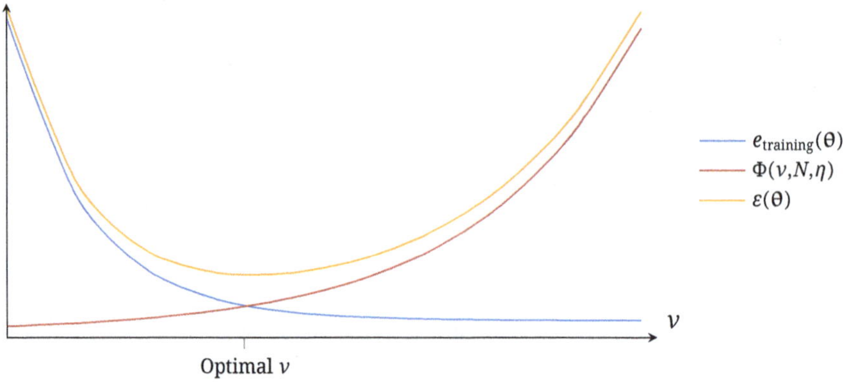

Fig. 9.3: Qualitative plot of the expected test error $\varepsilon(\Theta)$, the empirical training error $e_{\text{training}}(\Theta)$, and VC confidence $\Phi(\nu,N,\eta)$, against VC dimension ν.

Φ also increases, while the empirical training error decreases. The expected test error $\varepsilon(\Theta)$ first decreases, but increases again when the classifier begins to overfit the data. The optimal ν, that is, the model family with optimal complexity with respect to ε, is found where $e_{\text{training}}(\Theta) + \Phi(\nu,N,\eta)$ is minimal.

In general, an increasing VC dimension ν means that the underlying set of functions $\mathcal{K}$ becomes more and more “malleable”. With $\nu_1 < \nu_2$ and corresponding sets $\mathcal{K}_1, \mathcal{K}_2$, functions in $\mathcal{K}_2$ make fewer errors on the training set. Indeed, often (but not always) the set $\mathcal{K}_2$ includes all decision functions in $\mathcal{K}_1$ or $\mathcal{K}_1 \subset \mathcal{K}_2$. However, increasing the malleability of $\mathcal{K}$ also increases the risk that a classifier will overfit the training data.

Vapnik–Chervonenkis learning theory offers a theoretical framework for finding a compromise between a reduced training error and the generalizability of the model. A more detailed discussion and a formal definition of the concepts discussed above are found in Vapnik [1998].

The upper bound on the expected test error from Equation (9.1),

$$\begin{aligned}\varepsilon(\Theta) &\le e_{\text{training}}(\Theta) + \Phi(\nu,N,\eta)\\ &= e_{\text{training}}(\Theta) + \sqrt{\frac{\nu\left(\log\frac{2N}{\nu}+1\right) - \log\left(\frac{\eta}{4}\right)}{N}},\end{aligned} \tag{9.3}$$

and the Figure 9.3 give rise to the following observations:

1. If the number of training samples is vastly larger than the VC dimension, as $\frac{N}{\nu} \to \infty$, then $\Phi \to 0$ and in turn $\varepsilon \to e_{\text{training}}$. In this case, it is sufficient to minimize the training error, as a small training error guarantees a small test error. This is known as the empirical risk minimization (ERM) principle (Vapnik [1998]).
2. If, on the other hand, the ratio $\frac{N}{\nu}$ is small, then Φ dominates the bound. A small training error no longer guarantees a small test error. Here, e_{training} and Φ must be

minimized simultaneously instead. This principle is known as the structural risk minimization (SRM) principle (Vapnik [1998]).

Most practical applications of pattern recognition fall in the second case: there is a relatively small dataset on which to train the classifier, which means one should either choose a model with low VC dimension or train the model using SRM.

9.2 No-free-lunch theorem

Wolpert and Macready [1997] introduced the *no-free-lunch theorem* that deals with the question: which of the classification methods is the best if no a priori information of the task at hand is available? To answer this question, an appropriate evaluation measure has to be defined. For this, some terminology is introduced in the following.

Consider a binary classification problem with two classes ω_1 and ω_2, a training set $\mathcal{D} = \mathcal{D}^1 \cup \mathcal{D}^2$ with labelled data $\mathcal{D}^1 = \{\mathbf{m}^1\}$ and $\mathcal{D}^2 = \{\mathbf{m}^2\}$, as well as an objective function $\omega(\mathbf{m})$ to be learned that correctly classifies all data points: $\omega(\mathbf{m}^i) = \omega_i$. Then, for any learning method, there exists a set $\mathcal{H}$ of achievable hypotheses $k(\mathbf{m}) \in \mathcal{H}$ and the probability that a specific hypothesis k emerges as a result of the learning method depends on the training set: $P(k) = P(k|\mathcal{D})$. Note that a specific hypothesis $k(\mathbf{m})$ can be described, for example, with learned parameters Θ such that $k(\mathbf{m}) = k(\mathbf{m}|\Theta)$. In the optimal case, the hypothesis is equal to the objective function, i.e., $k(\mathbf{m}) = \omega(\mathbf{m})$, and the learned classifier makes no errors on the training set.

An appropriate evaluation measure for the quality of a learning method is the expected generalization error $\mathrm{E}\{\varepsilon|\mathcal{D}\}$ on unseen data $\mathbf{m} \notin \mathcal{D}$ given by

$$\mathrm{E}\{\varepsilon|\mathcal{D}\} = \sum_{k,\omega} \sum_{\mathbf{m}\notin\mathcal{D}} P(\mathbf{m})[1 - \delta_{k(\mathbf{m})}^{\omega(\mathbf{m})}]P(k|\mathcal{D})P(\omega|\mathcal{D}) \tag{9.4}$$

with $\delta_{k(\mathbf{m})}^{\omega(\mathbf{m})}$ being the Kronecker delta which becomes 1 if $\omega(\mathbf{m}) = k(\mathbf{m})$ (correct classification) and 0 otherwise. This generalization error and thus the quality of a learning method depends on the data distribution $P(\mathbf{m})$ and also on how good the learning method that produces hypotheses $P(k|\mathcal{D})$ matches to the actual distribution $P(\omega|\mathcal{D})$. Here, $P(k|\mathcal{D})$ denotes the probability that the learning method produces hypothesis k when trained on the dataset $\mathcal{D}$ and is zero everywhere except for a specific k if the learning method is deterministic (e.g., nearest neighbor classifier) or a broad distribution if the method is stochastic (e.g., neural networks with randomly initialized weights). A direct consequence of Equation (9.4) is that without prior knowledge about the problem at hand given by $P(\omega|\mathcal{D})$, little can be said about the performance of a specific classifier.

Given a specific objective function $\omega(\mathbf{m})$, the expected generalization error of a particular learning method $P_j(k(\mathbf{m})|\mathcal{D})$ can be calculated with

$$\mathrm{E}_j\{\varepsilon|\omega,n\} = \sum_{\mathbf{m}\notin\mathcal{D}} P(\mathbf{m})[1 - \delta_{k(\mathbf{m})}^{\omega(\mathbf{m})}]P_j(k(\mathbf{m})|\mathcal{D}) \tag{9.5}$$

where n is the number of test samples not included in the training set $\mathbf{m} \notin \mathcal{D}$.

With the aforementioned definitions, the no-free-lunch theorem makes inter alia the following statements:

$$\sum_{\omega}\sum_{\mathcal{D}} p(\mathcal{D}|\omega)(\mathrm{E}_1\{\varepsilon|\omega,n\} - \mathrm{E}_2\{\varepsilon|\omega,n\}) = 0, \tag{9.6}$$

$$\sum_{\omega} \mathrm{E}_1\{\varepsilon|\omega,\mathcal{D}\} - \mathrm{E}_2\{\varepsilon|\omega,\mathcal{D}\}) = 0. \tag{9.7}$$

Equation (9.6) says that the mean error when considering all possible objective functions ω and under the assumption that all of them are equally likely is identical for each learning method. According to Equation (9.7), this is even true if the training set $\mathcal{D}$ is known.

From the no-free-lunch theorem follows that no classifier is suited for all kind of problems (all objective functions) and that different classifiers should be applied dependent on the concrete problem at hand (relevant objective functions). To summarize, choosing an appropriate classifier and making the right assumptions about the problem (prior knowledge) are crucial for a good classification performance.

9.3 Empirical evaluation of classifier performance

The design phases of a pattern recognition system in Figure 1.4 listed the evaluation of the classifier as the last step. In the previous chapters, estimators of the error probability P_e were given for some specific classifiers, but this is not the only possibility for assessing the performance of a classifier. This section will fill that gap by introducing some common performance measures and techniques to validate classifiers with a finite test sample.

We will first focus on a binary classifier that decides between only $c = 2$ classes. Usually, one class is called the "negative class" and the other one the "positive class". Here, we associate ω_1 with the negative class and ω_2 with the positive class. Binary classifiers are always used if the goal is to detect the presence or absence of some quality, e.g., diseased vs. healthy fruit, defective vs. intact workpiece, etc. Moreover, every multi-class classification task can be solved using a combination of binary classifiers (see, for example, Section 7.1.1). Because binary classifiers play such an important role in pattern recognition, many technical terms have been invented to precisely describe their characteristics. In order to simplify the upcoming discussion, we restrict the scenario to a one-dimensional feature space ($d = 1$) and a linear classifier. This means the decision boundary is just a single point m^* and the decision regions are given by $\mathcal{R}_1 = \{m \in \mathcal{M}|m < m^*\}$ and $\mathcal{R}_2 = \{m \in \mathcal{M}|m > m^*\}$.

As already discussed in the previous chapters, the feature distributions usually overlap, which means that there is some minimum error probability that cannot be reduced any further. This situation is depicted in Figure 9.4. In both sub-figures, the decision boundary is marked by m^* and the optimal decision boundary according to Bayes is

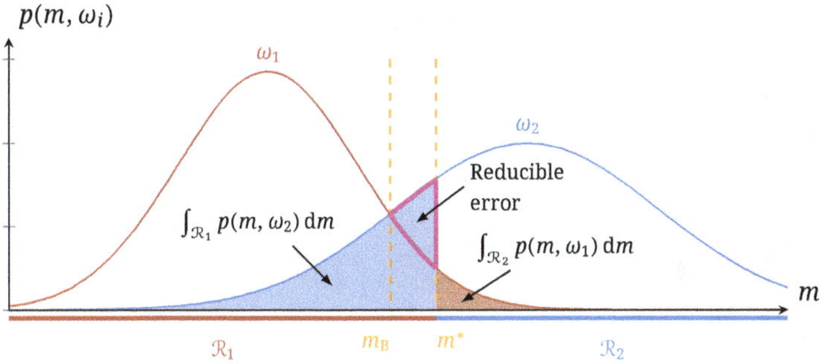

(a) Non-optimal choice of decision boundary

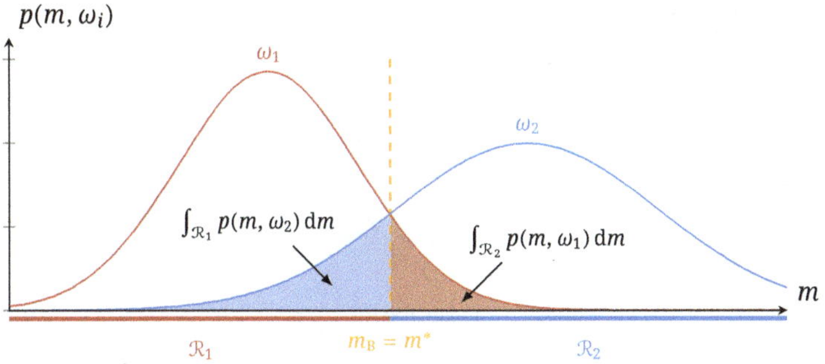

(b) Optimal choice of decision boundary

Fig. 9.4: Classification error probability. Even with an optimal decision boundary, there is a remaining, irreducible error probability.

labeled m_{B}. A classification error occurs if a feature falls in a different decision region than that to which the true class belongs. In this example, the overall error equals

$$P_e = P(\omega \neq \hat{\omega}) = \int\limits_{\mathcal{R}_2} P(\omega_1, \mathbf{m})\,\mathbf{dm} + \int\limits_{\mathcal{R}_1} P(\omega_2, \mathbf{m})\,\mathbf{dm}, \tag{9.8}$$

which is the sum of the red and blue areas in Figure 9.4. In Figure 9.4b, this sum attains its minimum, as the decision boundary equals the optimal boundary $m_{\mathrm{B}} = m^*$. In Figure 9.4a, the sum is larger, because the boundary introduces an additional reducible error (purple frame).

Depending on the true class ω and the classifier prediction $\hat{\omega}$, one can distinguish four cases (see Figure 9.5a):

Correct rejection True class and prediction are negative $\omega = \omega_1 = \hat{\omega}$.

Discovery (or hit) True class and prediction are positive $\omega = \omega_2 = \hat{\omega}$.

False alarm The true class is the negative class, $\omega = \omega_1$, but the classifier predicts the positive class, $\hat{\omega} = \omega_2$.

Slack The true class is the positive class, $\omega = \omega_2$, but the classifier predicts the negative class, $\hat{\omega} = \omega_1$.

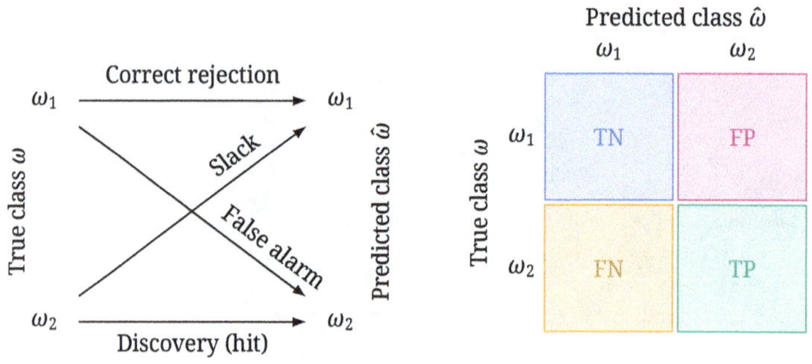

(a) Terms for classification outcomes **(b)** Confusion matrix

Fig. 9.5: Classification outcomes in a 2-class scenario.

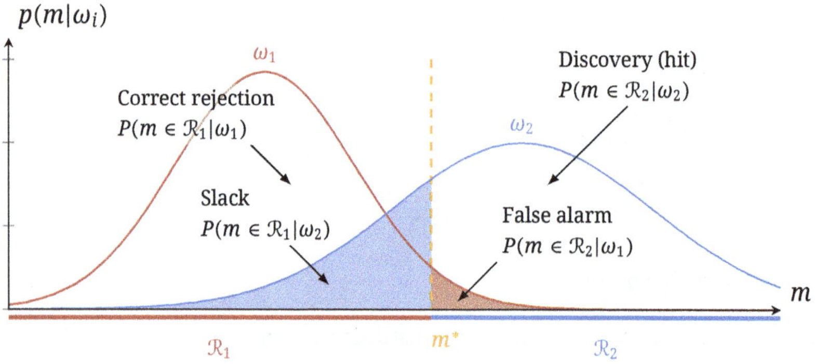

Fig. 9.6: Performance indicators for a binary classifier.

Figure 9.6 shows the class-conditional decision probabilities that correspond to the probabilities for each of the cases. Please note that these are *not* the same as the error probability, which is the *unconditional* probability that the classification is false.

When evaluating a classifier using a test set $\mathcal{T}$, the outcome of each case is counted and recorded in a confusion matrix, as shown in Figure 9.5b. In the two-class setting, the cells are often labeled as follows: *true negatives*, which counts the number of correct rejections, *false positives*, which counts the number of false alarms, *false negatives*, which is the number of falsely rejected samples, and *true positives*, which is the number of discoveries in the dataset. We denote the number of samples in each cell by TN, FP, FN, and TP, respectively. Using these quantities, one can compute higher order performance measures that characterize the classifier. These measures usually approximate a characteristic classification probability. A list of common measures is given in Table 9.1.

These measures are coupled in interesting ways. Changing a parameter to improve one measure will also change the values of the other measures. Unfortunately, the coupling is sometimes reversed: improving one measure may have a negative impact on another. Consider the example in Figure 9.7a, where the classifier has only one parame-

Tab. 9.1: Common binary classification performance measures derived from a confusion matrix.

Performance measure		Estimates
True positive rate (recall)	$\text{tpr} = \frac{\text{TP}}{\text{TP+FN}}$	$P(\mathbf{m} \in \mathcal{R}_2 \mid \omega_2)$
True negative rate	$\text{tnr} = \frac{\text{TN}}{\text{TN+FP}}$	$P(\mathbf{m} \in \mathcal{R}_1 \mid \omega_1)$
False positive rate (false alarm rate)	$\text{fpr} = \frac{\text{FP}}{\text{FP+TN}}$	$P(\mathbf{m} \in \mathcal{R}_2 \mid \omega_1)$
False negative rate	$\text{fnr} = \frac{\text{FN}}{\text{FN+TN}}$	$P(\mathbf{m} \in \mathcal{R}_1 \mid \omega_2)$
Precision	$\text{prec} = \frac{\text{TP}}{\text{TP+FP}}$	$P(\omega_2 \mid \mathbf{m} \in \mathcal{R}_2)$
Negative predictive value	$\text{npv} = \frac{\text{TN}}{\text{TN+FN}}$	$P(\omega_1 \mid \mathbf{m} \in \mathcal{R}_1)$
Accuracy	$\text{acc} = \frac{\text{TP+TN}}{\text{TP+TN+FP+FN}}$	$\Pr(\hat{\omega} = \omega)$

ter m^*. Increasing m^* will decrease the false positive rate (the false alarm rate), but the true positive rate (the recall) will also be decreased. Moving the decision boundary in the opposite direction will increase the recall, but also increase the false alarm rate.

9.3.1 Receiver operating characteristic

The receiver operating characteristic (ROC), originally developed to evaluate radar systems, is a tool to visualize the dependence between the false positive rate and the true positive rate. ROC curves are plots of the false positive rate against the recall for different parameter values. The shape of the curve depends on the underlying class-dependent feature distributions.

ROC curves for varying distances between class-specific feature distributions of the example in Figure 9.7a are shown in Figure 9.7b. If $d = 0$, then the feature is uninformative about the class. As the distributions overlap totally, the true and false positive rates will always be the same and the ROC curve will be a straight, diagonal line. ROC curves near the diagonal mean that the classifier is no better than random guessing. When the feature becomes more informative, i.e., if $d > 0$ grows, the classifier will become more and more capable of discriminating between the two classes. The recall will grow faster than the false alarm rate and the ROC curve will bend towards the upper left corner of the plot. Fully informative features, where the class-specific feature distributions do not overlap, will result in the ROC curve shown by the dashed line. In practice, this perfect ROC curve is never obtained.

A ROC curve should never lie below the diagonal. If it does, this means that the classifier misinterprets the features and should predict the opposite of what it does. Note that while ROC curves are used to evaluate a classifier, they are actually more indicative about the suitability of the features than about the suitability of the classifier.

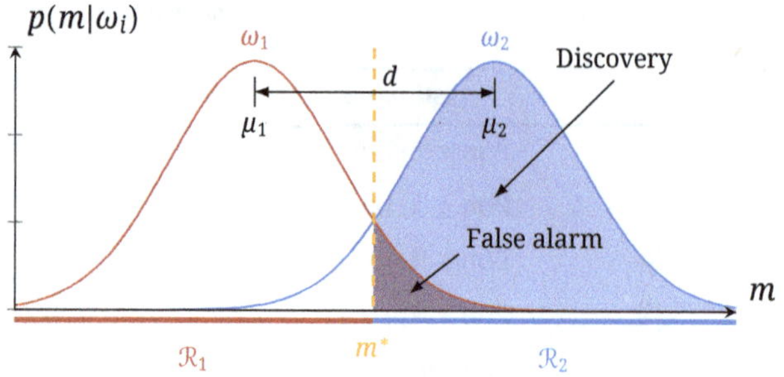

(a) Underlying class-specific feature distributions

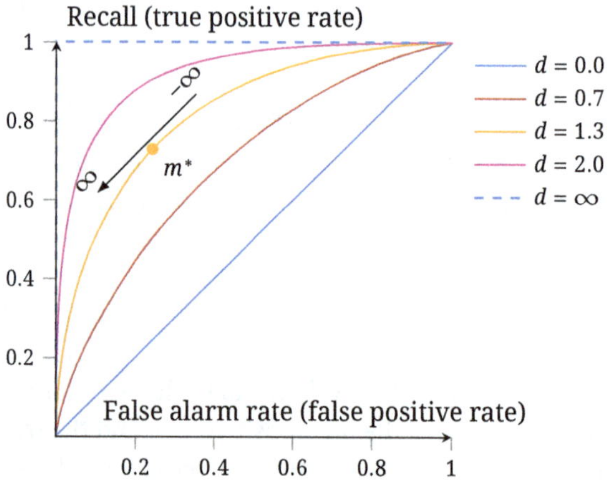

(b) Corresponding ROC curves for varying d

Fig. 9.7: Examples of ROC curves for two Gaussian feature distributions with variance $\sigma^2 = 1$ and distance d between the expectations μ_1 and μ_2.

9.3.2 Multi-class setting

We now turn our attention back to the generic multi-class setting. Generalizing the classification error probability P_e of the binary classifier (see Equation (9.8)) to the multi-class setting is straightforward:

$$P_e = \Pr(\omega \neq \hat{\omega}) = 1 - \sum_{i=1}^{c} P(\mathbf{m} \in \mathcal{R}_i, \omega_i) = 1 - \sum_{i=1}^{c} \int_{\mathcal{R}_i} p(\mathbf{m} \mid \omega_i) P(\omega_i) \, d\mathbf{m}. \tag{9.9}$$

Intuitively, the error probability P_e is the probability that the feature vector $\mathbf{m}$ does not fall in the correct decision region.

Other measures from the binary case, like precision and recall, cannot directly be applied in a multi-class setting. However, these measures can still be computed individually for each class: the c-class classification problem is treated as c separate binary

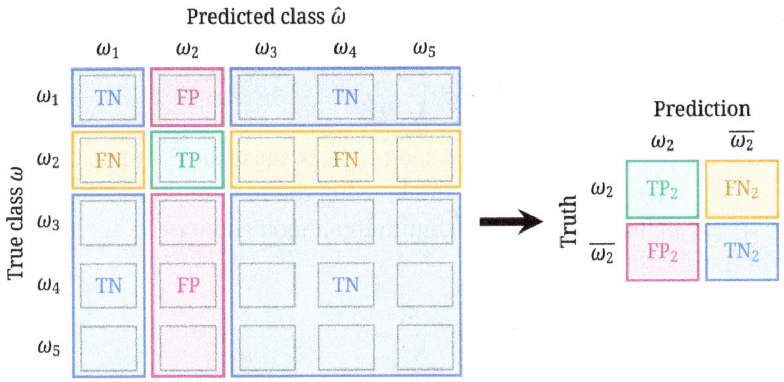

Fig. 9.8: Converting a multi-class confusion matrix to binary confusion matrices: a multi-class confusion matrix of c classes can be subsumed into c binary confusion matrices, one for each class ω_i. Example here: the reduced confusion matrix with respect to ω_2, i.e., $i = 2$.

classification problems, where the goal is to separate the target class ω_i from all other classes ω_k, $k \neq i$, or $\overline{\omega_i}$ for short.

Formally, let $C(\omega_j,\omega_k)$ denote the number of samples with true class $\omega = \omega_j$ that were classified as $\hat{\omega} = \omega_k$. The number of true positives, false positives, false negatives, and true negatives for class ω_i are then computed by (see Figure 9.8):

$$\mathrm{TP}_i = C(\omega_i,\omega_i) \tag{9.10}$$

$$\mathrm{FP}_i = \sum_{j \neq i} C(\omega_j,\omega_i) \tag{9.11}$$

$$\mathrm{FN}_i = \sum_{k \neq i} C(\omega_i,\omega_k) \tag{9.12}$$

$$\mathrm{TN}_i = \sum_{j,k \neq i} C(\omega_j,\omega_k). \tag{9.13}$$

Given these counts, the measures from Table 9.1 can be derived for each class individually. Sometimes, the class-wise measures are further averaged over all classes to grade the overall performance with a single number.

9.3.3 Theoretical bounds with finite test sets

In the following, assume that a classifier with c classes was trained on the training set $\mathcal{D}$ and the error probability will be estimated from the test set $\mathcal{T}$. Again, all elements in the training and test sets are i.i.d. The number of samples in $\mathcal{T}$ is denoted by $|\mathcal{T}| = N_{\mathcal{T}}$. Denote by $\mathcal{T}_j$ the set of the $|\mathcal{T}_j| = N_{\mathcal{T}j}$ test samples that belong to class ω_j. Let P_j denote the (unknown) error probability for class ω_j and let n_j denote the number of incorrectly classified samples in $\mathcal{T}_j$. The probability to incorrectly classifying n_j items of $\mathcal{T}$ is

$$\Pr(n_j \text{ items of } \mathcal{T}_j \text{ misclassified}) = \binom{N_{\mathcal{T}j}}{n_j} P_j^{n_j}(1-P_j)^{N_{\mathcal{T}j}-n_j}. \tag{9.14}$$

Note, again, that this error probability cannot be computed, because the P_j are not known. It can, however, be estimated from the test sample using the maximum likelihood estimate of P_j, $\hat{P}_j = \frac{n_j}{N_{\mathcal{T}j}}$, which leads to the maximum likelihood estimate of the overall error probability of the classifier,

$$\hat{P}_e = \sum_{j=1}^{c} P(\omega_j)\frac{n_j}{N_{\mathcal{T}j}}. \tag{9.15}$$

This estimator is unbiased, because with $\mathrm{E}\{n_j\} = N_{\mathcal{T}j}P_j$ it follows that

$$\mathrm{E}\{\hat{P}_e\} = \sum_{j=1}^{c} P(\omega_j)\frac{\mathrm{E}\{n_j\}}{N_{\mathcal{T}j}} = \sum_{j=1}^{c} P(\omega_j)P_j = P_e. \tag{9.16}$$

The variance is given by

$$\mathrm{Var}\{\hat{P}_e\} = \sum_{j=1}^{c} P^2(\omega_j)\frac{P_j(1-P_j)}{N_{\mathcal{T}j}}. \tag{9.17}$$

This highlights two problems. First, the variance of the estimator is inversely proportional to $N_{\mathcal{T}j}$, which means that small test subsets will result in a relatively high variance of the estimated error probability. Second, the comparison of two different classifiers with respect to their error probability is only valid if the difference in $\hat{P}_e$ is significant w.r.t. the stochastic error of the estimate $\sqrt{\mathrm{Var}\{\hat{P}_e\}}$.

The question, then, is how to choose a good test sample in order to get good estimates $\hat{P}_e$. General approaches to choosing adequate test sample sizes are described by Guyon et al. [1998]. Here, however, we are only interested in choosing the number of test samples $N_{\mathcal{T}}$ so that, with probability $(1-a)$, the true error rate P_e does not exceed the estimated error rate $\hat{P}_e$ by more than some small quantity $\varepsilon(N_{\mathcal{T}}, a)$:

$$\Pr\big(P_e \geq \hat{P}_e + \varepsilon(N_{\mathcal{T}},a)\big) \leq a \quad \text{with} \quad 0 \leq a \leq 1. \tag{9.18}$$

Defining $\varepsilon(N_{\mathcal{T}},a) := \beta P_e$ as a fraction of the true classification error, one can compute suitable test set sizes. For example, for $a = 0.05$ and $\beta = 0.2$, i.e., that there should be a 95 % probability that the true error P_e does not exceed the estimated error $\hat{P}_e$ by more than 20 %, it follows that the number of test samples should be $N_{\mathcal{T}} \approx \frac{100}{P_e}$ (Guyon et al. [1998]). Note that this result is independent of the number of classes c.

9.3.4 Dealing with small datasets

As discussed above, the performance evaluation of a classifier becomes more unreliable when fewer test samples are available. Unfortunately, it is often not feasible or even pos-

$$\hat{P}_e = \frac{1}{5}\sum_{f=1}^{5}\hat{P}_{ef} \qquad \text{Var}\{\hat{P}_e\} \approx \frac{1}{4}\sum_{f=1}^{5}\left(\hat{P}_{ef} - \hat{P}_e\right)^2$$

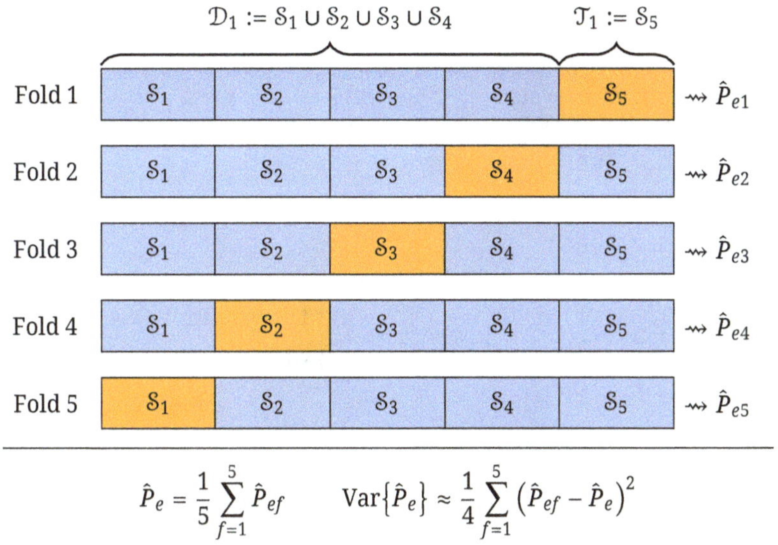

Fig. 9.9: Example of a five-fold cross-validation: The dataset S is partitioned into five equally sized subsets S_i. In each fold, one subset is used as the test set $\mathcal{T}$ and the training set $\mathcal{D}$ is the union of the remaining subsets.

sible to obtain more samples. In such situations, one can use a cross-validation scheme to perform the evaluation.

In an *m*-fold cross-validation, the dataset S is partitioned into m equally sized subsets $S = S_1 \cup S_2 \cup \ldots \cup S_m$ with $S_i \cap S_j = \emptyset$ for $i \neq j$ and $|S_1| \approx |S_2| \approx \ldots \approx |S_m|$.

In the *j*th of the m rounds (also called *folds*), the learning set $\mathcal{D} = S \setminus S_j$ is constructed from all but the *j*th subset, which is used as the test set $\mathcal{T} = S_j$. The classifier is trained using $\mathcal{D}$ and evaluated using $\mathcal{T}$ as usual, and the classification error—or any other performance measure—is recorded. The process is repeated for each $j = 1, \ldots, m$, so that every sample is used for testing once. The final estimate of the test error is taken as the arithmetic mean of the m test errors estimated in the folds. Cross-validation also makes it possible to estimate the variance of the test error estimate. Typical values for the number of folds are $m = 5, 10$. A schematic example of a five-fold cross-validation is shown in Figure 9.9.

A special case occurs if $m = N = |S|$, i.e., when the number of folds is equal to the number of samples. Here, the classifier is trained with all but one sample in S and evaluated on the sample that was left out. This scheme is therefore also known as the leave-one-out cross-validation. The estimates obtained with a leave-one-out cross-validation are typically more reliable than with m-fold cross-validation, but this increased precision is paid for by a much larger evaluation effort.

Algorithm 9.1: The AdaBoost algorithm (Freund and Schapire [1997]).

Data: Training set $\mathcal{D}$, number of iterations M

Result: Strong classifier $k(\mathbf{m}) = \operatorname{sign}\left(\sum_{j=1}^{M} a_j k^j(\mathbf{m})\right)$

$w_i \leftarrow \frac{1}{N}$ **for** $i = 1, \ldots, N$

for $j \leftarrow 1$ **to** M **do**

$k^j \leftarrow \text{train}(\mathcal{D}, \{w_1, \ldots, w_N\})$ `// Train classifier w.r.t. the` w_i

`// Calculate weighted training error`

$$\varepsilon_j \leftarrow \frac{1}{\sum_{i=1}^{N} w_i} \left(\sum_{i=1}^{N} w_i \, \delta_{[z_i \neq k^j(\mathbf{m}_i)]} \right) \text{ where } z_i := \begin{cases} 1 & \text{if } \omega(\mathbf{m}_i) = \omega_1 \\ -1 & \text{if } \omega(\mathbf{m}_i) = \omega_2 \end{cases}$$

$$a_j \leftarrow \log \frac{1 - \varepsilon_j}{\varepsilon_j}$$

$w_i \leftarrow w_i \exp\left(a_j \, \delta_{[z_i \neq k^j(\mathbf{m}_i)]}\right)$ **for** $i = 1, \ldots, N$

return $k(\mathbf{m}) = \operatorname{sign}\left(\sum_{j=1}^{M} a_j k^j(\mathbf{m})\right)$

9.4 Boosting

Boosting is a meta-technique to combine the outcomes of an ensemble of classifiers into a single classification. The underlying idea is that multiple weak classifiers, classifiers that perform marginally better than a random guess, can form a strong classifier. The decision function $k(\mathbf{m})$ of the strong classifier is a weighted sum of those of the weak classifiers $k^j(\mathbf{m})$,

$$k(\mathbf{m}) = \operatorname{sign}\left(\sum_{j=1}^{M} a_j k^j(\mathbf{m}) \right). \tag{9.19}$$

The weights a_j are adapted so that the weak classifiers $k^j(\mathbf{m})$ are weighted according to their classification performance on the training set. Note that in Section 8.2, we have discussed another instance of ensemble methods: random forests. Here, however, the base learners $k^j(\cdot)$ may be any type of classifier, not just decision trees. The only restriction is, again, that the $k^j(\cdot)$ must perform better than random guessing. Note that here we restrict ourselves to binary classification, hence the scalar decision function $k^j(\cdot)$ is used in place of the vectorial $\mathbf{k}^j(\cdot)$.

One of the best-known boosting algorithms is AdaBoost (short for *adaptive boosting*) of Freund and Schapire [1997]. The algorithm is very simple, yet produces very strong classifiers out of the box. The basic idea of AdaBoost is to iteratively generate weak classifiers that minimize the weighted training error on the training set. Initially, all training samples are weighted equally, but after each iteration more weight is put on misclassified samples. Consequently, the classifier chosen in the subsequent iterations

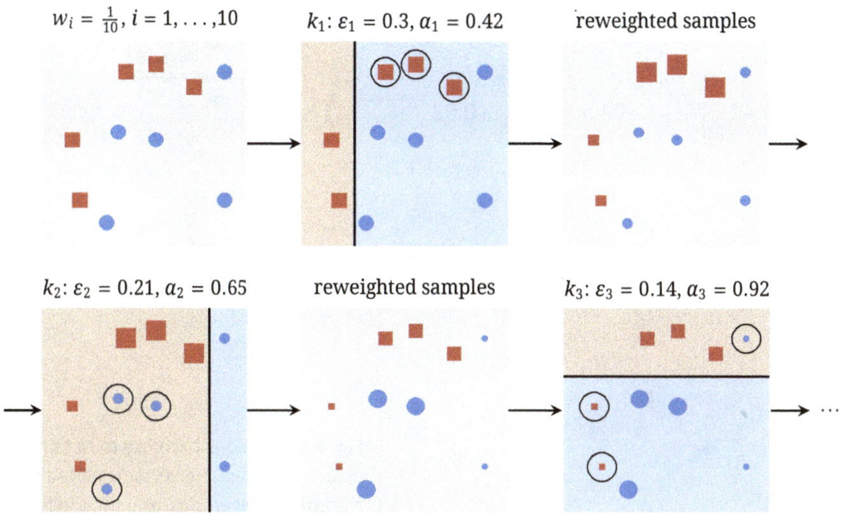

Fig. 9.10: Schematic example of AdaBoost training ($M = 3$) for two classes with the decision functions $k_i(\mathbf{m}) := \text{sign}(m_{d_i} - \tau_i)$ with $d_i \in 1,2$. The size of a mark indicates the magnitude of the associated weight. Example adapted from Freund and Schapire [2004].

will put more emphasis on correctly classifying samples that were often misclassified. Pseudo-code of the algorithm is given in algorithm 9.1. If the weak classifier is chosen from a set of base classifiers, line 3 of the algorithm is replaced with choosing the classifier that minimizes the weighted training error. A visual example of AdaBoost training and the resulting strong classifier is given in Figures 9.10 and 9.11.

Boosting algorithms are meta-algorithms that can be used with many different weak classifiers or even combinations of classifiers of different types. Popular choices are shallow decision trees, i.e., decision trees that are only a few levels deep, and linear SVMs. While simple, the approach is very powerful. Given that the weak classifiers are simple classifiers, the strong classifier is usually very fast to compute. At the same time, boosted classifiers typically generalize well and do not overfit easily. Weak classifiers that perform well on the training set will receive larger weights α_i than classifiers that do not.

The main design parameters are the types of the weak classifiers $k^j(\cdot)$ or a set of pre-selected classifiers to choose from, and the number of iterations M. There is no need for prior knowledge of the distribution of the features, but boosting typically needs large training sets to work well.

$$k(\mathbf{m}) = \text{sign}\left(0.42 \quad k_1 \quad +0.65 \quad k_2 \quad +0.92 \quad \overline{k_3} \quad \right) =$$

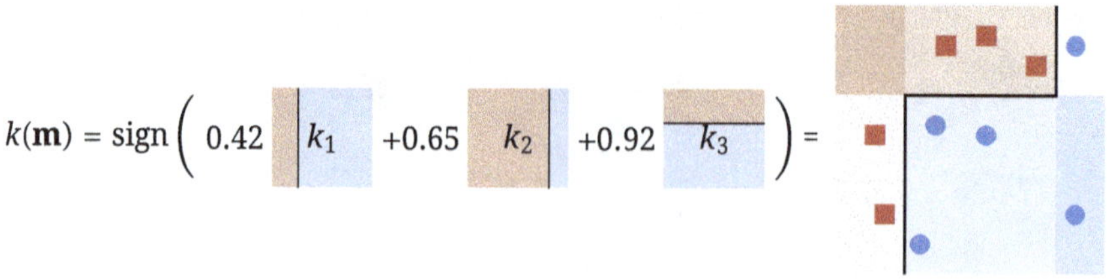

Fig. 9.11: Visual representation of the AdaBoost classifier obtained after three iterations in Figure 9.10.

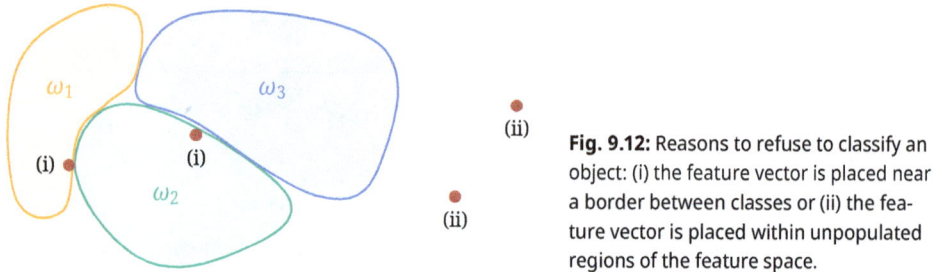

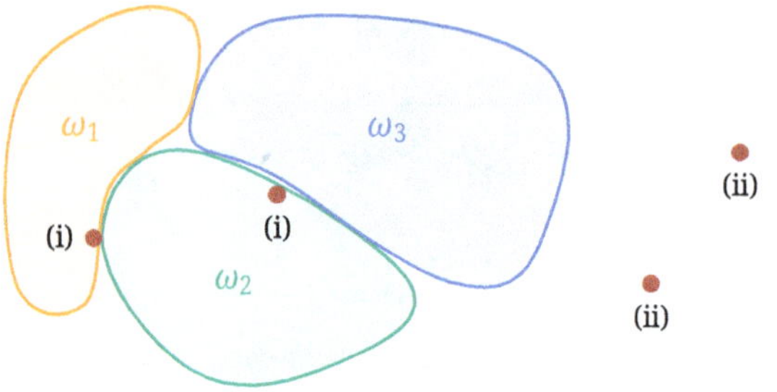

Fig. 9.12: Reasons to refuse to classify an object: (i) the feature vector is placed near a border between classes or (ii) the feature vector is placed within unpopulated regions of the feature space.

9.5 Rejection

In Chapter 1 and throughout the book so far, all objects were assumed to belong to one of the classes $\omega_1, \ldots, \omega_c$. In other words, the relevant part of the world Ω is partitioned into equivalence classes ω_i, $\Omega/\sim = \{\omega_1, \ldots, \omega_c\}$. The underlying (implicit) assumption that all classes are known at design time and that every object $o \in \Omega$ belongs to one of the classes ω_i is called the *closed world assumption*.

In practice, this assumption rarely holds, but it is often close enough to the truth to not cause any harm. An intelligent scale that classifies fruit and vegetables, for example, will work with a model of just these produce items of the supermarket. Other products in the store are not included in the model and will be misclassified when put on the scale, but the harm due to the misclassification is negligible.

In any case, there are at least two valid reasons to refuse a classification:
1. Ambiguous situation: the decision functions k_i do not exhibit a significant maximum.
2. Unknown object: the object lies outside the domain explained by $\Omega/\sim$.

The first case occurs when the feature vector of an object falls near a decision boundary. In other words, a sample should be rejected if the sample falls into a narrow strip around the decision boundaries. The second case occurs if the feature vector falls outside of the area occupied by the feature vectors of the known objects (see Figure 9.12). The intelligent scale, for example, may encounter a nashi pear that looks similar to an apple. Customers might also be tempted to put unrelated items, e.g., a loaf of bread, on the scale to see how the system reacts. In both cases, the task of classification should be declined.

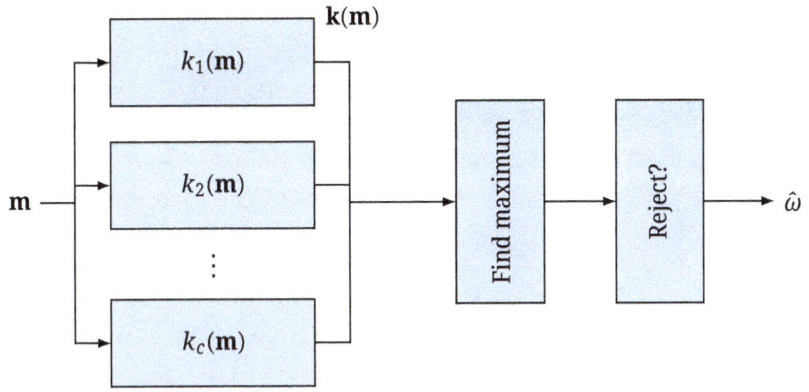

Fig. 9.13: Scheme of a classifier with rejection option. As the rejection option is a subsequent step after maximum search, it can be applied to any classifier.

To treat both cases, the workflow of a classifier (see Figure 3.3) is extended by a subsequent rejection stage, as shown in Figure 9.13. The rejection test might be inherent to the specific classifier, but there are four classifier-independent rejection criteria that work with any classifier that outputs some measure of confidence. These are (due to Schürmann [1996]):

- Maximum criterion: Reject if $\max\{k_i\} < \tau$, i.e., if the decision functions show no significant maximum. The corresponding rejection region in the decision space is shown in Figure 9.14a.
- Difference criterion: Reject if $\max\{k_i\} - \max\{\{k_j\} \setminus \max\{k_i\}\} < \tau$, i.e., the two top ranked decision options have a similar confidence. The corresponding rejection region in the decision space is shown in Figure 9.14b.
- Distance criterion: Reject if $\min\{\|\mathbf{k} - \boldsymbol{\omega}_i\|\} > \tau$, i.e., if the confidence of the closest class in the decision space is not large enough. The corresponding rejection region in the decision space is shown in Figure 9.14c.
- Minimum criterion: Reject if $\min\{k_i\} < \tau < 0$, i.e., if at least one decision function expresses high confidence that the object does *not* belong to any of the classes defined during design time. The corresponding rejection region in the decision space is shown in Figure 9.14d.

Formally, the rejection option can be treated as an additional class ω_0, with which the original class-partition of Ω is augmented (recall Section 3.3):

$$\Omega^0/\sim := \Omega/\sim \cup \{\omega_0\}. \tag{9.20}$$

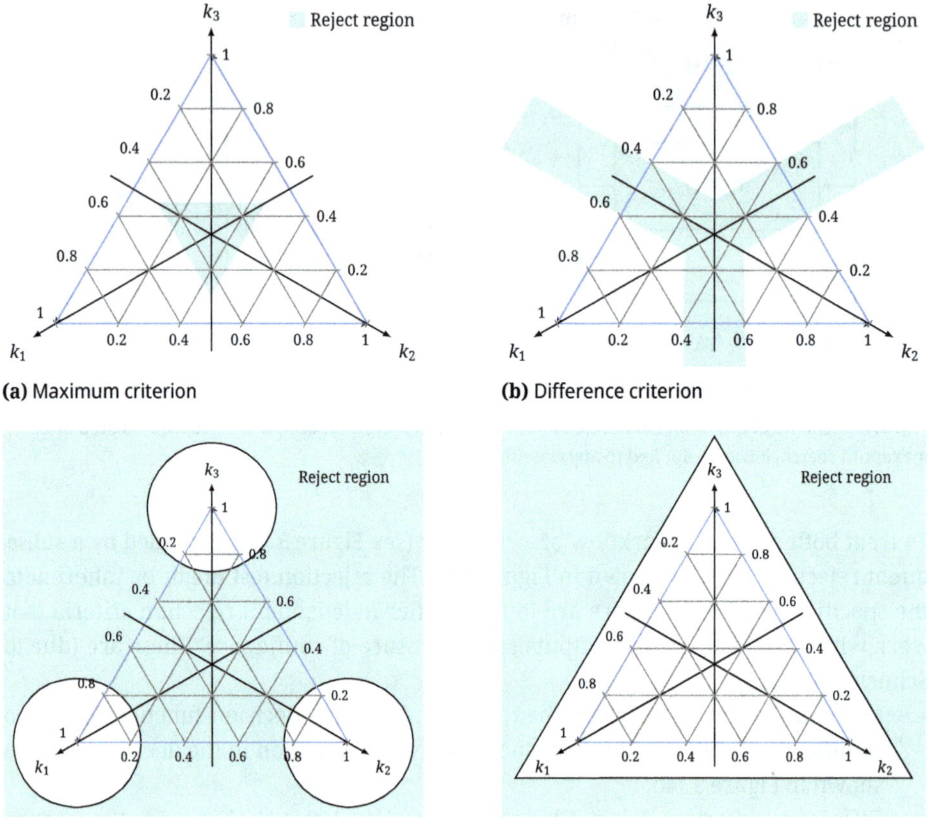

(a) Maximum criterion

(b) Difference criterion

(c) Distance criterion

(d) Minimum criterion

Fig. 9.14: Rejection criteria and the corresponding rejection regions in the decision space.

9.6 Exercises

(9.1) A binary classifier achieves the following results on a test set:

$$\omega_1: \quad N_{\mathcal{T}1} = 20, \quad k_1 = 14,$$
$$\omega_2: \quad N_{\mathcal{T}2} = 30, \quad k_2 = 12.$$

Here, $N_{\mathcal{T}i}$ denotes the number of test samples of class ω_i and k_i denotes the number of correctly classified samples of that class. The a priori probabilities are given by $P(\omega_1) = p$ and $P(\omega_2) = 2p$ for $p \in [0,1]$.

1. Give an estimate $\hat{P}_e$ of the error probability of the classifier.
2. In operation, an error rate of $P_e = 0.4$ is observed. Assuming that the class-dependent error rates are correct, what are the values of the true a priori probabilities?

(9.2) A test of a classifier with three classes $\omega_1, \omega_2, \omega_3$ results in the following confusion matrix:

	True class		
Prediction	ω_1	ω_2	ω_3
ω_1	120	6	3
ω_2	16	21	7
ω_3	8	9	26

Give an estimate $\hat{P}_e$ of the error probability of the classifier. Assume the following a priori probabilities:

$$P(\omega_1) = 0.5, \quad P(\omega_2) = 0.2, \quad P(\omega_3) = 0.3. \tag{9.21}$$

A Solutions to the exercises

A.1 Chapter 1

(1.1) The relation is an equivalence relation:
- Reflexivity: x attends the same class as x.
- Symmetry: If x attends the same class as y, then y also attends the same class as x.
- Transitivity: If x attends the same class as y and y attends the same class as z, then x and z also attend the same class.

(1.2) The relation is not transitive and hence not an equivalence relation, as seen in this family tree:

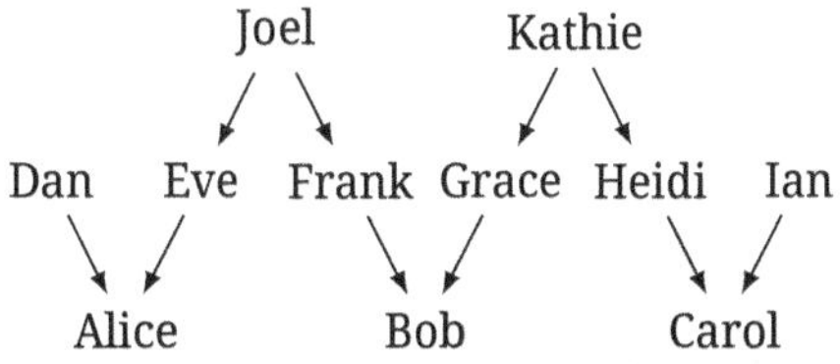

Alice and Bob share a grandparent (Joel), and Bob and Carol share a grandparent (Kathie), but Alice and Carol do not.

(1.3) The relation $\mathbf{x} \sim \mathbf{y} \Leftrightarrow \mathbf{x}^\mathsf{T}\mathbf{y} = 0$ is not reflexive for $\mathbf{x} \neq \mathbf{0}$ and hence not an equivalence relation.

(1.4) $\mathbf{x} \sim \mathbf{y} \Leftrightarrow \mathbf{x}^\mathsf{T}\mathbf{y} \geq 0$ is reflexive and symmetric, but not transitive and hence not an equivalence relation. Let $\mathbf{x} = (0, -1)^\mathsf{T}$, $\mathbf{y} = (1,0)^\mathsf{T}$, and $\mathbf{z} = (0,2)^\mathsf{T}$. Then:

$$\begin{aligned} \mathbf{x}^\mathsf{T}\mathbf{y} &= 0 &&\Rightarrow \mathbf{x} \sim \mathbf{y} \\ \mathbf{y}^\mathsf{T}\mathbf{z} &= 0 &&\Rightarrow \mathbf{y} \sim \mathbf{z} \\ \mathbf{x}^\mathsf{T}\mathbf{z} &= -2 &&\Rightarrow \mathbf{x} \nsim \mathbf{z}. \end{aligned}$$

(1.5) The relation is not an equivalence relation, because symmetry does not hold if $f(x) < f(y)$ (strictly smaller) for any $x,y \in \mathbb{N}$. In this case, $x \sim y$ holds, but $y \nsim x$, because $f(y) \nleq f(x)$.

(1.6) The relation is not symmetric and hence not an equivalence relation:
Let $r(X,n) \in \mathcal{O}(n^a)$ and $r(Y,n) \in \mathcal{O}(a^n)$ for $a > 1$. Then $r(X,n) \in \mathcal{O}(r(Y,n)) = \mathcal{O}(a^n)$ and therefore $X \sim Y$. However, $r(Y,n) \notin \mathcal{O}(r(X,n)) = \mathcal{O}(n^a)$ and thus $Y \nsim X$.

https://doi.org/10.1515/9783111339207-010

A.2 Chapter 2

(2.1) m_{n} allows any relabeling function, i. e., any injective function. m_{o} allows only functions that preserve the ordering. m_{r} allows only functions that also preserve relative distances, i. e., only linear functions. m_{r} allows only functions that also preserve the zero, i. e., only scaling functions. m_{a} allows only the identity mapping.

1. $f(m) = 3\,m + a$ is injective and strictly increasing and therefore allowed for both m_{n} and m_{o}. Since $3 > 0$, it can also be applied to m_{i}, but is only allowed for m_{r} if $a = 0$.
2. m_{n} and m_{o} both allow $f(m) = \mathrm{e}^m$, since the function is injective and strictly increasing. It is not linear and therefore not allowed with either m_{i} or m_{r}.
3. The function is only allowed with m_{n}, as it is injective, but not strictly increasing and not linear.

(2.2) The features are sorted in the scales as below:

- Nominal scale: car brands, genders, varieties of apple.
- Ordinal scale: grades, clothing sizes, places in a race.
- Interval scale: date of birth, motor temperature in °C, intelligence quotient.
- Ratio scale: area of the canvas of a sail, engine speed/revolution, height of body, optical magnification, account balance, electrical voltage, population density, annual income in EUR.
- Absolute scale: number of cows in a herd, display of a Geiger counter.

(2.3) The KL divergence between P_1 and P_2 is

$$\begin{aligned}
D_{\mathrm{KL}}(P_1 \| P_2) &= \sum_{x \in \operatorname{supp}(P_1)} P_1(x) \log \frac{P_1(x)}{P_2(x)} \\
&= P_1(a) \log \frac{P_1(a)}{P_2(a)} + P_1(b) \log \frac{P_1(b)}{P_2(b)} \\
&\quad + P_1(c) \log \frac{P_1(c)}{P_2(c)} + P_1(d) \log \frac{P_1(d)}{P_2(d)} \\
&= \frac{1}{3} \log \frac{1/3}{1/3} + 0 \log \frac{0}{1/6} + \frac{1}{3} \log \frac{1/3}{1/6} + \frac{1}{3} \log \frac{1/3}{1/3} \\
&= \frac{1}{3} \log 1 + \frac{1}{3} \log 2 + \frac{1}{3} \log 1 \\
&= \frac{1}{3} \log 2 \approx 0.23.
\end{aligned}$$

(2.4) The KL divergence between P_1 and P_2

$$D_{\mathrm{KL}}(P_1 \| P_2) = \sum_{i=1}^{3} P_1(i) \log \frac{P_1(i)}{P_2(i)}$$

$$= \frac{1}{3}\log\frac{1/3}{1/6} + \frac{1}{3}\log\frac{1/3}{a} + \frac{1}{3}\log\frac{1/3}{b}$$
$$= \frac{1}{3}(\log 2 - \log 3 - \log a - \log 3 - \log b)$$
$$= \frac{1}{3}(\log 2 - \log 3) - \frac{1}{3}(\log a + \log b)$$
$$= \text{const.} - \frac{1}{3}\log(ab)$$

assumes a minimum if $\log(ab)$ assumes a maximum. Since log is strictly monotonically increasing, it is sufficient to maximize ab. Furthermore, the following restrictions hold:

$$P_2(1) + P_2(2) + P_2(3) = 1$$
$$\Rightarrow a + b = \frac{5}{6} \quad \Leftrightarrow \quad \frac{5}{6} - a.$$

Therefore $ab = a\left(\frac{5}{6} - a\right) = \frac{5}{6}a - a^2$ needs to be maximized. Deriving by a and setting to zero yields

$$-2a + \frac{5}{6} \overset{!}{=} 0 \quad \Leftrightarrow \quad a = b = \frac{5}{12}.$$

(2.5) The distance from the center, e. g., $m' := \sqrt{x^2 + y^2}$, $m' := x^2 + y^2$, or $m' := |x| + |y|$. Other solutions are also possible.

(2.6) The area under the diagonal, i. e., $m' := m_1 - m_2$, or explicitly coded $m' := \delta_{[m_1 < m_2]}$. Other solutions are also possible.

(2.7) The feature m is ...
- ...invariant to scaling, because every term is divided by $|Z_1|$.
- ...not invariant to location, because the location coefficient Z_0 is a summand.
- ...not invariant to rotation, because the phase of the Z_i is not normalized.

(2.8) Translation invariance: $\alpha = 0$.
Scale invariance: $\beta = 2$ and $\gamma = 0$.

(2.9) The following patterns are equivalent under a feature invariant to ...
1. translation: 1 and 2; 3 and 4.
2. translation and rotation: 1 and 2; 3, 4 and 7.
3. translation, rotation and scaling: 1, 2, 5 and 6; 3, 4 and 7.

(2.10) A feature is invariant to translation if the location parameter $\mathbf{c}$ does not influence the feature. Similarly, a feature is invariant to rotation if its computation does not involve the rotation parameter φ. All other measurements—perimeter P, area

A, and axis lengths l_1 and l_2—are sensitive to scaling, but to different degrees; a scaling factor s affects P, l_1 and l_2 linearly, but A will grow quadratically in s. Therefore, one can derive the following invariances:

- m_1 is only invariant to rotation, but not translation or scaling.
- m_2 is invariant to translation, rotation and scaling.
- m_3 is invariant to translation and scaling, but not rotation.
- m_4 is invariant to translation, but not rotation or scaling.
- m_5 is invariant to neither translation, rotation nor scaling.

A.3 Chapter 3

(3.1) $P(a \mid b) + P(\overline{a} \mid b) = 1 \Rightarrow P(\overline{a} \mid b) = 1 - P(a \mid b) = 1 - P(a) = P(\overline{a})$.

(3.2)
1. If the a priori probability distribution is uninformative about the class, i. e., if $P(\omega_i)$ is the same for all classes ω_i.
2. If the class-specific feature distribution is the same for both features, i. e., $p(\mathbf{m}_1 \mid \omega_i) = p(\mathbf{m}_2 \mid \omega_i)$ for all ω_i.

(3.3)
1. ω_1 and ω_2 can not be separated using $\mathbf{m}$. The classification depends solely on the a priori probabilities.
2. ω_3 and ω_2 can be separated without error. Since the class-specific feature distributions for ω_1 and ω_2 are the same, ω_3 and ω_1 can also be perfectly separated.
3. Class-dependent error probabilities:

$$\begin{aligned}
\omega(\mathbf{m}) &= \omega_3\colon \quad P(\hat{\omega} \neq \omega_3 \mid \omega_3) = 0 \\
\omega(\mathbf{m}) &= \omega_2\colon \quad P(\hat{\omega} \neq \omega_2 \mid \omega_2) = 0 \\
\omega(\mathbf{m}) &= \omega_1\colon \quad P(\hat{\omega} \neq \omega_1 \mid \omega_1) = 1
\end{aligned} \Bigg\} \text{ since the classifier always decides on } \omega_2$$

$$\Rightarrow P_e = 0.1 \cdot 1 + 0.6 \cdot 0 + 0.3 \cdot 0 = 0.1.$$

(3.4)
1. Sketch of the feature distributions $p(m \mid \omega)$ and the decision boundaries for parts (2) and (3):

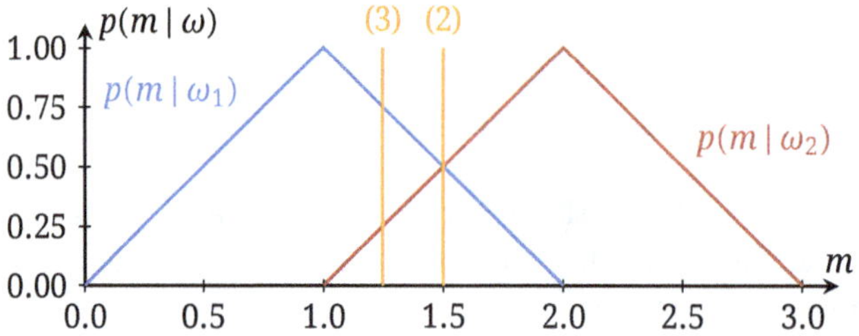

2. With equal a priori probabilities $P(\omega_1) = P(\omega_2)$, the decision boundary is at the point $p(m \mid \omega_1) = p(m \mid \omega_2)$. The supports of the densities overlap in the range $m \in [1,2]$, which gives

$$p(m \mid \omega_1) \overset{!}{=} p(m \mid \omega_2) \;\Leftrightarrow\; 2 - m \overset{!}{=} m - 1 \;\Rightarrow\; m = \frac{3}{2} = 1.5.$$

3. With unequal a priori probabilities, it holds that

$$\begin{aligned} \hat{\omega} = \omega_1 \;&\Leftrightarrow\; P(\omega_1 \mid m) > P(\omega_2 \mid m) \\ &\Leftrightarrow p(m \mid \omega_1)P(\omega_1) > p(m \mid \omega_2)P(\omega_2). \end{aligned}$$

The decision boundary is therefore at

$$\begin{aligned} & p(m \mid \omega_1)P(\omega_1) = p(m \mid \omega_2)P(\omega_2)) \\ \Leftrightarrow\; & \frac{1}{4}(2 - m) = \frac{3}{4}(m - 1) \\ \Leftrightarrow\; & \frac{2}{4} - \frac{m}{4} = \frac{3m}{4} - \frac{3}{4} \;\Rightarrow\; m = \frac{5}{4} = 1.25. \end{aligned}$$

4. The error probability with two classes is

$$P_e = \int_{\mathcal{R}_1} P(\omega_2)p(m \mid \omega_2)\,\mathrm{d}m + \int_{\mathcal{R}_2} P(\omega_1)p(m \mid \omega_1)\,\mathrm{d}m.$$

For the first boundary (part 2)

$$\begin{aligned} P_e &= \int_0^{\frac{3}{2}} P(\omega_2)p(m \mid \omega_2)\,\mathrm{d}m + \int_{\frac{3}{2}}^{3} P(\omega_1)p(m \mid \omega_1)\,\mathrm{d}m \\ &= \int_1^{\frac{3}{2}} \frac{1}{2}(m - 1)\,\mathrm{d}m + \int_{\frac{3}{2}}^{2} \frac{1}{2}(2 - m)\,\mathrm{d}m \\ &= \frac{1}{2}\left(\frac{1}{8} + \frac{1}{8}\right) = \frac{1}{8}. \end{aligned}$$

For the second boundary (part 3)

$$\begin{aligned} P_e &= \int_0^{\frac{5}{4}} P(\omega_2)p(m \mid \omega_2)\,\mathrm{d}m + \int_{\frac{5}{4}}^{3} P(\omega_1)p(m \mid \omega_1)\,\mathrm{d}m \\ &= \int_1^{\frac{5}{4}} \frac{3}{4}(m - 1)\,\mathrm{d}m + \int_{\frac{5}{4}}^{2} \frac{1}{4}(2 - m)\,\mathrm{d}m \\ &= \frac{3}{4} \cdot \frac{1}{32} + \frac{1}{4} \cdot \frac{9}{32} = \frac{3}{32}. \end{aligned}$$

(3.5) 1. $P(\omega_1) = P(\omega_2) = \frac{1}{2}(1 - P(\omega_3) - P(\omega_4)) = 0.1$.
2. Class ω_4 will be chosen, since it has highest a priori probability of all classes.
3. ω_1 and ω_4 have the highest class-specific feature densities at m_1. Since $P(\omega_4) > P(\omega_1)$,

$$p(m_1 \mid \omega_1)P(\omega_1) < p(m_1 \mid \omega_4)P(\omega_4) \Leftrightarrow P(\omega_1 \mid m_1) < P(\omega_4 \mid m_1)$$

and hence ω_4 is chosen again.
4. ω_1 and ω_4 as well as ω_2 and ω_4 are best separated using only m_2. ω_1 and ω_2 cannot be separated, since the a priori probability and the class-specific feature distributions, and hence the a posteriori probabilities, are the same for these classes.
5. Since ω_1 and ω_2 can not be separated using m_2, but they can be separated using m_1, one should use m_1 instead of m_2.

(3.6) As m_1 and m_2 are stochastically independent,

$$p(\mathbf{m}) = p(m_1,m_2) = p(m_1)p(m_2) \quad \text{and}$$
$$p(m_1 \mid m_2,\omega) = p(m_1 \mid \omega).$$

Therefore

$$\begin{aligned} P(\omega \mid \mathbf{m}) &= \frac{p(\mathbf{m} \mid \omega)P(\omega)}{p(\mathbf{m})} = \frac{p(m_1,m_2 \mid \omega)P(\omega)}{p(m_1,m_2)} \\ &= \frac{p(m_1 \mid m_2,\omega)p(m_2 \mid \omega)P(\omega)}{p(m_1)p(m_2)} = \frac{p(m_1 \mid \omega)p(m_2 \mid \omega)P(\omega)}{p(m_1)p(m_2)}. \end{aligned}$$

(3.7) 1. Cost matrix $\mathbf{L}$:

$$\mathbf{L} = \begin{pmatrix} l(\omega_1,\omega_1) & l(\omega_1,\omega_2) & l(\omega_1,\omega_3) \\ l(\omega_2,\omega_1) & l(\omega_2,\omega_2) & l(\omega_2,\omega_3) \\ l(\omega_3,\omega_1) & l(\omega_3,\omega_2) & l(\omega_3,\omega_3) \end{pmatrix} = \begin{pmatrix} 0 & 1 & \frac{1}{100} \\ \frac{1}{100\,000} & 0 & 0 \\ \frac{1}{100\,000} & 0 & 0 \end{pmatrix}.$$

The costs $l(\omega_1,\omega_2)$ and $l(\omega_1,\omega_3)$ are directly taken from the description. The costs $l(\omega_2,\omega_1)$ and $l(\omega_3,\omega_1)$ are inferred from the value of one grain: 1/100 000.
2. According to the description, the a priori probabilities are

$$P(\omega_1) = \frac{97}{100}, \qquad P(\omega_2) = \frac{2}{100}, \qquad P(\omega_3) = \frac{1}{100}.$$

The a posteriori probabilities $\mathbf{p}$ are therefore

$$\begin{aligned} p_1 &= \Pr(\omega_1 \mid \text{length} < 7\,\text{mm}) = \frac{90}{100} \cdot \frac{97}{100} \cdot \frac{10\,000}{90 \cdot 97 + 3 \cdot 2 + 5 \cdot 1}, \\ p_2 &= \Pr(\omega_2 \mid \text{length} < 7\,\text{mm}) = \frac{3}{100} \cdot \frac{2}{100} \cdot \frac{10\,000}{90 \cdot 97 + 3 \cdot 2 + 5 \cdot 1}, \\ p_3 &= \Pr(\omega_3 \mid \text{length} < 7\,\text{mm}) = \frac{5}{100} \cdot \frac{1}{100} \cdot \underbrace{\frac{10\,000}{90 \cdot 97 + 3 \cdot 2 + 5 \cdot 1}}_{\text{normalization constant}}. \end{aligned}$$

The classifier will choose class ω_1.

3. The a posteriori risk is computed to be $\mathbf{r} = (r_1, r_2, r_3)^{\mathsf{T}} = \mathbf{L}\mathbf{p}$. Now

$$\begin{aligned}
r_1 &= \frac{10\,000}{90 \cdot 97 + 2 \cdot 3 + 5} \cdot \left(1 \cdot \frac{3}{100} \cdot \frac{2}{100} + \frac{1}{100} \cdot \frac{5}{100} \cdot \frac{1}{100}\right) \\
&= \frac{10\,000}{90 \cdot 97 + 2 \cdot 3 + 5} \cdot \frac{602}{1\,000\,000}, \\
r_2 &= \frac{10\,000}{90 \cdot 97 + 2 \cdot 3 + 5} \cdot \frac{1}{100\,000} \cdot \frac{90}{100} \cdot \frac{97}{100} \\
&= \frac{10\,000}{90 \cdot 97 + 2 \cdot 3 + 5} \cdot \frac{8730}{1\,000\,000\,000}, \\
r_3 &= \frac{10\,000}{90 \cdot 97 + 2 \cdot 3 + 5} \cdot \frac{1}{100\,000} \cdot \frac{90}{100} \cdot \frac{97}{100} \\
&= \frac{10\,000}{90 \cdot 97 + 2 \cdot 3 + 5} \cdot \frac{8730}{1\,000\,000\,000}.
\end{aligned}$$

It is easy to see that $r_1 > r_2 = r_3$. The classifier will choose class ω_2 or ω_3—the opposite result of the maximum a posteriori classification!

A.4 Chapter 4

(4.1) Estimator $\hat{m}_1$ has $\mathrm{E}\{\hat{m}_1\} = 15$ and $\mathrm{Var}\{\hat{m}_1\} = 0$. Therefore

$$\mathrm{MSE}(\hat{m}_1) = (\mathrm{E}\{\hat{m}_1\} - m)^2 + \mathrm{Var}\{\hat{m}_1\} = (15 - m)^2,$$
$$\mathrm{MSE}(\hat{m}_2) = (\mathrm{E}\{\hat{m}_2\} - m)^2 + \mathrm{Var}\{\hat{m}_2\} = 36.$$

Since the weight m is in the interval [10,20] and the maximum of $\mathrm{MSE}(\hat{m}_1)$ in this interval is 25 (at $m = 10$ and $m = 20$), the estimator $\hat{m}_1$ has lower mean squared error than $\hat{m}_2$.

(4.2) To find the maximum likelihood estimator, we maximize the log-likelihood function:

$$\begin{aligned}
l(\mu) &= \log \prod_{i=1}^{N} \frac{1}{2\sigma} \exp\left(-\frac{|m_i - \mu|}{\sigma}\right) = \sum_{i=1}^{N} \left(\log \frac{1}{2\sigma} - \frac{|m_i - \mu|}{\sigma}\right) \\
&= -N \log(2\sigma) - \frac{1}{\sigma} \sum_{i=1}^{N} |m_i - \mu| \\
\Rightarrow \mu_{\mathrm{ML}} &= \arg\max_{\mu} l(\mu) = \arg\max_{\mu} \left\{-N \log(2\sigma) - \frac{1}{\sigma} \sum_{i=1}^{N} |m_i - \mu|\right\} \\
&= \arg\max_{\mu} \left\{-\frac{1}{\sigma} \sum_{i=1}^{N} |m_i - \mu|\right\} = \arg\max_{\mu} \left\{-\sum_{i=1}^{N} |m_i - \mu|\right\} \\
&= \arg\min \mu \left\{\sum_{i=1}^{N} |m_i - \mu|\right\} = \hat{\mu}.
\end{aligned}$$

(4.3) Again, we need to maximize the log-likelihood

$$l(\theta) = \log \prod_{i=1}^{N} \theta e^{-\theta x_i} = \sum_{i=1}^{N} (\log\theta - \theta x_i)$$
$$= N \log\theta - \theta \sum_{i=1}^{N} x_i$$

to find the maximum likelihood estimator $\hat{\theta}_{\mathrm{ML}}$. Differentiating $l(\theta)$ by θ and setting to zero yields

$$\frac{\mathrm{d}}{\mathrm{d}\theta} l(\theta) = \frac{N}{\theta} - \sum_{i=1}^{N} x_i \overset{!}{=} 0$$
$$\Leftrightarrow \frac{N}{\theta} = \sum_{i=1}^{N} x_i \Rightarrow \hat{\theta}_{\mathrm{ML}} = \frac{N}{\sum_{i=1}^{N} x_i}.$$

(4.4) 1. $\hat{\mu}$ is unbiased iff $\mathrm{E}\{\hat{\mu}\} = \mu$:

$$\mathrm{E}\{\hat{\mu}\} = \mathrm{E}\left\{\frac{1}{N-4}\sum_{i=3}^{N-2} x_i\right\} = \frac{1}{N-4}\sum_{i=3}^{N-2} \mathrm{E}\{x_i\}$$
$$= \frac{1}{N-4}(N-4)\,\mu = \mu.$$

2. Both estimators are unbiased, but the variance of $\hat{\mu}$ is larger than the variance of $\hat{\mu}_{\mathrm{ML}}$ and therefore $\hat{\mu}_{\mathrm{ML}}$ is a better estimator. This can be shown using Equation (4.23):

$$\mathrm{Var}\{\hat{\mu}\} = \mathrm{Var}\left\{\frac{1}{N-4}\sum_{i=3}^{N-2} x_i\right\} = \frac{1}{(N-4)^2}\sum_{i=3}^{N-2} \mathrm{E}\left\{(x_i-\mu)^2\right\}$$
$$= \frac{1}{(N-4)^2}\sum_{i=3}^{N-2} \sigma^2 = \frac{1}{N-4}\sigma^2 > \frac{1}{N}\sigma_2 = \mathrm{Var}\{\hat{\mu}_{\mathrm{ML}}\}.$$

(4.5) The expected value of the estimator $\widehat{\sigma^2}$ is

$$\mathrm{E}\left\{\widehat{\sigma^2}\right\} = \frac{1}{a-N}\sum_{i=1}^{N} \mathrm{E}\left\{(m_i-\mu)^2\right\} \qquad (\mathrm{E}\{\cdot\}\text{ linear})$$
$$= \frac{1}{a-N}\sum_{i=1}^{N} \mathrm{E}\left\{(m-\mu)^2\right\} \qquad (m_i\text{ i. i. d.})$$
$$= \frac{1}{a-N}\cdot N \cdot \mathrm{Var}\{m\} = \frac{N}{a-N}\cdot\sigma^2.$$

In order for the estimator to be unbiased, the expected value must be equal to the true value, which gives

$$\mathrm{E}\left\{\widehat{\sigma^2}\right\} = \sigma^2 \Leftrightarrow \frac{N}{a-N} = 1 \Leftrightarrow a = 2N.$$

(4.6) The expected value of the estimator $\hat{\mu}$ is

$$\mathrm{E}\{\hat{\mu}\} = \mathrm{E}\left\{\frac{N}{N-\alpha}\sum_{i=1}^{N} f(m_i)\right\} = \frac{N}{N-\alpha}\sum_{i=1}^{N}\mathrm{E}\{f(m_i)\}$$
$$= \frac{N}{N-\alpha}\sum_{i=1}^{N}\mathrm{E}\{f(m)\} = \frac{N^2\,\mu}{N-\alpha}.$$

It is unbiased if

$$\mathrm{E}\{\hat{\mu}\} \overset{!}{=} \mu \;\Leftrightarrow\; \frac{N^2\,\mu}{N-\alpha} = \mu \;\Rightarrow\; N^2 = N-\alpha \;\Leftrightarrow\; \alpha = N - N^2.$$

A.5 Chapter 5

(5.1) The inverse mapping $x = A^{-1}(y) = \pm\sqrt{y}$ is not unique, therefore the inference from y to x is ill-posed.

(5.2) The system of equations to solve this problem is over-determined, since there are more data points ($N > 2$) than degrees of freedom (a,b). This means that in general there is no solution that interpolates all data points. Therefore, the problem is ill-posed.
The following variation is well-posed.

Find the parameters $a,b \in \mathbb{R}$ of a straight line $y = a\,x + b$ that minimizes the distance between the line and the data points, i. e., find the parameters $a,b \in \mathbb{R}$ that minimize

$$\sum_{i=1}^{N} d(y_i, a\,x_i + b)$$

for a suitable distance measure, e. g., $d(u,v) = (u-v)^2$.

(5.3) Parzen window estimation:

$$\hat{p}(x) = \frac{1}{N}\sum_{i=1}^{N}\frac{1}{V_N}\varphi\left(\frac{x-x_i}{h_N}\right) \quad \text{where} \quad \varphi(y) := \begin{cases} 1 & \text{if } |y| \le 0.5 \\ 0 & \text{else} \end{cases}.$$

Here, $V_N = h_N^1 = 1$, and hence

$$\hat{p}(6) = \frac{1}{10}\cdot 0 = 0$$
$$\hat{p}(8) = \frac{1}{10}\cdot 2 = 0.2$$
$$\hat{p}(10) = \frac{1}{10}\cdot 3 = 0.3$$
$$\hat{p}(12) = \frac{1}{10}\cdot 1 = 0.1$$
$$\hat{p}(14) = \frac{1}{10}\cdot 0 = 0.$$

(5.4) The dataset can be sorted to quickly find the nearest neighbors to a given m: $\mathcal{D} = \{7.1, 7.6, 8.0, 8.5, 9.3, 9.7, 10.0, 10.5, 12.2, 14.9\}$.
The density is estimated according to $\hat{p}(m) = \frac{k}{NV}$. The volume V depends on the position of the neighbors, $V(m) = 2 \cdot |n_3(m) - m|$, where $n_3(m)$ denotes the third-closest neighbor of m). Putting everything together,

$$\hat{p}(m=6) = \frac{3}{10 \cdot 2 \cdot |8-6|} = \frac{3}{40}$$
$$\hat{p}(m=8) = \frac{3}{10 \cdot 2 \cdot |8.5-8|} = \frac{3}{10}$$
$$\hat{p}(m=10) = \frac{3}{10 \cdot 2 \cdot |10.5-10|} = \frac{3}{10}$$
$$\hat{p}(m=12) = \frac{3}{10 \cdot 2 \cdot |10-12|} = \frac{3}{40}$$
$$\hat{p}(m=14) = \frac{3}{10 \cdot 2 \cdot |10.5-14|} = \frac{3}{70}.$$

(5.5) The feature space, sample, decision boundary, and samples to classify, are shown in the diagram below:

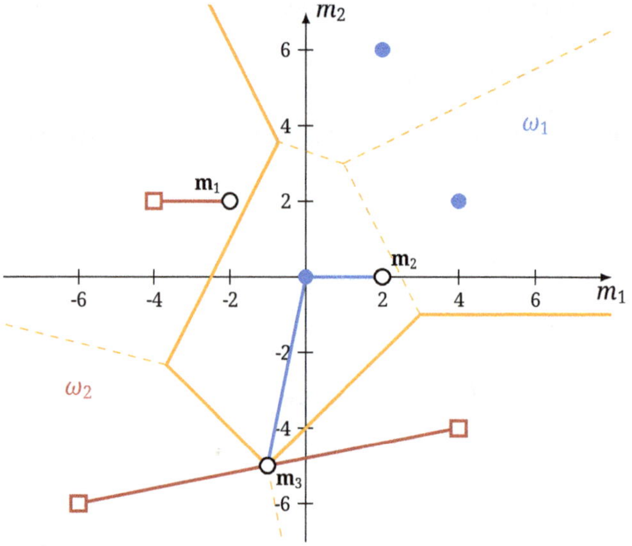

$$\Rightarrow \hat{\omega}(\mathbf{m}_1) = \omega_2; \quad \hat{\omega}(\mathbf{m}_2) = \omega_1; \quad \hat{\omega}(\mathbf{m}_3) = \omega_1 \text{ or } \hat{\omega}(\mathbf{m}_3) = \omega_2.$$

A.6 Chapter 6

(6.1) Each of the three Gaussian components $g_k(\mathbf{m})$ is parametrized by a mean $\boldsymbol{\mu}_k$ and a covariance matrix $\boldsymbol{\Sigma}_k$. The mean requires five (5) parameters since the feature space is five-dimensional. The covariance matrix is symmetric and requires estimating fifteen (15) parameters. Two parameters are needed to estimate the α_k, since $\alpha_1 + \alpha_2 + \alpha_3 = 1$. In all, there are $3 \cdot (5 + 15) + 2 = 62$ parameters to estimate for the Gaussian mixture.
The Parzen window method does not require estimating any parameters. The meta-parameters (window type, window size) are chosen beforehand.

(6.2) The linear classifier needs to estimate four (4) parameters: three for the normal of the hyperplane $\mathbf{w}$ and one for the distance to the origin b.
For the Gaussian classifier, four (4) parameters are needed for each mean and ten (10) parameters are needed for the covariance matrices, so there are $2 \cdot (4 + 10) = 28$ parameters to estimate in all. The a priori probabilities are not estimated from the sample.

(6.3)
1. A linear classifier requires d parameters to define the hyperplane in a d-dimensional feature space. Here, $d = 6$, and hence $\lfloor 256/6 \rfloor = 42$ classifiers can be saved on the device. $256 - 42 \cdot 6 = 4$ parameters remain for other use.
2. A multivariate Gaussian distribution requires d parameters for the mean and $\frac{d(d+1)}{2}$ parameters for the covariance matrix. In addition, the a priori probabilities have to be stored, but since $\sum_{i=1}^{c} P(\omega_i) = 1$, only $c-1$ parameters are needed for c classes. In all, the required number of parameters is $c \cdot \left(d + \frac{d(d+1)}{2} + 1\right) - 1$, which yields the inequality

$$c \cdot \left(6 + \frac{6 \cdot 7}{2} + 1\right) - 1 = 28\,c - 1 < 256 \;\Leftrightarrow\; c < \frac{257}{28} = 9 + \frac{5}{28}.$$

All in all, $c = 9$ classes can be separated, at a maximum, using this device. $256 - 9 \cdot 28 + 1 = 5$ parameters remain unused.

(6.4) The probability that the feature m_i lies in the interval $[-2,5]$ is

$$\Pr\left(m_i \in [-2,5]\right) = \frac{5 - (-2)}{11 - (-10)} = \frac{7}{21} = \frac{1}{3}.$$

Since the m_i are stochastically independent, the probability that $\mathbf{m} \notin [-2,5]^d$ is

$$\Pr\left(\mathbf{m} \notin [-2,5]^d\right) = 1 - \Pr\left(\mathbf{m} \in [-2,5]^d\right) = 1 - \left(\frac{1}{3}\right)^d =: P_d.$$

Plugging in different values for d yields

$$d = 1 \Rightarrow P_1 = 1 - \frac{1}{3} = \frac{2}{3} \ngtr \frac{9}{10},$$
$$d = 2 \Rightarrow P_2 = 1 - \frac{1}{9} = \frac{8}{9} \ngtr \frac{9}{10},$$
$$d = 3 \Rightarrow P_3 = 1 - \frac{1}{27} = \frac{26}{27} > \frac{9}{10}.$$

Therefore, more than 90 % of the probability mass is outside the hypercube $[-2,5]^d$ when the dimensionality of the feature space is at least $d = 3$.

A.7 Chapter 7

(7.1) $K(\mathbf{m},\mathbf{m}') = \|\mathbf{m}\|_2$ does not depend on $\mathbf{m}'$, hence it is not symmetric and not a kernel function.

(7.2) $K(\mathbf{m},\mathbf{m}') = 4\mathbf{m}^\top\mathbf{m}' - \mathbf{m}^\top\mathbf{m} - (\mathbf{m}')^\top\mathbf{m}'$ is symmetric, but not positive definite, and therefore not a kernel function:

$$K\left(\begin{pmatrix}1\\0\end{pmatrix}, \begin{pmatrix}0\\1\end{pmatrix}\right) = 4 \cdot 0 - 1 - 1 = -2 < 0$$
$$K\left(\begin{pmatrix}1\\0\end{pmatrix}, \begin{pmatrix}1\\1\end{pmatrix}\right) = 4 - 1 - 2 = 1 > 0.$$

(7.3) The kernel function is the scalar product of the lifted features:

$$K(\mathbf{p},\mathbf{q}) = \langle \varphi(\mathbf{p}), \varphi(\mathbf{q}) \rangle = (p_1\, p_2)(q_1\, q_2) + (p_2\, p_3)(q_2\, q_3) + (p_3\, p_1)(q_3\, q_1) + (p_1 + p_2 + p_3)^3(q_1 + q_2 + q_3)^3.$$

(7.4) 1. Sketch of the features and decision boundary. Note that there are no feature vectors inside the margin, since this is a hard margin SVM:

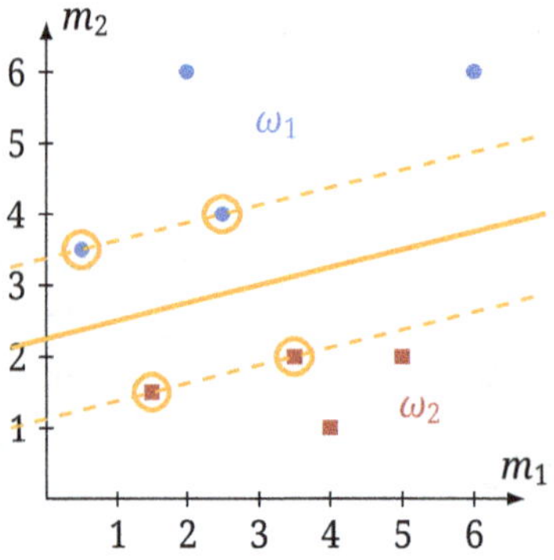

2. Line equation from the diagram:

$$
\begin{aligned}
& m_2 = \frac{1}{4} m_1 + 2.25 \;\Leftrightarrow\; 4\,m_2 = m_1 + 9 \\
\Leftrightarrow\;& m_1 - 4\,m_2 + 9 = 0 \;\Leftrightarrow\; \begin{pmatrix} 1 & -4 \end{pmatrix} \begin{pmatrix} m_1 \\ m_2 \end{pmatrix} + 9 = 0 \\
\Rightarrow\;& \mathbf{w} = \begin{pmatrix} 1 \\ -4 \end{pmatrix}, b = 9.
\end{aligned}
$$

Note: Any solution $\alpha(\mathbf{w}^\top \mathbf{m} - b) = \alpha \mathbf{w}^\top \mathbf{m} - \alpha b = 0$ with $\alpha \in \mathbb{R} \setminus \{0\}$ is correct.

3. $\hat{P}_e = \frac{|\mathcal{SV}|}{N} = \frac{4}{8} = 0.5$.
4. The number of support vectors stays the same, but $N = 20$, hence $\hat{P}_e = \frac{4}{20} = 0.2$.

A.8 Chapter 8

(8.1) One possible solution is the following tree:

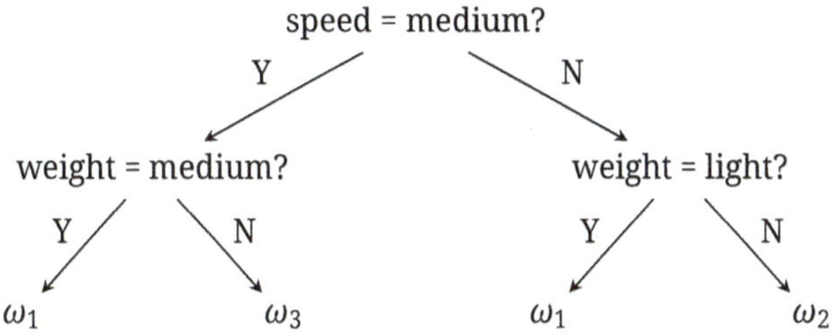

(8.2) The decision boundary is shown in the following sketch:

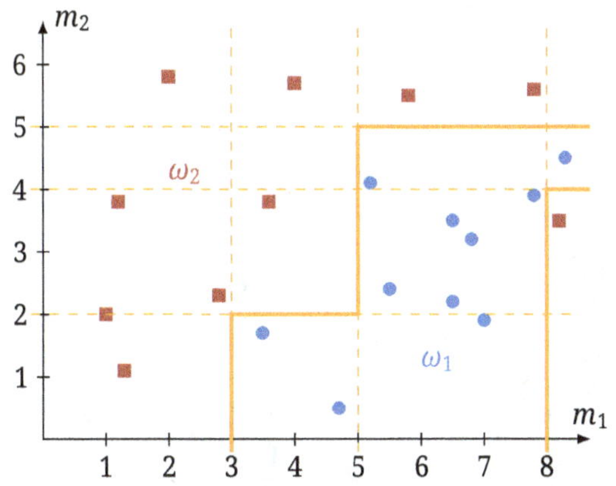

Overfitting occurs in the region $m_1 > 8$, i. e., in the partial tree reached by N→Y→N: here, both leaves contain only one sample.
This overfitting is eliminated by replacing the partial tree reached by N→Y with a leaf node ω_1, yielding the following decision tree:

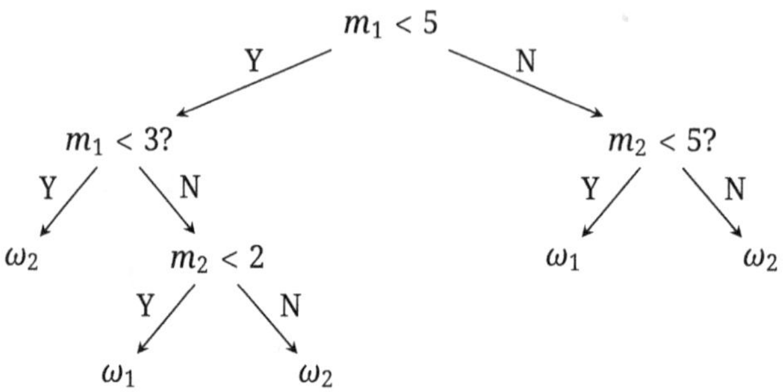

This tree misclassifies one of the training samples.

A.9 Chapter 9

(9.1) 1. $P(\omega_1) + P(\omega_2) \stackrel{!}{=} 1 \Rightarrow P(\omega_1) = \frac{1}{3}, P(\omega_2) = \frac{2}{3}$. From Equation (9.16) it follows:

$$\hat{P}_e = \sum_{i=1}^{2} P(\omega_i) \cdot \frac{N_{\mathcal{T}i} - k_i}{N_{\mathcal{T}i}} = \frac{1}{3} \cdot \frac{6}{20} + \frac{2}{3} \cdot \frac{18}{30} = \frac{1}{2}.$$

2. Approach: calculate $P(\omega_1)$ (and hence $P(\omega_2)$) from the observed error rate $P_e = P(\omega_1)\frac{n_1}{N_{\mathcal{T}1}} + (1 - P(\omega_1))\frac{n_2}{N_{\mathcal{T}2}}$:

$$\begin{aligned} & P(\omega_1) \cdot \frac{6}{20} + (1 - P(\omega_1)) \cdot \frac{18}{30} \stackrel{!}{=} \frac{4}{10} \\ \Leftrightarrow \quad & \frac{6}{10} - \frac{3}{10} \cdot P(\omega_1) \stackrel{!}{=} \frac{4}{10} \\ \Rightarrow \quad & P(\omega_1) = \frac{2}{3} \quad \Rightarrow \quad P(\omega_2) = \frac{1}{3}. \end{aligned}$$

(9.2) The numbers of testing samples per class are $N_{\mathcal{T}1} = 144$, $N_{\mathcal{T}2} = 36$, and $N_{\mathcal{T}3} = 36$. The class-dependent error probabilities are estimated as

$$\hat{P}(\overline{\omega_1} \mid \omega_1) = \frac{16 + 8}{144}, \quad \hat{P}(\overline{\omega_2} \mid \omega_2) = \frac{6 + 9}{36}, \quad \hat{P}(\overline{\omega_3} \mid \omega_3) = \frac{3 + 7}{36}.$$

Putting both together yields

$$\begin{aligned} \hat{P}_e &= P(\omega_1)\, \hat{P}(\overline{\omega_1} \mid \omega_1) + P(\omega_2)\, \hat{P}(\overline{\omega_2} \mid \omega_2) + P(\omega_3)\, \hat{P}(\overline{\omega_3} \mid \omega_3) \\ &= \frac{1}{2} \cdot \frac{24}{144} + \frac{1}{5} \cdot \frac{15}{36} + \frac{3}{10} \cdot \frac{10}{36} = \frac{1}{12} + \frac{1}{12} + \frac{1}{12} = \frac{1}{4} = 0.25. \end{aligned}$$

B A primer on Lie theory

The tangential distance in Section 2.4.8 as well as the construction of invariant features in Section 2.6.3 used concepts from Lie theory, but gave no formal introduction of these concepts. This section will give a concise introduction to the postponed mathematical details.

Definition B.1 (Topological manifold). Let Π be a Hausdorff space, i. e., a set of points with a system of open sets such that for each pair of two distinct points the points can be placed in two disjoint open sets.

1. A *chart* (or *coordinate chart* or *coordinate map*) is a pair $(\mathfrak{U}, \varphi)$ of an open subset $\mathfrak{U} \subseteq \Pi$ and a corresponding injective map $\varphi : \mathfrak{U} \to \mathcal{V}$ to an open subset $\mathcal{V} \subseteq \mathbb{R}^d$ of Euclidean space such that φ is a *homeomorphism*. To be a homeomorphism means that φ and φ^{-1} are both continuous, i. e., the pre-image of an open set is an open set.
2. Let $(\mathfrak{U}_i, \varphi_i), (\mathfrak{U}_j, \varphi_j), i \neq j$ be two charts with a nonempty intersection $\mathfrak{U}_{ij} = \mathfrak{U}_i \cap \mathfrak{U}_j \neq \emptyset$ and let $\overline{\varphi}_i$ and $\overline{\varphi}_j$ denote the restrictions of φ_i and φ_j to the intersection $\mathfrak{U}_{ij}$. The map
$$\tau = \overline{\varphi}_i \circ \overline{\varphi}_j^{-1} : \overline{\varphi}_i(\mathfrak{U}_{ij}) \to \overline{\varphi}_j(\mathfrak{U}_{ij}) \tag{B.1}$$
is called a *transition map*. As φ_i and φ_j are homeomorphisms, τ is also a homeomorphism on $\mathbb{R}^d$.
3. A system $\mathfrak{A} = \{(\mathfrak{U}, \varphi) | \mathfrak{U} \subseteq \Pi \text{ open}\}$ is called an *atlas* of Π iff there is a chart $(\mathfrak{U}, \varphi) \in \mathfrak{A}$ for every point $\mathbf{p} \in \Pi$ such that $\mathbf{p} \in \mathfrak{U}$.
4. A Haussdorff space Π together with an atlas $\mathfrak{A}$ is called a (topological) *manifold*.

The above definition looks rather complicated, but can be understood intuitively—the phrases “coordinate map” and “atlas” were not chosen by chance. A (topological) manifold is a set that locally looks like Euclidean space.

The canonical example is a globe: a sphere and a plane have different global geometries, but they look the same when you zoom in close enough. It is possible to choose an open neighborhood on the sphere and map it to the Euclidean plane the same way as one can flatten pieces of an orange peel. Otherwise, it would not be possible to transfer a map printed on a flat sheet of paper onto a spherical globe.

On a grander scale, the earth looks flat (apart from the occasional hill or valley) from a human perspective, but from outer space, it is clear that it is (approximately) a sphere.

Most of the above definition is necessary to ensure that every point of the manifold is on some map and that the same point on the manifold looks similar on different maps. For example, two maps of Western and Middle Europe both include Germany, but in different places. Still, the border of the country should look approximately the same on both maps.

As a manifold locally looks like Euclidean space, many properties of Euclidean space can be transfered to manifolds by virtue of the charts. One example is the idea of a tangent space (used in Section 2.4.8), which requires some notion of differentiation.

https://doi.org/10.1515/9783111339207-011

Definition B.2 (Smooth manifold, diffeomorphism). Let Π be a (topological) manifold and $\mathfrak{A} = \{(\mathcal{U}, \varphi) | \mathcal{U} \subseteq \Pi \text{ open}\}$ its atlas.

1. Π is a *smooth manifold* iff the transition maps τ_{ij} of each pair of chart functions φ_i, φ_j are smooth, i. e., infinitely differentiable.
2. Let Π and $\widetilde{\Pi}$ be two smooth manifolds. A function $F : \Pi \to \widetilde{\Pi}$ is *differentiable* at $\mathbf{x} \in \Pi$ if there are maps $\varphi : \mathcal{U} \to \mathbb{R}^d$ and $\widetilde{\varphi} : \widetilde{\mathcal{U}} \to \mathbb{R}^d$ with $\mathbf{x} \in \mathcal{U} \subseteq \Pi$ and $F(\mathbf{x}) \in \widetilde{\mathcal{U}} \subseteq \widetilde{\Pi}$ such that
$$f = \widetilde{\varphi} \circ F \circ \varphi^{-1} : \varphi(\mathcal{U}) \subset \mathbb{R}^d \to \mathbb{R}^d \tag{B.2}$$
is differentiable at $\varphi(\mathbf{x})$.
3. A function $F : \Pi \to \widetilde{\Pi}$ is a *diffeomorphism* if it is bijective and F as well as F^{-1} are both differentiable.

Note that in the first item, differentiability is defined, because τ_{ij} are functions from and to Euclidean space, where this concept already exists. In the second item, differentiability is independent of the choice of the chart functions φ and $\widetilde{\varphi}$, because the transition function τ between different charts is required to be differentiable. Hence F is either differentiable with respect to every chart or not differentiable at all.

The definition ensures that the manifold has no "edges" or "corners" but rather looks smooth, as the name suggests. From this definition, it can be seen that the previous example for a topological manifold—the globe—is also a smooth manifold.

For the purposes of this book, this short introduction to manifolds will suffice. We will now turn our attention to a different mathematical field, group theory, but come back to manifolds at the end of this section.

Definition B.3 (Group). A *group* $(\mathcal{G}, \odot)$ is a set $\mathcal{G}$ with a binary composition $\odot : \mathcal{G} \times \mathcal{G} \to \mathcal{G}$ with the following properties:

1. Associativity: $(g_1 \odot g_2) \odot g_3 = g_1 \odot (g_2 \odot g_3)$.
2. Neutral element: there exists $e \in \mathcal{G}$ with $g \odot e = e \odot g = g$ for all $g \in \mathcal{G}$.
3. Inverse element: for every $g \in \mathcal{G}$, there is a $g^{-1} \in \mathcal{G}$ with $g \odot g^{-1} = g^{-1} \odot g = e$.

Definition B.4 (Group action). Let $\mathcal{G}$ be a group and $\mathcal{S}$ an arbitrary set. A (left) *group action* of $\mathcal{G}$ on $\mathcal{S}$ is a function $A\colon \mathcal{G} \times \mathcal{S} \to \mathcal{S}$ such that:

1. for all $g_1, g_2 \in \mathcal{G}$ and $s \in \mathcal{S}$
$$A\left(g_2, A\left(g_1, s\right)\right) = A\left(g_2 \odot g_1, s\right), \tag{B.3}$$
2. for all $s \in \mathcal{S}$ and with e denoting the neutral element of $\mathcal{G}$,
$$A\left(e, s\right) = s. \tag{B.4}$$

Both items of the definition are reasonable when put into words: The first item requires that the result of composing two group actions will be the same as the group action on the composition of the elements. The second item requires that the neutral element of the group has no effect with respect to the group action. One sometimes writes $g \odot s$ instead of $A(g, s)$, i. e., the group composition $\odot$ is also used to indicate the group action.

In the following two definitions, let $\mathcal{G}$ be a group, $\mathcal{S}$ a set, A a group action of $\mathcal{G}$ on $\mathcal{S}$, and $s \in \mathcal{S}$.

Definition B.5 (Group orbit). The set $\mathcal{G}s = \{A(g, s) | g \in \mathcal{G}\} \subseteq \mathcal{S}$ is called the *orbit* of s.

Definition B.6 (Stabilizer). The subgroup $\mathcal{G}_s = \{g \in \mathcal{G} | A(g, s) = s\}$ is called the *stabilizer* of s.

Although the definitions look similar and the notations differ only in the use of a concatenation vs. the use of a subscript, the semantics are quite different. The group orbit $\mathcal{G}s$ is a subset of $\mathcal{S}$ and contains all the points that can be reached from s by a transformation. The stabilizer $\mathcal{G}_s$ is a subset of $\mathcal{G}$ and contains all the group elements that do not affect s.

To see the difference, consider the group of all rotation matrices of two-dimensional Euclidean space,

$$\mathcal{G} = \left\{ \begin{pmatrix} \cos\alpha & \sin\alpha \\ -\sin\alpha & \cos\alpha \end{pmatrix} \middle| \alpha \in \mathbb{R} \right\}, \tag{B.5}$$

where the group composition $\odot$ is given by the usual matrix multiplication. This group is also called the two-dimensional special orthogonal group SO(2).

It is easy to see that $\mathcal{G}$ is indeed a group. Associativity follows by the associativity of matrix multiplication and the concatenation of two rotations by α and β gives the same result as one rotation by $\alpha + \beta$. The inverse of a rotation is the rotation by the negative of its angle and the neutral element is a rotation by 0.

Now, let $\mathcal{S} = \mathbb{R}^2$ be the Euclidean plane. The group action of $\mathcal{G}$ on $\mathcal{S}$ is given by the usual multiplication of a vector by a matrix. Then the orbit of an arbitrary point $\mathbf{s} \in \mathbb{R}^2$ consists of all points with the same distance from the origin:

$$\begin{aligned} \mathcal{G}\mathbf{s} &= \{A(g, \mathbf{s}) | g \in \mathcal{G}\} = \left\{ \begin{pmatrix} \cos\alpha & \sin\alpha \\ -\sin\alpha & \cos\alpha \end{pmatrix} \mathbf{s} \middle| \alpha \in \mathbb{R} \right\} \\ &= \left\{ \mathbf{s}' \in \mathbb{R}^2 \middle| \|\mathbf{s}'\| = \|\mathbf{s}\| \right\}. \end{aligned} \tag{B.6}$$

This means that these orbits are circles around the origin—hence the name. For $\mathbf{s} = \mathbf{0}$, the orbit is a degenerate circle with radius $\mathbf{0}$, i. e., a point. Hence, the stabilizer is given by

$$\mathcal{G}_0 = \mathcal{G} \qquad \mathcal{G}_{\mathbf{s}} = \{\mathbf{I}\} \text{ for } \mathbf{s} \neq \mathbf{0}. \tag{B.7}$$

It is easy to see that any rotation maps the origin to the origin, which means that the stabilizer $\mathcal{G}_0$ is the whole group. All other points are rotated along a circle around the origin, hence the only stabilizer is the neutral element $\mathbf{I}$.

The special orthogonal group shares an important property that links the world of algebra with manifolds: each rotation matrix can be decomposed into smaller and smaller rotations. In particular, a rotation can be decomposed into

$$\begin{pmatrix} \cos\alpha & \sin\alpha \\ -\sin\alpha & \cos\alpha \end{pmatrix} = \begin{pmatrix} \cos\frac{\alpha}{n} & \sin\frac{\alpha}{n} \\ -\sin\frac{\alpha}{n} & \cos\frac{\alpha}{n} \end{pmatrix}^n \tag{B.8}$$

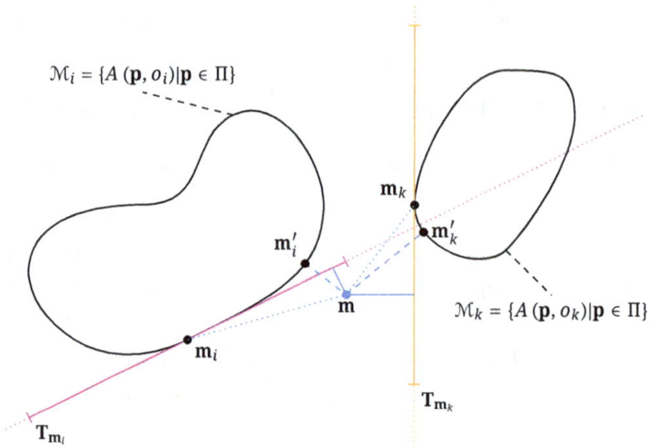

Fig. B.1: Tangential distance measure, reproduced from Figure 2.12.

for every $n \in \mathbb{N}$. Indeed, any rotation can be decomposed into infinitely many, infinitesimally small rotations.

This observation leads to the following definition.

Definition B.7 (Lie group). A group Π that is also a smooth manifold such that the group operation $p_1 \odot p_2^{-1}$ is differentiable is a *Lie group*.

The two-dimensional special orthogonal group SO(2) is a Lie group and a one-dimensional smooth manifold, which can be seen if one considers that a suitable chart function maps a rotation matrix to its angle $\alpha \in \mathbb{R}$.

The definition of a Lie group only makes a statement about the group operation, but not about group actions. The combination yields the following definition.

Definition B.8 (Lie transformation group). Let $\mathcal{M}$ be a smooth manifold, Π a Lie group, and $A : \Pi \times \mathcal{M} \to \mathcal{M}$ a group action of Π on $\mathcal{M}$. Π is called a *Lie transformation group* with respect to $\mathcal{M}$ if A is differentiable.

Note that this definition uses some kind of differentiability three times: the group Π is a Lie group and therefore equipped with a differentiable structure; the space $\mathcal{M}$ is a smooth manifold, i. e., it possesses a differentiable structure; and the map A is required to be differentiable, too.

To conclude, consider Figure 2.12 (reproduced in Figure B.1 for convenience) in light of the new concepts. The feature space $\mathcal{M}$ is assumed to be a smooth manifold and the disturbances are modeled as a Lie transformation group Π that acts on the feature space. Then the orbit of a feature vector $\mathbf{m}_i$ under the group action are smooth sub-manifolds. Although this section skipped a mathematical definition of the *tangent space*, it should be intuitively clear that a tangent exists at each point of each sub-manifold, because all involved maps are differentiable.

C Random processes

The following will give a brief overview of random processes. This overview is by no means meant to be comprehensive, but should be sufficient to understand the concepts in Chapter 2.

A random process $\underline{g}$ is a random variable that is also a function of some arguments. Here, we will focus only on two-dimensional random processes over the real numbers, i. e., random functions of the form

$$\underline{g} : \mathbb{R}^2 \to \mathbb{R}. \tag{C.1}$$

From one perspective, $\underline{g}$ is a random function that is evaluated at any point $\mathbf{x} \in \mathbb{R}^2$: depending on the realization of $\underline{g}$, a different result will be obtained. Hence, $\underline{g}$ can be described by a probability distribution over the space of all possible functions $g : \mathbb{R}^2 \to \mathbb{R}$ that map from $\mathbb{R}^2$ to $\mathbb{R}$. In particular, one can find an expectation μ and a variance σ^2 for $\underline{g}$,

$$\mu = \mathrm{E}\{\underline{g}\} \quad \text{with} \quad \mu : \mathbb{R}^2 \to \mathbb{R} \qquad \text{and} \tag{C.2}$$

$$\sigma^2 = \mathrm{Var}\{\underline{g}\} \quad \text{with} \quad \sigma^2 : \mathbb{R}^2 \to \mathbb{R}^+. \tag{C.3}$$

Note that both μ and σ^2 are themselves functions. The covariance is a function of two points $\mathbf{x},\mathbf{y} \in \mathbb{R}^2$,

$$\mathrm{Cov}\{\underline{g}\} : \mathbb{R}^2 \times \mathbb{R}^2 \to \mathbb{R}. \tag{C.4}$$

A second interpretation considers $\underline{g}$ as a (deterministic) function that maps $\mathbf{x}$ to a random variable $\underline{r}_{\mathbf{x}}$ on $\mathbb{R}$. From this perspective, there is an infinite (and uncountable) set $\{\underline{r}_{\mathbf{x}} \mid \mathbf{x} \in \mathbb{R}^2\}$ of random variables. Assume that a probability density function $p(\mathbf{x}_1, \ldots, \mathbf{x}_k)$ exists for each finite subset with k arbitrary points $\mathbf{x}_1, \ldots, \mathbf{x}_k \in \mathbb{R}^2$. The corresponding distributions are called finite-dimensional marginal distributions (fidis). Moreover, we require that

$$\mu(\mathbf{x}) = \left(\mathrm{E}\{\underline{g}\}\right)(\mathbf{x}) \stackrel{!}{=} \mathrm{E}\{\underline{g}(\mathbf{x})\} = \mathrm{E}\{\underline{r}_{\mathbf{x}}\}, \tag{C.5}$$

$$\sigma^2(\mathbf{x}) = \left(\mathrm{Var}\{\underline{g}\}\right)(\mathbf{x}) \stackrel{!}{=} \mathrm{Var}\{\underline{g}(\mathbf{x})\} = \mathrm{Var}\{\underline{r}_{\mathbf{x}}\} \tag{C.6}$$

for all $\mathbf{x}$ and for all other moments, in case they exist. Note the subtle mathematical difference: on the left of the equations, the stochastic moment of a random function is calculated first and then the resulting (non-random) function is evaluated at the point $\mathbf{x}$; on the right, the function is evaluated at $\mathbf{x}$ to a random variable first and then the stochastic moment of that random variable is calculated.

With these notations at hand it is possible to introduce two properties of stochastic processes.

Definition C.1 ((Strictly) stationary process (of order m)). Let $\underline{g}$ be a random process, $k \in \mathbb{N}$ be a finite dimension, $\mathbf{x}_1, \ldots, \mathbf{x}_k \in \mathbb{R}^2$ arbitrary points, and $\boldsymbol{\tau} \in \mathbb{R}^2$ a translation vector.

https://doi.org/10.1515/9783111339207-012

1. If
$$p(\mathbf{x}_1, \ldots, \mathbf{x}_k) = p(\mathbf{x}_1 + \boldsymbol{\tau}, \ldots, \mathbf{x}_k + \boldsymbol{\tau}) \tag{C.7}$$
holds for all valid choices of k, $\mathbf{x}_1, \ldots, \mathbf{x}_k$, and $\boldsymbol{\tau}$, the process $\underline{g}$ is called *(strictly) stationary*. This means that all fidis are invariant under translation.
2. A stochastic process is called *(strictly) stationary of order m*, if the above holds for all $k \le m$.

Definition C.2 (Homogeneity (of order m)). A stochastic process is called *homogeneous of order m* if
$$\mathrm{E}\{\underline{g}^\nu\}(\mathbf{x}) = \text{const} \qquad \forall \nu \in \{1, \ldots, m\}. \tag{C.8}$$

This means the first m moments do not depend on the point $\mathbf{x}$. Obviously, stationarity is much stronger than homogeneity. A stationary process is always homogeneous (up to the same order) but not vice versa.

Definition C.3 ((Two-dimensional) weakly stationary process). Let $\underline{g}$ denote a two-dimensional random process, $\mathbf{x}, \mathbf{y} \in \mathbb{R}^2$ two points, and $\boldsymbol{\tau} \in \mathbb{R}^2$ a translation vector. The process is a *weakly stationary process* if for all $\mathbf{x}, \mathbf{y}, \boldsymbol{\tau}$,
$$\mathrm{E}\{\underline{g}\}(\mathbf{x}) = \mu, \tag{C.9}$$
$$\mathrm{Cov}\{\underline{g}, \underline{g}\}(\mathbf{x},\mathbf{y}) = \mathrm{Cov}\{\underline{g}, \underline{g}\}(\mathbf{x} + \boldsymbol{\tau}, \mathbf{y} + \boldsymbol{\tau}). \tag{C.10}$$

This means the expectation is constant for every point $\mathbf{x}$ and the covariance is also constant in the sense that its value only depends on the relative position of $\mathbf{x}$ and $\mathbf{y}$ but is not affected by a translation. Especially for $\mathbf{x} = \mathbf{y}$, this implies $\mathrm{Var}\{\underline{g}\}(\mathbf{x}) = \sigma^2$ is constant for all $\mathbf{x} \in \mathbb{R}^2$.

The condition of weak stationarity is more restrictive than a homogeneity of order two, but less restrictive than stationarity of order two. For a process to be homogeneous, it is only required that its expectation and variance be constant: this does not say anything about its covariance. In contrast, to be a stationary process of order two, it is required that all two-dimensional marginal distributions be identical. The latter is much stronger than having only identical covariances.

Definition C.4 (Expectation-free (two-dimensional) weakly stationary process). Let $\underline{g}$ be a two-dimensional weakly stationary process. The process is called *expectation free* if for all $\mathbf{x} \in \mathbb{R}^2$
$$\mathrm{E}\{\underline{g}\}(\mathbf{x}) = 0. \tag{C.11}$$

Note that the term “expectation free” is a bit misleading: an expectation free random process is not free of having an expectation. It has an expectation: 0.

Definition C.5 ((Two-dimensional) white noise). A two-dimensional random process $\underline{e}_{mn}$ is *white noise* if it is weakly stationary and fulfills the additional requirements

$$\mathrm{E}\{\underline{e}\} = 0, \tag{C.12}$$

$$\mathrm{Cov}\{\underline{e}\}(\mathbf{x}, \mathbf{x} + \boldsymbol{\tau}) = \begin{cases} \sigma^2 & \boldsymbol{\tau} = \mathbf{0} \\ 0 & \text{else} \end{cases}. \tag{C.13}$$

Actually, both requirements already ensure that it is a weakly stationary process, but demand much more. Especially, the last requirement implies that any two states are uncorrelated with each other.

Lastly, we consider a certain assumption about random processes that makes reasoning about them easier in many circumstances: *ergodicity*. Informally, in an ergodic process, a reasonably large sample from that process is representative of the process as a whole. Formally, let $\mathcal{E}$ denote a probability space and let $e \in \mathcal{E}$ be an elementary event. Moreover, let $g(\mathbf{x}) = \underline{g}(\mathbf{x}, e)$ denote the realization of the random process $\underline{g}(\mathbf{x})$ with respect to the elementary event e.

Definition C.6 (Ergodic process). Let $\underline{g}$ be a stationary process and let $\mu(\mathbf{x}) = \mathrm{E}\{\underline{g}\}(\mathbf{x})$ denote the expectation of $\underline{g}$. This means that the expectation $\mu(\mathbf{x}) = \mu$ is constant for all $\mathbf{x} \in \mathbb{R}^2$. The process $\underline{g}$ is said to be *ergodic* if for all events $e \in \mathcal{E}$ and all $\mathbf{y} \in \mathbb{R}^2$,

$$\lim_{\substack{w\to\infty \\ h\to\infty}} \frac{1}{wh} \int_{-\frac{w}{2}}^{\frac{w}{2}} \int_{-\frac{h}{2}}^{\frac{h}{2}} \underline{g}(\mathbf{x}, e)\, \mathrm{d}\mathbf{x} = \mu = \mathrm{E}\{\underline{g}\}(\mathbf{y}). \tag{C.14}$$

On the right side of Equation (C.14), one arbitrary point $\mathbf{y}$ is fixed and the average over all possible realizations g of $\underline{g}$ is calculated. On the left, one realization $g(\mathbf{x}) = \underline{g}(\mathbf{x}, e)$ is fixed and the average over all points $\mathbf{x} \in \mathbb{R}^2$ is calculated. Hence, under the assumption that $\underline{g}$ is ergodic, one can determine the unknown expectation and variance of $\underline{g}$ by taking the average over all points of only one single realization.

Bibliography

R. Aster, B. Borchers, and C. Thurber. *Parameter Estimation and Inverse Problems*. International Geophysics Series. Academic Press, 2013.

Y. Bengio, P. Lamblin, D. Popovici, and H. Larochelle. Greedy Layer-Wise Training of Deep Networks. In *Advances in Neural Information Processing Systems*, pages 153–160, 2007.

J. Beyerer. *Analyse von Riefentexturen*. PhD thesis, Universität Karlsruhe, 1994.

J. Beyerer, F. Puente León, and C. Frese. *Machine Vision*. Springer, 2016.

P. Billingsley. *Probability and Measure*. Wiley, 3 edition, 1995.

C. Bishop. Mixture Density Networks. 1994.

C. Bishop. *Pattern Recognition and Machine Learning*. Springer, 2007.

B. Boser, I. Guyon, and V. Vapnik. A Training Algorithm for Optimal Margin Classifiers. In *Proceedings of the Fifth Annual Workshop on Computational Learning Theory*, pages 144–152, 1992.

L. Breiman. Random Forests. *Machine Learning*, 45(1):5–32, 2001.

J. Bromley, J. Bentz, L. Bottou, I. Guyon, Y. LeCun, C. Moore, E. Säckinger, and R. Shah. Signature Verification using a “Siamese” Time Delay Neural Network. *International Journal of Pattern Recognition and Artificial Intelligence*, 7(4):669–688, 1993.

K. Cho, B. van Merrienboer, Ç. Gülçehre, D. Bahdanau, F. Bougares, H. Schwenk, and Y. Bengio. Learning Phrase Representations using RNN Encoder–Decoder for Statistical Machine Translation. In *Proceedings of the Conference on Empirical Methods in Natural Language Processing*, pages 1724–1734, 2014.

C. Cortes and V. Vapnik. Support-Vector Networks. *Machine Learning*, 20(3):273–297, 1995.

T. Cover and P. Hart. Nearest Neighbor Pattern Classification. *IEEE Transactions on Information Theory*, 13(1): 21–27, 1967.

K. Crammer and Y. Singer. On the Algorithmic Implementation of Multiclass Kernel-based Vector Machines. *Journal of Machine Learning Research*, 2(Dec):265–292, 2001.

A. Criminisi, J. Shotton, and E. Konukoglu. Decision Forests: A Unified Framework for Classification, Regression, Density Estimation, Manifold Learning and Semi-Supervised Learning. *Foundations and Trends in Computer Graphics and Vision*, 7(2–3):81–227, 2012.

N. Cristianini and J. Shawe-Taylor. *An Introduction to Support Vector Machines and other Kernel-based Learning Methods*. Cambridge University Press, 2000.

G. Csurka, C. Dance, L. Fan, J. Willamowski, and C. Bray. Visual Categorization with Bags of Keypoints. In *Workshop on Statistical Learning in Computer Vision, European Conference on Computer Vision*, pages 1–22, 2004.

A. Dempster, N. Laird, and D. Rubin. Maximum Likelihood from Incomplete Data via the EM Algorithm. *Journal of the Royal Statistical Society: Series B*, 39:1–38, 1977.

J. Deng, W. Dong, R. Socher, L. Li, K. Li, and L. Fei-Fei. ImageNet: A Large-Scale Hierarchical Image Database. In *Proceedings of the IEEE Conference on Computer Vision and Pattern Recognition*, pages 248–255, 2009.

R. Duda, P. Hart, and D. Stork. *Pattern Classification*. Wiley, 2 edition, 2001.

B. Efron and T. Hastie. *Computer Age Statistical Inference: Algorithms, Evidence, and Data Science*. Cambridge University Press, 2016.

G. Fink. *Mustererkennung mit Markov-Modellen*. Vieweg+Teubner Verlag, 2003.

G. Forney. The Viterbi Algorithm. *Proceedings of the IEEE*, 61(3):268–278, 1973.

Y. Freund and R. Schapire. A Decision-Theoretic Generalization of On-Line Learning and an Application to Boosting. *Journal of Computer and System Sciences*, 55(1):119–139, 1997.

Y. Freund and R. Schapire. A Tutorial on Boosting, 2004.

A. Georghiades, P. Belhumeur, and D. Kriegman. From Few to Many: Illumination Cone Models for Face Recognition under Variable Lighting and Pose. *IEEE Transactions on Pattern Analysis and Machine Intelligence*, 23(6):643–660, 2001.

https://doi.org/10.1515/9783111339207-013

I. Goodfellow, J. Pouget-Abadie, M. Mirza, B. Xu, D. Warde-Farley, S. Ozair, A. Courville, and Y. Bengio. Generative Adversarial Nets. In *Advances in Neural Information Processing Systems*, 2014.

I. Guyon, J. Makhoul, R. Schwartz, and V. Vapnik. What Size Test Set Gives Good Error Rate Estimates? *IEEE Transactions on Pattern Analysis and Machine Intelligence*, 20(1):52–64, 1998.

T. Hastie, R. Tibshirani, and J. Friedman. *The Elements of Statistical Learning.* Springer, 2001.

K. He, X. Zhang, S. Ren, and J. Sun. Deep Residual Learning for Image Recognition. In *Proceedings of the IEEE Conference on Computer Vision and Pattern Recognition*, pages 770–778, 2016.

C. Herrmann, D. Willersinn, and J. Beyerer. Low-Resolution Convolutional Neural Networks for Video Face Recognition. In *Proceedings of the 13th IEEE International Conference on Advanced Video and Signal Based Surveillance*, 2016.

T. Ho. Random Decision Forests. In *Proceedings of the Third International Conference on Document Analysis and Recognition*, pages 278–282, 1995.

S. Hochreiter and J. Schmidhuber. Long Short-Term Memory. *Neural Computation*, 9(8):1735–1780, 1997.

R. Hoffmann. Signalanalyse und -erkennung. *Springer*, 1998.

A. Hyvärinen, J. Karhunen, and E. Oja. *Independent Component Analysis.* Wiley, 2004.

A. Jaglom and I. Jaglom. *Wahrscheinlichkeit und Information.* Deutscher Verlag der Wissenschaften, 1960.

J. Jensen. Sur les fonctions convexes et les inégalités entre les valeurs moyennes. *Acta Mathematica*, 30: 175–193, 1906.

A. Kolmogorov. On the representation of continuous functions of many variables by superposition of continuous functions of one variable and addition. *American Mathematical Society Translation*, 28(2): 55–59, 1963.

D. Krahe and J. Beyerer. A Parametric Method to Quantify the Balance of Groove Sets of Honed Cylinder Bores. In *Intelligent Systems and Advanced Manufacturing*, pages 192–201, 1997.

A. Krizhevsky, I. Sutskever, and G. Hinton. Imagenet Classification with Deep Convolutional Neural Networks. In *Advances in Neural Information Processing Systems*, pages 1097–1105, 2012.

K. Küpfmüller. Die Entropie der deutschen Sprache. *Fernmeldetechnische Zeitung*, 7(6):265–272, 1954.

A. Laubenheimer. *Automatische Registrierung adaptiver Modelle zur Typerkennung technischer Objekte.* PhD thesis, Universität Karlsruhe, 2004.

Y. LeCun, L. Bottou, Y. Bengio, and P. Haffner. Gradient-Based Learning Applied to Document Recognition. *Proceedings of the IEEE*, 86(11):2278–2324, 1998.

A. Lipkus. A proof of the triangle inequality for the Tanimoto distance. *Journal of Mathematical Chemistry*, 26 (1–3):263–265, 1999.

K. Liu and G. Mattyus. Fast Multiclass Vehicle Detection on Aerial Images. *IEEE Geoscience and Remote Sensing Letters*, 12(9):1938–1942, 2015.

D. Loftsgaarden and C. Quesenberry. A Nonparametric Estimate of a Multivariate Density Function. *The Annals of Mathematical Statistics*, 36(3):1049–1051, 1965.

D. Lowe. Distinctive Image Features from Scale-Invariant Keypoints. *International Journal of Computer Vision*, 60(2):91–110, 2004.

V. Maz'ya and G. Schmidt. On Approximate Approximations Using Gaussian Kernels. *IMA Journal of Numerical Analysis*, 16(1):13–29, 1996.

J. Mercer. Functions of Positive and Negative Type, and their Connection with the Theory of Integral Equations. *Philosophical Transactions of the Royal Society of London*, 209:415–446, 1909.

O. Mogren. C-RNN-GAN: A Continuous Recurrent Neural Network with Adversarial Training. In *Constructive Machine Learning Workshop, Advances in neural information processing systems*, 2016.

T. Moon and W. Stirling. *Mathematical Methods and Algorithms for Signal Processing.* Prentice Hall, 2000.

J. Neyman and E. Pearson. On the Problem of the Most Efficient Tests of Statistical Hypotheses. In *Breakthroughs in Statistics*, pages 73–108. Springer, 1992.

A. Novikoff. On Convergence Proofs on Perceptrons. *Proceedings of the Symposium on the Mathematical Theory of Automata*, 12:615–622, 1962.

C. Rasmussen and C. Williams. *Gaussian Processes for Machine Learning*. Adaptative Computation and Machine Learning Series. MIT Press, 2006.

M. Richter, T. Längle, and J. Beyerer. Knowing When You Don't: Bag of Visual Words with Reject Option for Automatic Visual Inspection of Bulk Materials. In *Proceedings of the 23rd International Conference on Pattern Recognition*, 2016.

A. Rieder. *Keine Probleme mit Inversen Problemen: Eine Einführung in ihre stabile Lösung*. Vieweg+Teubner Verlag, 2003.

H. Ritter, T. Martinetz, and K. Schulten. Neuronale Netze. *Addison-Wesley*, 1990.

C. Robert. A Comparison of the Bayesian and Frequentist Approaches to Estimation by Francisco J. Samaniego. *International Statistical Review*, 79(1):117–118, 2011.

L. Rokach. *Pattern Classification Using Ensemble Methods*. World Scientific Publishing, 2010.

F. Rosenblatt. *The Perceptron: A Perceiving and Recognizing Automaton*. Cornell Aeronautical Laboratory, 1957.

F. Rosenblatt. *Principles of Neurodynamics: Perceptrons and the Theory of Brain Mechanisms*. Spartan, 1962.

D. Rumelhart, G. Hinton, and R. Williams. *Learning Internal Representations by Error Propagation*. MIT Press, 1986.

J. Schmidhuber. Deep Learning in Neural Networks: An Overview. *Neural Networks*, 61:85–117, 2015.

B. Schölkopf and C. Burges. *Advances in Kernel Methods: Support Vector Learning*. MIT press, 1999.

B. Schölkopf, A. Smola, and K. Müller. Kernel Principal Component Analysis. In *International Conference on Artificial Neural Networks*, pages 583–588, 1997.

J. Schürmann. *Pattern Classification: A Unified View of Statistical and Neural Approaches*. Wiley, 1996.

C. Shannon. A Mathematical Theory of Communication. *The Bell System Technical Journal*, 27:379–423, 1948.

L. Sommer, T. Schuchert, and J. Beyerer. Deep learning based multi-category object detection in aerial images. In *Automatic Target Recognition XXVII*, 2017.

N. Srivastava, G. Hinton, A. Krizhevsky, I. Sutskever, and R. Salakhutdinov. Dropout: A Simple Way to Prevent Neural Networks from Overfitting. *Journal of Machine Learning Research*, 15(1):1929–1958, 2014.

S. Stevens. On the Theory of Scales of Measurement. *Science*, 103(2684):677–680, 1946.

L. Valiant. A Theory of the Learnable. *Communications of the ACM*, 27(11):1134–1142, 1984.

L. van der Maaten and G. Hinton. Visualizing Data Using t-SNE . *Journal of Machine Learning Research*, 9: 2579–2605, 2008.

V. Vapnik. *Statistical Learning Theory*. Wiley, 1998.

A. Vaswani, N. Shazeer, N. Parmar, J. Uszkoreit, L. Jones, A. Gomez, L. Kaiser, and I. Polosukhin. Attention is All You Need. In *Proceedings of the 31st International Conference on Neural Information Processing Systems*, pages 6000–6010, 2017.

A. Viterbi. Error Bounds for Convolutional Codes and an Asymptotically Optimum Decoding Algorithm. *IEEE Transactions on Information Theory*, 13(2):260–269, 1967.

P. Werbos. Backpropagation Through Time: What It Does and How to Do It. *Proceedings of the IEEE*, 78(10): 1550–1560, 1990.

D. Wolpert and W. Macready. No Free Lunch Theorems for Optimization. *IEEE Transactions on Evolutionary Computation*, 1(1):67–82, 1997.

M. Zeiler and R. Fergus. Visualizing and Understanding Convolutional Networks. In *European Conference on Computer Vision*, pages 818–833, 2014.

Z. Zhou. *Ensemble Methods: Foundations and Algorithms*. CRC press, 2012.

Glossary

a posteriori distribution The distribution of the classes with respect to a fixed feature
a priori distribution The distribution of the classes without knowledge of the features
absolute norm A special type of Minkowski norm
absolute scale Scale of measurement for counting quantities
ANN Artificial neural network
AR model Autoregressive signal model
artificial neural network Model of a biological neural network that is a type of artificial intelligence
attention mechanism A computational mechanism in neural networks that allows the model to focus on and relate between specific parts of the input sequence when making predictions
autoencoder Neural network architecture that compresses (or encodes) input data into a lower-dimensional representation and then reconstructs the original input
autoregressive signal model Representation of a type of random process

backpropagation Supervised learning algorithm used for training artificial neural networks by performing a gradient descent on a loss function
bagging Bootstrap aggregating
Baum–Welch algorithm An algorithm used to estimate the unknown parameters of a hidden Markov model, instance of the EM algorithm
Bayes' law A fundamental result in probability theory
Bayesian classifier A special classifier that uses all the ingredients of the Bayesian framework and is optimal with respect to the risk
bias In parameter estimation: error of an estimator that is not due to chance
binary classifier A classifier that only decides between two classes
boosting Meta-method that combines several weak classifiers into one strong classifier
bootstrap aggregating Ensemble method in which several classifiers are trained on random subsets of the same training data

central limit theorem A central theorem in probability theory
Chebyshev norm A special type of Minkowski norm
class A subset of the world grouping similar objects
class-specific feature distribution The distribution of the features given a class
classifier A (mathematical) method of assigning an object to an equivalence class based on features
CNN Convolutional neural network
conditional distribution The distribution of a random quantity if another quantity from a joint probability space is kept fixed
confusion matrix Table that compares the ground truth with the classifier's prediction on a validation set
consistent estimator An estimator that converges almost surely to the true value
convolutional neural network Type of deep learning architecture suitable for multidimensional data with repeating local structure
cosine distance An angle-based distance measure (not metric) between two (often high-dimensional) vectors
cost function A function that describes the costs of assigning a class with respect to the true class
CR-efficient estimator A special estimator that has minimum variance
Cramér–Rao bound A lower bound on the variance of an unbiased estimator
CRB Cramér–Rao bound
cross-validation Technique to estimate the performance of a classifier with a small dataset

https://doi.org/10.1515/9783111339207-014

dataset The set of all objects that were collected to define, validate, and test a pattern recognition system
decision boundary The boundary of a decision region, the entirety of boundaries is an equivalent description of a classifier
decision function A function that maps a feature vector to one component of the decision space
decision region A partition in the feature space
decision space An intermediate space to unify the mathematical description of the classes
decision tree Tree-structured classifier where the inner nodes correspond to tests, the edges correspond to the outcomes of the tests, and the leaf nodes govern the class decision
decision vector The vector of decision functions of all classes
deep learning A field of machine learning that focuses on neural networks with a large amount of layers
Dirac sequence A sequence of probability distributions that converges to the Dirac distribution
discrepancy A function that quantifies the similarity between two (mathematical) objects that lacks some properties of a metric
distance function Usually a synonym for metric, usage may vary depending on context
distribution Mathematical object that encapsulates the properties of random variables
divergence A discrepancy between probability distributions

EM Expectation maximization
emission probability In hidden Markov models: probability of seeing an observable given a chain of states
empirical operation Mathematical operation that corresponds to an experiment, e.g., addition of the masses of two objects by putting both on a scale at the same time
empirical relation Mathematical relation that emerges from an experiment, e.g., by comparing the weight of two objects
empirical risk minimization From statistical learning theory: minimization of the average loss on a training set
empiricism Philosophy of science that emphasizes evidence and experiments
entropy measure Impurity measure corresponding to the entropy of the empirical class distribution of a dataset
ERM Empirical risk minimization
estimator A measurable function from the space of all finite datasets into the parameter space of a parametric distribution assumption
Euclidean norm A special type of Minkowski norm
expectation maximization An iterative optimization technique used to estimate parameters of Gaussian mixture models or other statistical models involving latent variables by maximizing the likelihood function

false alarm The event that a binary classifier incorrectly decides for “positive” although the sample is negative
false negative rate The probability of a binary classifier of deciding for “negative” although the sample actually belongs to the positive class
false positive rate The probability of a binary classifier of deciding for “positive” although the sample actually belongs to the negative class
feature A mathematical quantity that describes the characteristics of an object
feature space The set of all possible features
Fisher information The variance of the score

GAN Generative adversarial network
Gaussian mixture See Gaussian mixture model
Gaussian mixture model A random variable whose density is a convex combination of Gaussian densities

generalization Ability of a classifier to perform well on unseen data
generative adversarial network Neural network architecture that consists of a generator and a discriminator trained simultaneously through adversarial learning
Gini impurity Impurity measure corresponding to the expected error probability of random class assignment on a dataset
GMM Gaussian mixture model
grammar A set of rules and structures that defines the syntax of a language, can be used to classify sequences

hard margin SVM Traditional SVM that finds the decision boundary with maximum margin between linearly separable data
hidden Markov model Markov model where the states and state transitions are hidden and can only be inferred from observations
HMM Hidden Markov model
homogeneous process A random process whose moments do not depend on the point of evaluation
hyperparameter Parameter that governs a classifier but is not estimated from the training set

ICA Independent component analysis
impurity measure Measure that assesses the class distribution in a dataset
independent component analysis A statistical method to split features into components that are stochastically independent
interval scale Scale of measurement for measuring intervals but lacking a natural zero

joint distribution The distribution of several random quantities in a joint probability space

KL divergence Kullback–Leibler divergence
k-nearest neighbor method A parameter-free technique to define a density given a number of finite samples, see also Parzen window method
Kullback–Leibler divergence Measure (but not a metric) of the difference between probability distributions

leave-one-out cross-validation Cross-validation where only one sample is used for evaluation and the rest are used to train the classifier
likelihood function A function of the parameters of a statistical model for a given dataset
likelihood ratio The ratio between two likelihood functions with different models, used in hypothesis testing
linear discriminant A basic classifier that draws hyperplanes between classes in the feature space
log-likelihood function The logarithm of the likelihood function
long short-term memory Type of deep learning architecture suitable for sequential data
loss function A mathematical function that measures the difference between the predicted values and the actual values in a machine learning model
LSTM Long short-term memory

Mahalanobis norm Norm of a vector with respect to some positive definite matrix
Manhattan metric Metric deduced from the absolute norm, also: taxicab metric
MAP classifier maximum a posteriori classifier
marginal distribution The projection of a joint distribution onto one of the axes
Markov model Probabilistic model of states and transitions between states with certain restrictions
matched filter A template-matching based signal processing filter designed to maximize the signal-to-noise ratio when detecting a known object in the presence of noise

maximum a posteriori classifier A classifier that decides for the class with the highest a posteriori probability with respect to a given feature
maximum likelihood estimator An estimator that chooses the parameter that makes the given observation most likely under the model
maximum norm A special type of Minkowski norm
MDA Multiple discriminant analysis
MDN Mixture density network
mean squared error Mean of the squared derivations of an estimator to the target variable
median The middle entry in a sorted list of items
metric A function that defines a distance
metric space A set with a distance measure
minimax classifier A special type of classifier that estimates the class such that the maximal risk with respect to any a priori distribution is minimized, see also classifier
Minkowski norm A parametrized norm for real vector spaces
misclassification measure Impurity measure corresponding to the empirical error probability of the dominant class in a dataset
mixture density network A type of neural network that models probability distributions using a mixture of multiple (Gaussian) probability density functions
ML estimator Maximum likelihood estimator
mode In statistics: the global maximum of a probability mass or probability density, i.e., the most probable value
multiple discriminant analysis A statistical technique used for dimensionality reduction that accounts for the classes in the problem

naive Bayes assumption See naive Bayes classifier
naive Bayes classifier A simple Bayesian classifier with the "naive" assumption of independence among features
natural language processing A field of artificial intelligence that focuses on the interaction between computers and human language
nearest neighbor classifier A classifier that assigns an object the same class as the nearest sample (in the feature space) of the training set
NLP Natural language processing
nominal scale Scale of measurement made up of labels
norm Function to measure the length of a vector

ordinal scale Scale of measurement with an ordering
overfitting Phenomenon where a classifier performs well on the training set but very poorly on unseen data

parameter space The (vector) space of all quantities that define a classifier
parameter vector A point in the parameter space
Parzen window method A parameter-free technique to define a density given a number of finite samples, see also k-nearest neighbor method
pattern The raw data from a sensor
pattern space The set of all possible patterns
PCA Principal component analysis
perceptron The simplest form of a neural network consisting of a single layer of artificial neurons with binary outputs, central component to build more complex network architectures
permutation metric A metric for features on the ordinal scale

principal component analysis A method for finding a lower-dimensional subspace such that the projection of the dataset has a minimal squared reconstruction error
probability simplex A subset in the decision space

quantile Summary statistic to describe the location within an ordered sample

random forest An ensemble of decision trees
random process Mathematical description of a (time-ordered) series of random events
ratio scale Scale of measurement for measuring ratios
recall The event that a binary classifier correctly decides for "positive", the probability of the recall is the true positive rate
receiver operating characteristic Plot of the false positive rate against the true positive rate of a binary classifier
rectified linear unit Activation function used in deep learning
recurrent neural network Neural network architecture designed to handle sequential data by maintaining a hidden state that captures information about previous inputs
ReLU Rectified linear unit
residual network A type of deep neural network architecture that introduces residual connections, allowing the model to skip some layers during training
ResNet Residual network
risk The expected cost of the decisions of a classifier, see also cost function
RNN Recurrent neural network
ROC Receiver operating characteristic
rubber-sheeting Distortion of a surface to allow seamless joins

scale of measurement Defines certain types of variables and permissible operations on the variables of a given type
score In statistics: measure of how much a parameter influences the density of a random variable
sensitivity True positive rate
SGD Stochastic gradient descent
slack The event that a binary classifier incorrectly decides for "negative" although the sample is positive
slack variable In SVMs: variables associated with the training samples to measure the violation of the maximum margin constraint
soft margin SVM Extension of the hard margin SVM that allows misclassifications to account for noisy data
specificity True negative rate
SRM Structural risk minimization
state transition probability In Markov models: probability to switch between states
stationary process A random process that does not change the joint distribution of a derived time series when shifted in time
stochastic gradient descent Randomized version of the gradient descent optimization algorithm
string matching Class of algorithm to find occurrences of a substring (pattern) within a larger sequence
structural risk minimization From statistical learning theory: joint minimization of the average loss on a training set and the model complexity
supervised learning Learning when the classes of the training samples are known, e.g., classification
support vector Data point that influences the placement and orientation of the decision boundary by maximizing the margin between classes
support vector machine A linear classifier that maximizes the margin between the decision boundary and the training samples
SVM Support vector machine

target vector A unit vector in the decision space and a corner of the probability simplex
taxicab metric Metric deduced from the absolute norm, also: Manhattan metric
***t*-distributed stochastic neighbor embedding** A dimensionality reduction technique for visualization that emphasizes preserving local relationships between data points
template matching Technique in signal processing that finds small parts (templates) within a noisy signal, see matched filter
test set A special subset of the dataset that is used to test the performance of a classifier
training set A special subset of the dataset that is used to define the parameters of a classifier
transformer Neural network architecture designed to process sequential data in parallel, often used for NLP
true negative rate The probability of a binary classifier of deciding for "negative" if the sample actually belongs to the negative class
true positive rate The probability of a binary classifier of deciding for "positive" if the sample actually belongs to the positive class
***t*-SNE** *t*-distributed stochastic neighbor embedding

unbiased estimator A special estimator whose expectation value equals the parameter being estimated if considered as a random variable on its own
unbiasedness See unbiased estimator
unsupervised learning Learning when the classes of the training samples are not known or not needed, e.g., clustering, density estimation, etc.

VAE Variational autoencoder
validation set A special subset of the dataset that is used to define the design parameters of a classifier
Vapnik–Chervonenkis dimension Measure of complexity of a given family of classifiers
variational autoencoder A type of autoencoder that learns a probability distribution as its lower-dimensional representation and then samples from that distribution to capture the inherent uncertainty in the data
Viterbi algorithm A dynamic programming algorithm used for finding the most likely sequence of hidden states in a hidden Markov model

weak classifier A classifier that performs only marginally better than random guessing
weakly stationary process A random process whose expectation and covariance are constant at every point
window function A function that is nonzero only in some interval, often used to assign a weight according to some distance, e.g., in the Parzen window method

Index

https://doi.org/10.1515/9783111339207-015

www.ingramcontent.com/pod-product-compliance
Lightning Source LLC
Chambersburg PA
CBHW080909220326
41598CB00034B/5524
* 9 7 8 3 1 1 1 3 3 9 1 9 1 *